# The Rocket Lab

# THE FOUNDERS SERIES

The Founders Series publishes books on and about Purdue University, whether the physical campus, the University's impact on the region and world, or the many visionaries who attended or worked at the University.

OTHER TITLES IN THIS SERIES

*Planting the Seeds of Hope: Indiana County Extension Agents During the Great Depression and World War II*
Frederick Whitford

*Queen of American Agriculture: A Biography of Virginia Claypool Meredith*
Frederick Whitford, Andrew G. Martin and Phyllis Mattheis

*WBAA: 100 Years as the Voice of Purdue*
Angie Klink

*Pledge and Promise: Celebrating the Bond and Heritage of Fraternity, Sorority, and Cooperative Life at Purdue University*
Angie Klink

*Wings of Their Dreams: Purdue in Flight, Second Edition*
John Norberg

*Ever True: 150 Years of Giant Leaps at Purdue University*
John Norberg

*Purdue at 150: A Visual History of Student Life*
David M. Hovde, Adriana Harmeyer, Neal Harmeyer, and Sammie L. Morris

*Memories of Life on the Farm: Through the Lens of Pioneer Photographer J. C. Allen*
Frederick Whitford and Neal Harmeyer

*A Purdue Icon: Creation, Life, and Legacy*
James L. Mullins (Ed.)

*Scattering the Seeds of Knowledge: The Words and Works of Indiana's Pioneer County Extension Agents*
Frederick Whitford

*Enriching the Hoosier Farm Family: A Photo History of Indiana's Early County Extension Agents*
Frederick Whitford, Neal Harmeyer, and David M. Hovde

*The Deans' Bible: Five Purdue Women and Their Quest for Equality*
Angie Klink

# The Rocket Lab

## Maurice Zucrow, Purdue University, and America's Race to Space

Michael G. Smith

Purdue University Press • West Lafayette, Indiana

Printed in the United States of America.

Cataloging-in-Publication Data is available from the Library of Congress.
978-1-61249-840-9 (hardback)
978-1-61249-841-6 (paperback)
978-1-61249-842-3 (epub)
978-1-61249-843-0 (epdf)

The front cover image is a painting by Don Trembath titled *Space Trajectories*, which also served as the December 1961 cover for *Astronautics*, the journal of the American Rocket Society.

*For Melinda (MZ)*

# CONTENTS

# ACKNOWLEDGMENTS

FOR ITS GENEROUS RESEARCH SUPPORT, I THANK THE COLLEGE OF LIBERAL ARTS and Department of History at Purdue University. I also appreciate the crucial assistance of the directors and staff of the Maurice J. Zucrow Laboratories, especially Steve Heister, Robert Lucht, Scott Meyer, and Jennifer Ulutas. I was able to research the files still in storage at the labs in 2015 and 2016.

I am grateful to Marc Cohen, who kindly donated Dr. Zucrow's papers to Purdue University Archives and Special Collections in 2017, and to Jonathan Eisenburg, who donated a wide variety of Dr. Zucrow's genealogical materials and who shared important family contexts. I thank Sammie Morris, Tracy Grimm, Richard Bernier, Adriana Harmeyer, Stephanie Schmitz, and the Archives and Special Collections staff for processing the papers and assisting me with access to the full range of related papers in its collections. I thank Neil Harmeyer and R. Allen Bol for their expertise with the illustrations. I am also grateful to Justin Race and the staff of Purdue University Press and to senior production editor Kelley Kimm.

Around the country, for their help and advice, I thank all archivists, especially Douglas Bicknese at the National Archives (Chicago), Ray Ortensie at the Air Force Materiel Command Office of History, and Robert Arrighi at the NASA Glenn Research Center. I also offer special thanks to Steve Heister and Nicholas Sambaluk for reviewing the full manuscript and offering valuable improvements.

# ABBREVIATIONS

AAF Army Air Forces
AAS American Astronautical Society
ABM antiballistic missile
ABMA Army Ballistic Missile Agency
AE aeronautical engineering
AEC Atomic Energy Commission
AERA Association pour l'encouragement a la recherche aéronautique
AERL Aircraft Engine Research Laboratory (NACA Cleveland)
AIAA American Institute of Aeronautics and Astronautics
AFB Air Force base
AFHRA Air Force Historical Research Agency
AMC Air Materiel Command
APL Applied Physics Lab
ARDC Air Research and Development Command
ARGMA Army Rocket and Guided Missile Agency
ARPA Advanced Research Projects Agency
ARS American Rocket Society
ASEE American Society for Engineering Education
ASME American Society of Mechanical Engineers
ATSC Air Technical Service Command
BSAE bachelor of science in aeronautical engineering
BSAES bachelor of science in aeronautical and engineering science
BSME bachelor of science in mechanical engineering
BSEE bachelor of science in electrical engineering
BuAer Bureau of Aeronautics (Navy)
BuOrd Bureau of Ordnance (Navy)
Caltech California Institute of Technology (CIT)
CIA Central Intelligence Agency
DEW distant early warning
DOD Department of Defense
EES Engineering Experiment Station
ESMWT Engineering, Science, and Management War Training
GALCIT Guggenheim Aeronautical Laboratory at Caltech

| | |
|---|---|
| GMC | Guided Missiles Committee |
| GPO | Government Printing Office |
| HUAC | House Un-American Activities Committee |
| IAS | Institute of the Aeronautical (Aerospace) Sciences |
| ICBM | intercontinental ballistic missile |
| JCS | Joint Chiefs of Staff |
| JNW[E] | Joint Committee on New Weapons [and Equipment] |
| JP | jet propellant |
| JPL | Jet Propulsion Laboratory |
| JRDB | Joint Research and Development Board |
| LARS | Laboratory for Applications of Remote Sensing (Purdue) |
| LAW | light antitank weapon |
| ME | mechanical engineering |
| MICOM | Missile Command (Army) |
| MIT | Massachusetts Institute of Technology |
| MPF | Missiles Propulsion and Fuels (Panel) |
| MPH | miles per hour |
| MSAE | master of science in aeronautical engineering |
| MSAES | master of science in aeronautical and engineering science |
| MSME | master of science in mechanical engineering |
| NAA | North American Aviation |
| NACA | National Advisory Committee for Aeronautics |
| NAE | National Academy of Engineering |
| NARA | National Archives and Records Administration |
| NAS | National Academy of Sciences |
| NASA | National Aeronautics and Space Administration |
| NRL | Naval Research Laboratory |
| NSC | National Security Council |
| NSF | National Science Foundation |
| NYU | New York University |
| OART | Office of Advanced Research and Technology (NASA) |
| ONR | Office of Naval Research |
| ORD | Ordnance (Army) |
| ORI | Office of Research and Inventions (USN) |
| OSU | Ohio State University |
| PERT | Project [Program] Evaluation and Review Technique |
| PRF | Purdue Research Foundation |
| psi | pounds per square inch |
| R&D | research and development |

| | |
|---|---|
| RAND | Research and Development Corporation |
| RDB | Research and Development Board |
| RFNA | red fuming nitric acid |
| RG | Record Group |
| ROTC | Reserve Officers' Training Corps |
| SAC | Strategic Air Command |
| SAE | Society of Automotive Engineers |
| SAGE | Semi-Automatic Ground Environment |
| SATC | Student Army Training Corps |
| TEG | Technical Evaluation Group |
| TOW | tube-launched, optically tracked, wire-guided missile |
| T-2 | Technology-2 Branch (ATSC) |
| USAF | US Air Force |
| USN | US Navy |
| USSR | Union of Soviet Socialist Republics |
| WAC | without attitude control |
| WADC | Wright Air Development Center |
| WFNA | white fuming nitric acid |
| WPAFB | Wright-Patterson Air Force Base |
| WSEG | Weapons Systems Evaluation Group |

# INTRODUCTION

## A Study in Professional Biography

*"Arts of public use, as fortification, making of engines, and other instruments of war; because they confer to defence and victory, are power: and though the true mother of them be science, namely the mathematics; yet, because they are brought into the light by the hand of the artificer, they be esteemed . . . as his issue."*

THOMAS HOBBES, *LEVIATHAN*

PURDUE UNIVERSITY, SET ON A PLATEAU IN CENTRAL INDIANA, JUST ABOVE THE banks of the Wabash River, is framed by its aerospace history. At one corner of the main campus are the Maurice J. Zucrow Laboratories, with some of the country's premier chemical rocket test stands; at another corner, the Neil Armstrong Hall of Engineering, filled with cutting-edge laboratories and even a Moon rock. Each place represents a small chapter in the history of the space age. Both began as smaller, less glamorous buildings. Dr. Maurice Zucrow established what he called his Rocket Laboratory in 1948 next to the Purdue airport, in a farmer's field at the far southwest edge of campus. At the start, it was little more than a series of fortified cinder block garages, serving as test cells and offices for his liquid-propellant rocket experiments. Armstrong Hall, located at the northeast corner of the university, used to be a site filled with sheet metal Quonset huts from the Second World War era, originally serving as classrooms for chemistry and physics.

To make sense of the history, we need to return to these humbler beginnings. As we look back, this book offers a history of aerospace engineering in the US during its golden age, the twenty-some years between the first guided missiles of 1946 and the successful Saturn V Moon rocket of 1969. I center this story on the life and times of Maurice "Doc" Zucrow, who founded Purdue's Rocket Laboratory as a signal contribution to the American guided missile and spaceflight programs.

Dr. Zucrow has never been well known among the American public, beyond a small if significant readership of engineers. He remains relatively unknown today. He was not one of the named "men of space" at the dawn of spaceflight. He was not one of the "crowned heads of missiledom" through the early Cold War. He has not been an inductee in the National Aviation Hall of Fame, or one of the honorary US Air Force (USAF) Space and Missile Pioneers, or even a member of the International Space Hall of Fame. He is not included in David Darling's internet encyclopedia of rocket engineers and space scientists; nor has he received a tribute from the National Academy of Engineering.[1] This study is a small effort to rediscover this person, so lost to history, along with the people closest around him. I make the case that both he and they still matter.[2]

Zucrow was a true pioneer of rocket engineering, author of a number of impressive firsts. He was the first person to graduate from Harvard University's new engineering program magna cum laude. He was the first earned PhD from Purdue; and he was its first Distinguished Professor of Engineering. He taught the country's first university courses in jet propulsion during the Second World War and was the first leading engineer afterward to confirm the possibilities of rocket travel beyond Earth. He established the first university laboratory in the US dedicated to education and research in jet propulsion and rocketry He published the world's first textbook on jet propulsion and rocketry, *Principles of Jet Propulsion and Gas Turbines* (1948).

Zucrow pioneered the laboratory approach to liquid-propellant rocketry. As a traveling consultant, he mentored researchers at Ohio State University, Wright-Patterson Air Force Base, and Lewis Flight Propulsion Laboratory of the National Advisory Committee on Aeronautics (NACA). In 1952, after five years of serving on the Pentagon's Panel on Propulsion and Fuels, he joined an official committee to advocate for the Atlas intercontinental ballistic missile (ICBM). He also joined with colleagues in NACA and the American Rocket Society (ARS) to lobby for the peaceful exploration of outer space. He was a national spokesperson for spaceflight through the early 1960s and was one of the founding directors of the American Institute for Astronautics and Aeronautics (AIAA). Beyond all this, he mentored 27 PhD students, and just over 250 MS degrees, most with the thesis option. He was the author of project contracts totaling above $4 million, and the builder of a $2.5 million physical plant.[3] Not bad for an immigrant born in Kyiv, Ukraine, raised in London and Boston, and naturalized as a US citizen in 1921.

Each of these firsts is less a singular prize than a sign of multiples. Each is a reference point that bridges Zucrow's life with other lives. His story is not so much about priority as significance. He built these achievements upon the shared contributions of his immigrant family, his teachers at Harvard and Purdue, and his talented bands of engineering colleagues and graduate students. This is not a story of one life but of intersecting lives. His 1948 textbook, in an interesting turn of events, changed history. A young Neil Armstrong, as we shall see, chose Purdue University to study engineering because of it. Zucrow taught there, and that was enough. What matters, then, is biography through history, a life in context.

Zucrow's was a classic American story. During his second academic career at Purdue (1946–1966), the country experienced phenomenal growth in consumer goods and services, in prosperity and mobility, and in scientific and technological applications.[4] These were also years of extraordinary aerospace successes. The US and USSR deployed supersonic jets, guided missiles, and ICBMs and initiated their first spaceflight programs. Popular science magazines of the era commonly represented all of this progress in the steep inclines of exponential S curves. J. R. Van Pelt, a pioneer of science and technology education, called this a "generalized curve of progress" in social and engineering achievements, stating, "If we plot efficiency or state of development as ordinates against time as abscissas, the curve may rise slowly at first, then become much steeper, and then flatten out as a high level of efficiency is reached. It will remain on that plateau until some new and unconventional approach clears the way for further progress."[5] Historical progress spanned social, scientific, and technological developments. It was a function of spikes and plateaus, leaps, and gaps, if ever rising. Kenneth Mees, vice president of research at the Eastman Kodak Company, translated the curve into a "helix of progressive change," a coil of horizontal circles and vertical spirals representing moments of increasing lift and upsurge. This was a sign of the times. At select points around the world, especially in the US and Europe, people were living better and smarter, faster and farther lives. The middle decades of the twentieth century were all about a developmental boost phase in science, technology, and living standards. For Mees and like-minded thinkers, rising science empowered the economic and social takeoff of modernization.[6]

Jet propulsion, as a revolutionary technical development, was a perfect metaphor for the S curve of progress. Marvin McFarland, Guggenheim Chair of Aeronautics at the Library of Congress, celebrated both the "scientific-technological-military revolution" of the mid-twentieth century and "the rocket that has become its symbol." As he wrote, "It is in the nature of revolution, once unleashed, to accelerate at an ever-increasing rate until at some point it collapses, whether from exhaustion or decay, or is stopped by a counteracting force." Here was evidence of Newton's second law of nature at work in history: "For every action there is a corresponding reaction."[7] The rocket was a metaphor for the rapid pace and promise of modernization. Politicians and engineers shared the view. Congressman Overton Brooks, chair of the Science and Astronautics Committee in the US House of Representatives, called the early 1960s "this era of explosive technology," especially referencing the new "million-pound thrust single chamber rocket" that became the Saturn booster. Clifford Furnas, aerospace engineer and chancellor at the University of Buffalo, spoke of the twentieth century's "speeding up of technology," jumping from the automobile age to the space age in a generation, in part because "the results of science are cumulative—like the sustained thrust of a rocket motor. Each increment builds on what happened before, so the farther it goes the faster it goes."[8]

All this meant a fascinating process of accelerated competition and "creative destruction," in Joseph Schumpeter's terms: creating machines to break speed and altitude, distance, and accuracy records, and once accomplished, to create them better once again.[9] Engineering constantly relinquished the old to give way to the new. Engineers were the mostly silent stars in this drama of designing and redesigning, all in a research and development (R&D) race with each other, and with the nation's enemies in the Cold War. As one witticism of the time held, "If it works, it's obsolete." Achievement simply sent the engineers back to the research drawing board to make a better product, to extend the upward curve.[10] Change was the lot of engineers. Improvement their guiding principle, for reliability and use.

Zucrow participated in this story, helped make the revolution that was rocketry, if in measured ways. Writing of the turbojet engine as applied to aviation after the Second World War, he dramatized the steep vertical "slope of the curve of flight speed as a function of years."[11] Aircraft were flying faster and faster. He and his engineering colleagues experienced these accelerations in the machines they made and flew, and in their busy daily lives, driven by the competitions and pressures of the Cold War. Time seemed to be compressing all around them. For the goal of the rocket and guided missile was to achieve high speeds over prescribed distances. The engineer's task, in turn, was to reduce distance in space and time with the quickest and best results, with accelerated lead times and crash programs.[12]

These efforts aligned with the elaboration of the American "project system" between the 1940s and 1960s, framed by the peak achievements of the Manhattan Project that created the atomic bomb and Project Apollo that explored the Moon. The term is not new. It had its humble origins in early initiatives of the US Department of Agriculture, conceived to pay universities and educational foundations for applied research to improve farming. Clarence Danhof has surveyed the project system in the greatest breadth and depth, culminating thirty years of advances in atomic power, radar, jet aircraft, spaceflight, and computers. David Hart has explored its networks as the liberal associative state. Aaron Friedberg has explored its nuances and democratic checks and balances.[13] The scientists and engineers who belonged to the project system understood its methods: to mobilize expert talent and achieve results by way of the sponsored research contract. That was the fundamental procedure of the Cold War and Space Race.

Ever since President Dwight Eisenhower's famous remarks in 1961, the negative cliché "military–industrial complex" has overwhelmed the intricacies and achievements of the project system. Yet projects were ubiquitous and definitive in these years, an apogee of American engineering. The project was adept at "accelerating the accumulation of new knowledge and promoting new technologies," in Danhof's terms. It was so powerful because it "intricately intermingled the interests and activities" of three great forces: the power politics of the government and military, the profit motives of business and industry, and the academic freedoms of the university. The project system represented a unique separation and balance of powers. By 1968, individual research projects accounted for 72 percent of all US federal government academic R&D monies.[14] Zucrow's story empowers us to see them again anew, hidden all these years in plain view. His story also helps us to see the military–industrial–academic network as a natural outgrowth of the American economic and political systems, a network that engineering expertise helped to define and refine.

The contributions of Maurice Zucrow to this new era of jet propulsion and rocketry may seem rather modest. In scientific and technical terms, they amounted to several dozen peer-reviewed publications and several hundred credentialed graduate students. All of this was in line with Zucrow's cautious engineering views, sober laboratory methods, and sound business sense. He and his colleagues called this a stepwise approach in engineering, one at the heart of the project system. Engineering was about small and measured steps along the project way, between basic and applied science, moving from experiment to design, and on to the testing and development of a finished product. At times, Zucrow leapt forward into speculation about space travel, or encouraged his students to do so, something still risky in the 1940s and early 1950s. For the most part, though, he devoted his life to details, to the hard work of testing his jet and rocket engines, and of mentoring his students. He intentionally kept his laboratories small, supported by a wide range of limited military, government, and industrial projects. This was his preferred means of avoiding the statist excesses of Big Science.

The Purdue Rocket Lab was small but significant. As Zucrow expressed it, "I work in a little niche in a tremendous field."[15] It was a niche in the vast Cold War network of new aerospace institutions, but one whose influence and legacy far outweighed its smallness: like the very rocket science that Zucrow helped to create. According to fellow aerospace engineer John Sloop, "Large changes in range result from small changes in engine performance." Sloop was referring to the truth of the rocket equation, the mathematical explanation for its performance at launch and acceleration. He meant that small changes in values, as for example more efficient high-energy propellants, or chamber pressures, created large payoffs in results, the very carrying capacity of rockets for military or civilian payloads.[16] The measure of the rocket engineer was in the incremental.

Zucrow's labs never built actual rockets, at least not how the media portrayed them, with their sleek, bullet-shaped exteriors and downward plumes of exhaust. His labs only ever built and tested liquid-propellant rocket engines, messy and noisy affairs, gangly complexes of wiring and valves, pumps and combustion chambers, all fixed within their boxy horizontal test stands. The delicate experiments and measurements that Zucrow and his teams made on these engines were precise, focused on improving specific impulse, the key measure of a vehicle's exhaust thrust and its launching power into air and space. Zucrow developed a fine rocket science of slight improvements in the design and performance of the rocket motor, centered on higher chamber pressures, in order to extract more power from its propellants and improve the thrust-to-weight ratio of missiles and rockets. His quest for higher pressures inspired a major line of R&D in liquid-propellant rocketry. In the words of one of his PhD students, H. Doyle Thompson, Zucrow's combustion methods were "to spit fire, and produce thrust, and it was high pressure."[17] The trouble was that these higher pressures and temperatures created higher heat loads on the combustion chamber's walls. This demanded new methods of motor cooling, as well as solutions to problems like carbon deposits, ignition lags, and combustion oscillations. Zucrow's first improvisation was film cooling, or as he defined it, "Interposing a film of liquid between a wall that must be protected and hot gas (5,000° to 8,000°F.) flowing past the wall."[18]

Maurice Zucrow and his teams found meaning in small things. They were focused more on the technical than the technological. They inhabited the realm of the technological, of course. They designed and operated actual machines. They were part of the country's R&D apparatus, centered on the design and manufacture of things (technologies) as industrial or consumer products, especially as weapons. Yet their focus was on the details of actual technical parts and processes, mostly in order to create and impart knowledge about them to college students. This is a distinction worth teasing out here, as that publicists and historians tend to spotlight the mythical dimensions of the technological rather than the engineering facets of the technical. We give "big technology" an "unprecedentedly high cultural standing," even though the term can really mean anything. For as Paul Forman has further written, *technology* is "simply the collective noun for all the many things that are in fact done and made."[19] We entertain technology's vast evolutionary transformations.[20] Yet we neglect the less recognizable techniques and procedures that informed them, and the people who made them, in all their intricacies and combinations. Zucrow's story aligns more with the smaller spaces of engineering science, the delicate human what and the how of their profession, as Walter Vincenti has defined it. This meant the elaboration of mathematical tools and quantitative analyses for successful product designs, along a pathway of stepwise improvements.[21] Engineering as both knowledge and practice, premise and approach, method and application. Engineers as the often silent, hidden figures who make history; never alone, always in teams.

Among the leading rocket pioneers of the twentieth century, Zucrow was a consummate engineer, or as he called himself, a "rocket engineer."[22] Most of the other pioneers trained as credentialed physicists and only later reeducated themselves as engineers. Among the physicists were Robert H. Goddard, Hermann Oberth, Theodore von Kármán, Wernher von Braun, Fritz Zwicky, and Martin Summerfield. They all became self-trained engineers. Zucrow trained, from the start, and considered himself to the end, as a mechanical engineer, as a pioneer of the new engineering science. This meant not simply pure or basic research, but also applied research, creating new designs and products.[23] From the start, he inhabited the worlds of both the theoretical and the practical. This was thanks to the ways his mentors taught the fundamentals, the mathematical and scientific foundations of engineering. This was also thanks to his experiential learning: first in the mills and factories of Massachusetts, later in the industrial plants and research labs of Chicago and Pittsburgh, eventually at the pioneering Aerojet Corporation, and later in the deep networks of the Pentagon's Research and Development Board.

In these places, Zucrow was often the rover and fixer, moving about his organizations, or the whole country, rooted in first principles and ready for any challenge, true to the ideal type of rocket engineer. Abe Silverstein, a leading administrator at the National Aeronautics and Space Administration (NASA), defined this type as a "good technical man" with all the "fundamentals," but always self-educating and "expanding their horizons" and supremely "versatile." James B. Jones, a Zucrow student and professor of mechanical engineering at Virginia Polytechnic Institute, called his teacher a "complete engineer," meaning "one who practices superbly as teacher *and* researcher *and* administrator *and* counsellor *and* consultant."[24] The ideal had its limits. Zucrow was not an easy person. He was often direct and brash. One of his students, Elliott Katz, described him as "a very, very hard man who would remind you of Curtis LeMay except he had a sense of humor, but a very demanding individual."[25]

Zucrow was often a small part of big things. But sometimes he was a big part of big things too. He cultivated a unique skill set for jet propulsion. This included fluency in the German and French languages; higher mathematics; hydraulics, gas turbines, and power plant engineering; fluid mechanics and dynamics; instrumentation and control; and vibration theory. Zucrow mastered these fields with intensive readings and reworkings. His personal library, filling the basement walls of his home in West Lafayette, comprised "the great and noble engineering books of his time," remembered Bruce Reese, one of Zucrow's first graduate students. These were books that "he had read and studied and knew," filled with his notations and edits.[26] He and his teams of colleagues and students contributed propulsion systems for an impressive series of rockets in the US arsenal, including jet-assisted takeoff units (JATOs); the Corporal, Nike, and Atlas missiles; and the Saturn boosters, to name a few. In all these ways, Zucrow was one of the twentieth century's original project engineers, guiding his teams and students in the design and manufacturing of new machineries. The accent here was on contribution. Not the act of any one individual, but of many in common. The engineering way.

Besides the technical, Zucrow and his closest colleagues also valued the humanities. This was the result of their own traditional educations, which included balanced readings in the sciences and liberal arts. It was a result of their own passions and interests. Zucrow even started out at Harvard University planning to be a historian. This appreciation for the humanities was also the result of lived experience: the pace of technological progress through the twentieth century, the real and potential devastations of its economic crises and world wars, and the failure of modern civilization to manage its own survival and good order. Engineers lived through these challenges and crises. They recognized their own responsibility to understand and temper the political and social costs of their designs and creations.[27] We will meet many of them through Zucrow's biography, a band of American engineers schooled in the R&D of the Second World War. T. A. Heppenheimer has coined them as the "main line of American liquid rocketry."[28] They were largely shaped by the fight against Nazi Germany, which of course included the Peenemünde rocketry teams of Wernher von Braun. The von Braun teams were enemies who became colleagues in the Cold War, but this study offers evidence and interpretation as to how at least some of these American engineers never fully integrated or trusted them.

Zucrow owed his dynamic engineering approaches to his Jewish upbringing in London and Boston; to a stint in the US Army and a year at Tufts University; and to a BS and MS at Harvard, his alma mater. Above all, he owed his creative edge to Purdue, his second alma mater, where he earned another MS and his PhD in mechanical engineering. There he found mentors like Purdue's president Edward Elliott and dean of engineering Andrey Potter. There he found close patrons like professor of mechanical engineering Harry Huebotter, the university's chief business officer R. B. Stewart, its director of research G. Stanley Meikle, and the new president, Frederick Hovde. There he found students who became his colleagues, even surpassed him professionally, like George Hawkins, expert in thermodynamics and later dean of engineering. These were notable campus leaders, but they were also persons of high national prominence. With Zucrow, they helped to shape the engineering architectures of the Second World War and Cold War. It is quite a story.

These were also fiercely independent actors. It was an independence bordering on intransigence, something of a local trademark. Purdue University became the country's largest engineering school by the 1950s, thanks to their hard work. It was a university oversized but underrated. Purdue was something of an outlier on the national scene, halfway across the country between the glamour of the Massachusetts Institute of Technology (MIT) and Princeton University in the East, the California Institute of Technology (Caltech) and Stanford University in the West. These schools were the bearers of so much federal government research patronage between the 1940s and 1960s, at the expense of Purdue and other regional universities like it. Yet Purdue enjoyed the rare advantage of perspective, the poise that came with being set apart, something of an underdog. It was aspirational, focused on rising to the highest standards of higher education. Purdue was also fiercely pragmatic, like other land-grant schools, focused on economic and social improvements. Thus fortified, this rather disparate band of academics placed their accents not on privilege and prestige but on education, the hardscrabble education of young people in engineering and agriculture and science, and in time the humanities and social sciences too. Self-determination was their ideal. It was an ideal illustrated in the university's logo for most of the twentieth century: a roaring dragon, carrying the motto "Science, Technology, Agriculture," wielding a pointed tail and spitting fire.[29]

Higher education, both local and national, plays a central role in this story. The local is my priority, how Purdue made its way through the radical and largely rewarding transformations in the twenty years after the Second World War. Its settings and characters count.[30] But I also draw in other universities, each with its own valuable traditions and personalities, successes and failures. Each was a part of a national fabric, a partnership and consensus to accelerate American R&D. This worked, at least until the compact broke down in the late 1960s era of youth and anti-war protests.[31] The project system also exhausted itself by that time, aerospace engineering defunded and demoted on the national stage.

In the Purdue tradition, Zucrow devoted his career to what he called humanity's "prime mover." He meant the power plant, the engine that put people and things into motion, the machine that converted chemical energy into motive power. Potter experimented with steam engines; Huebotter with internal combustion engines; and Hawkins with guns, machines of a different order. Zucrow worked on steam and internal combustion engines while at Purdue. His dissertation was one of the first in engineering science to apply the aerodynamic insights of the airplane to the thermodynamics of the internal combustion engine, focused on the applied physics of the carburetor jet. By 1943, he devoted himself to two altogether new prime movers: the turbojet and the rocket engine. These were complex systems in terms of their structures and guidance and control. Yet they both began as power plants. As Zucrow expressed it, in a perfect axiom for his day, "Propulsion paces progress."[32] The power plant defined the initial ways into the stratosphere and outer space. The rocket, in its early days, was in fact a synonym for either the warhead or satellite, or for the interplanetary probe or spacecraft it propelled.[33] They were all rockets. The one thing meant all the other things. Zucrow's contributions in jet propulsion helped to pace these advances. His mandates were to help create the fundamentals of the new science, engineer sophisticated engines, and thereby prepare the ways for human spaceflight, all from his experiments with rockets along the Wabash.

# 1

# MOISHE ZUCROW

## An Emigrant's Life

MAURICE ZUCROW'S LIFE, LIKE THOSE OF HIS CLOSE FRIENDS AND COLleagues, represented a trajectory both outward and upward. His early itineraries carried him and his family in Ukraine from the small town of Bila Tserkva to Stavyshche, and from Stavyshche to Kyiv, then from Kyiv to London, and eventually from London to Boston. These were twisted turns of mostly good fortune. The trajectory followed the classic pathways of the Jewish diaspora in Eastern Europe at the turn of the twentieth century. Its migrants came from the infamous Pale of Settlement, the territories that initially segregated Russia's Jews within the Tsarist Empire. But they were eventually drawn toward the bigger cities of Russia, and eventually to the English-speaking worlds of the UK and US. Zucrow and his family found themselves in Jewish enclaves in each case, but also found ways to break out and adapt to the new cultures at large. Solomon moved his family during waves of rising in-migration. He was one of thousands of village Jews (*yeshuvnikes*) who came to the big city of Kyiv. He later became one of the "greeners" who settled in London. In each case he found gradual success and upward mobility. The trajectory had consequences. The family turned its enclaves into horizons. At each turn they prospered. The father rose as a teacher and scholar. The son and his sisters received fine educations.

The trajectory had a deeper symbolic meaning, marking the family as "people of the air" (*luftmensch*). This term had mostly negative connotations, thanks to Max Nordau, who coined it to mean the growing numbers of unemployed Jewish men who seemed to roam aimlessly across the landscapes of Eastern Europe. They were vagrants, without jobs or steady incomes. The term also referred to Jews as rootless and despairing wanderers. The stories of Sholem Aleichem featured the *luftmensch* as pathetic figures, without a homeland, or often stuck between religious tradition and secular modernity. Yet even for Nordau and his devotees, the term had positive overtones, meaning those mobile people who lived in "the free air," who lived "by wonders and miraculous chances." These were Jews who might even become "muscular" and modern. Nordau's drive for a "muscular Judaism" meant that they needed strength of character and livelihood, embodied in physical and mental health. He and his progressives encouraged the Jews of Europe and the Americas, so elevated and free, to thereby participate in the new utopian age of science and machines, to become aspirational and progressive with it.[1]

One of the leading proselytizers of these trends, Israel Zangwill, certainly understood the nuances of the *luftmensch*. In a short story of 1907, he featured the character Nehemiah as a man with a "ridiculous" courage. He "was not an earth-man in gross contact with solidities. He was an air-man, floating on facile wings through the ether." His was a "pessimism that was merely optimism in disguise," filled with a "robust faith in life, his belief in God, man, himself." He was a man with an "invincible resilience." Zangwill turned the negative "aerial man" *luftmensch* into something freer and positive.[2] In this, he was part of a wider literary trend in European intellectual culture, making a play upon the "superman" (*übermensch*) of Friedrich Nietzsche's fame, about the new man breaking free from moral conventions and even earthly gravity. For some European Jews, this trend culminated in the appeal to become a new "race of fliers," in fulfillment of the old warrior "spirit of the Maccabees," now steeled and sped along by metal airplanes and rockets, free flight into the atmosphere and even into outer space, ultimate marks of the muscular ideal.[3] In these strains of Jewish utopianism, Maurice Zucrow found a pathway for his future career.

Maurice's father, Solomon, may have been inspired by Zangwill's writings. Their families certainly shared the same life paths. Like the Zucrows, Zangwill's family fled from the Pale, settling in London, where he graduated from the Jews' Free School and University of London, to become one of the country's premier writers. *The Melting Pot*, his play staged in England to popular acclaim between 1908 and 1910, projected a future for the young Solomon. It celebrated the power of American society to free humanity from ethnic and religious divisions, to overcome the hatreds and oppressions of old Europe.[4] Here was a powerful secular ideal, tugging at Solomon's devout religious faith, summarizing the hopes for himself and especially his children to move on toward freer, better lives.

## FLIGHT OUT OF EGYPT

Maurice Zucrow was born near Kyiv, Ukraine, named Moishe Yossel (Moses Joseph), son to a family of migrants. They had not come far. They were not done moving. His father, Solomon Zucrow (1870–1932) was born in a shtetl at Bila Tserkva, about seventy-five kilometers south of Kyiv, the son of a rabbi, Moishe Zucrow, who died when he was three. He and his brother were raised by a Hassidic Orthodox single mother. Solomon likely named his own son, Moishe, after his father. But the name was also a presentiment, a sign of things to come: the Moses who would lead the family to a promised land.

Both Bila Tserkva (and Stavyshche, where Solomon married) were in Volyn', a part of the old Polish Commonwealth, which had taken over these lands after 1569. The Polish magnates and nobles, those who received land grants and moved into these territories, brought Jewish migrants with them to assume managerial posts, as stewards in their agricultural enterprises. More migrants followed, settling in the shtetl neighborhoods near marketplaces and bazaars. The wealthier Jews were traders in cattle, lumber, and grain, or enjoyed the privileges to run distilleries and the liquor trade. They owned taverns where travelers and their horses found rest and replenishment. To the south, on the lands adjacent to the Black Sea, as at Kherson, they were farmers. Most Jewish subjects of the Commonwealth were middling or poor: tailors, tinsmiths, and blacksmiths; or rope makers and itinerant peddlers.[5]

Between 1772 and 1795, as the Russian, Austrian, and Prussian states partitioned the Polish state out of existence, Solomon's birthplace, Bila Tserkva, became part of the Russian Empire. In coming decades, Christians and Jews lived in relative peace, prosperity, and mutual respect. The shtetl even gave rise to a cultural Jewishness (*Yiddishkeit*) centered on Eastern European Yiddish. It had its high manifestations, as in the art of Marc Chagall, or the stage designs of Boris Aronson, later applied to the popular Broadway show and film *Fiddler on the Roof*. It had its lower manifestations too, just as significant, in the entertainment and musical routines that producers and performers, like Louis Mayer, Sam Goldwyn, and the Warner brothers (Harry, Albert, Sam, and Jack), brought to vaudeville and Hollywood.[6] This was a culture that Maurice, at least in part, made his own, in the Yiddish proverbs and humor of which he was so fond.

There were interethnic strains. In Solomon's day, shtetl life was becoming overpopulated and stagnant, able-bodied men forced to join the ranks of the restless unemployed, especially with the economic downturns of the 1890s. The new global economy and coming of the railroads meant that they lost their jobs as tradespeople, small merchants, and wagon drivers. Shtetl lands were also crossroads of violence. Some were the sites of major anti-Jewish pogroms, one of which Solomon would have remembered from 1881, when criminal bands reacted in anger and bloodshed to the assassination of Tsar Alexander II. More happened shortly after he emigrated, in the 1903–1906 revolutionary events, encompassing the major cities and towns, Kyiv included, especially during spring and summer 1905. With Berdychiv to the east and Uman' to the south, Bila Tserkva and Stavyshche were among the few towns in the Russian Empire where Jews counted as half or more of the local population. This made them targets for more pogroms to come through the years of war and revolution.[7]

According to family accounts, Solomon was a gifted young man with dreams of taking up a learned profession and achieving career success. At six years old he had already read all of the Bible; at fourteen the Talmud; and by age seven he had learned Russian.

He was the talk of the town in Bila Tserkva, this budding professor bound for the gymnasium and university. That was his hope, anyhow, one that he suppressed in deference to his overbearing mother, who wanted him to become a Hasidic rabbi. Caught between tradition and modernity, Solomon chose a middle path, becoming a Hebrew and Talmud teacher. "Thus, without a murmur," as Solomon's daughter later wrote, "out of filial devotion to a puritanical woman, he sacrificed both his childhood and his future."[8] As a religious scholar, he was a sympathetic to the Jewish enlightenment movement (*Haskalah*), for Hebrew-based rational education as opposed to Talmud scholasticism. He was like-minded with the secular and assimilationist reformers (*maskilim*), those advocating for Western dress and mores, in tandem with their faith. Yet he was also fiercely independent, free of any school or cult.

Solomon married Dova Smushkin in 1894. He was twenty-four. She was twenty-one, the "natural blue-eyed, blond-haired" daughter of Shalom Smushkin, a prosperous flour merchant, and who also operated an inn and stables at Stavyshche. According to one census, he was the "richest man" in the town. Wealth and power at Stavyshche were centered in the estates of the famed Count Francis Xavier Branicki, eighteenth-century Polish noble, whose descendants now prospered from their grain harvests, sugar beet fields, and cattle herds. Thousands of people came to Stavyshche from surrounding farms and communities for the Tuesday fair and Sunday market. Peasants sold produce, livestock, or their knitted goods and woodwork. They bought dry goods (clothes and shoes) and finished foods from Jewish shops, filled with barrels of pickles and sauerkraut, goose fat and cracklings for sale. There were two distilleries (for wine and beer) and two flour mills, an apothecary, and a hospital. There was no train station, so fleets of wagons took produce by the local highways to the nearby towns of Berdychiv and Uman'. Trade reached as far as Kyiv to the north and Odesa to the south. Stavyshche was a crossroads of faiths and cultures. The Eastern Orthodox, Jewish, and Roman Catholic traditions mingled there; as did the Russian, Ukrainian, Yiddish, Hebrew, and Polish languages. The town itself, according to several reminiscences, was a place of contrasts: centered by a cobblestone boulevard leading to the Branicki estate, with tall pines and poplars; and framed by the poorer Jewish section with mud streets with mud-floored homes. The outer landscapes were enchanted places of deep pine woods, ponds and lakes, and grassy fields dotted by lilac bushes and park benches. Those Jews who grew up in Stavyshche, eventually to emigrate, came to remember the place with a special nostalgia. They recalled their vibrant shtetl as a "holy community" of "love and faith." There were six synagogues and prayer houses, along with both Jewish and Russian primary schools. The older and able boys went to high school in Kyiv. Jewish newspapers arrived in town from Poland in either Hebrew or Yiddish. Orthodox believers mixed with the reformers and liberals, as for example the Zionists in favor of a Palestine homeland, or the Jewish Bundists in favor of a socialist revolution.[9]

According to traditional practice, Shalom's money financed Solomon's marriage to his daughter Dova, along with his education as a religious scholar. The new family took him from one shtetl to another, Bila Tserkva to Stavyshche, only fifty kilometers apart. Dova sometimes derided the arrangement, and the trying journeys to come in their married life. Several family stories relate how she spoke of her husband as the "scholar-poor man," how she was "embittered" by their life of poverty and want in London, how she later resented all those "people who eat off gold plates" in Boston. These were snatches of conversation, from her and others who remembered them, either in Yiddish or her broken English.[10] They were complaints, but also marks of her courage during all the hardships and moves to come.

After their marriage in 1894, Solomon and Dova first moved from Stavyshche to Kyiv. For Jews of the Pale of Settlement, Kyiv was Egypt (*Yehupets*). It was a place of wonders and worries, enchantment and mobility, perhaps even some security and prosperity. But it was also a place of bondage to an alien culture, with all its allures of secularization and acculturation. Russian Jews migrated there between 1881 and 1917 under civil registration regimes that were at times lax, at other times severe. The tsarist regime required that Jews have residency permits, although benign neglect, or the occasional payment of bribes, helped many of them to stay in place. Police dragnets and the two major pogroms, in 1881 and 1905, displaced them. Yet the migrants made Kyiv Russia's most Jewish of cities: Petersburg was 1.8 percent Jewish in 1910, Moscow below 1 percent in 1912. By 1917, the Jewish population of Kyiv was at a high of 15 percent.

Kyiv was a city of its own dramatic contrasts. The poor concentrated in the outlying districts of Plosskii, Lybedskoi, and the Podol. Wealthy Jewish merchants, some of Russia's wealthiest, lived in fine central enclaves like the Pechersk. These included the elites of the Brodskii sugar conglomerate, the Margolin shipping concerns, and the Lev Ginzburg building company. Their money helped to build two prominent synagogues in the city center, blocks from some of the holiest Orthodox religious sites. Their philanthropy also funded the various portals of Jewish religious education, as well as the vibrant market reported in Yiddish newspapers and books. Jewish monies helped establish the Kyiv Polytechnic Institute, founded in 1898.[11] Solomon and his family occupied a lower-middling place in this world, attuned to all its nuances and contrasts. With support from his father-in-law, he continued his studies, training as a teacher of Hebrew and the Talmud, two foundations of the faith, used by rabbis in their sacred rites, and taught by *melamdim* (elementary school teachers) to the young. He navigated the half-legal world of Talmud Torah schools. The couple did not stay long, but long enough to grow their family, beginning with daughters Bess and Nancy. Maurice was born on 15 December 1899 "in a little town near Kyiv," as he recounted, probably Stavyshche. He was the family's Moses, a sign to flee out of Egypt. Their journey westward began the next year, when Maurice was around six months old. To London, where Solomon traveled first.[12]

## LONDON'S EAST END

Millions of Jews migrated out of Eastern Europe at the turn of the twentieth century. About 2 million went to North America. Only about 120,000 ever got to England. Among them were the Zucrows, who arrived at the peak of the arriving migrants. One family story held that a visiting English lecturer heard Solomon speak in Kyiv and invited him to London with promises of support. But the sponsor never made good on the deal. Betrayed by his supposed benefactor, and then cheated of his monies by scoundrels, Solomon at first lived in a "dark gloomy attic." He ate a "scanty diet of peanuts and bread." He even once pawned his pillow to pay for food. Yet he taught himself English by reading books in the free public library, receiving some charity from the Jewish community. He was befriended by Israel Abrahams, reader in Rabbinics at Cambridge University, with the promise of a university education. Instead, Solomon scraped together enough of a living as a religion teacher and tutor to eventually send for his family.[13]

Migrant journeys out of the Pale of Settlement were often harrowing, beset by incidents of crime and corruption, robberies and bribes. Young men and women from the shtetls and towns of the Pale, often with a handful of children of varying ages, set off by foot or train, and eventually steamer, by way of Berlin and Bremen, on their way to England and points westward. The sea journey to London took two or three days, probably the shortest leg of the trip. Those with enough money, and an awaiting American family, continued on to New York. Those without the money, or with relatives awaiting there, stayed on at London. Their arrivals were often cinematic, framed by the Tower Bridge and the city's smokestacks on all sides. Disembarking at the wharfs, a short walk took them to the East End, the neighborhoods of Whitechapel Road.[14] The trip was worth the risks and costs. Had they stayed in the ghettos or shtetls of Russia, the Zucrow girls would not have received any formal education. For boys, there was little opportunity for higher education. For the family, London opened new vistas. And for infant Maurice, the world.

Their first home was in Stepney, called the East End or the "Ghetto of London," an "area of about a square mile." It was east of the Liverpool Street train station; north of the Thames River, the Tower of London, and London Docks; centered at the great triangle formed by Whitechapel and Commercial Roads. These were the neighborhoods of the Jewish garment industry and its workers, a population constantly replenished at the turn of the century by new arrivals, in-migrants from Eastern Europe. They came, as one worried commentator had it, like the "slow rising of a flood." All faces were now of the "Hebrew stamp." As Jewish synagogues replaced Christian parishes, the Hebrew script and Yiddish replaced English on storefronts and in newspapers.[15]

The East End was a busy place. William Booth founded the Salvation Army here in 1865. One of Jack the Ripper's victims was also found here in 1888. The neighborhoods saw a frenzy of political movements and events. The family was exposed to all variety of political influences. Theodor Herzl roused the ghetto in 1896 with Zionist calls for a mass return to Palestine, the ancient homeland. Young Zionist men took on the clean-shaven, "robust good looks of the open-air pioneer." Given the pace of urban life and work, others broke their Sabbath devotions and Kosher diets for acculturated ways.[16] Socialists and anarchists fought for the workers' loyalties. Vladimir Lenin lived and visited here in 1902 and 1907. The family might have passed him on the street, maybe even with a nod or the touch of the cap. They might have encountered all manner of Mensheviks and Bolsheviks, as for example during the Fifth Congress of the Russian Social Democratic Labor Party (1907), when their delegates stayed in the East End. These included revolutionaries with such exotic names as Rosa Luxemburg, Leon Trotsky, and Joseph Stalin. In 1911, the East End also saw the Sidney Street siege, known as the Battle of Stepney, when local police and the army guardsmen surrounded two Latvian political activists and criminals for half a day, filled with exchanges of gunshots, until the culprits were burned to death by a fire. Winston Churchill, as home secretary, helped to direct the operation. These were just some of the local news stories for this intrepid family. Or as one memoir had it: "Our environment made us tough. Tough not in body, but in mind. Aware of evil, depravity, wickedness, hunger, sorrow, misery; aware too of the innate goodness of the people around us, even the wrong types."[17]

Synagogues remained the center of Jewish devotional life. They could be as small as storefronts or upper rooms, as large as the cathedral of the synagogue at Great Portland Street. Here the faithful found prayer and fellowship. They heard the colorful Yiddish sermons of visiting preachers (*maggidim*), sometimes hours long, paced by Lithuanian or Ukrainian, Polish or Russian accents. In the most Orthodox neighborhoods, prayer shawls filled the streets on holy days. The Sabbath was the heart of weekly life, with steam baths on Fridays to prepare, and the Friday night meal and Kiddush prayer. Children learned a bit of Hebrew, along with Bible stories and rabbinic tales, in their religious lessons and Talmud Torah schools.[18]

As the numbers of synagogues and schools rose in the first decades of the twentieth century, so did Solomon's prospects as a teacher and tutor.[19] He was "no 'go-getter,'" as his daughter Lillian remembered. He was a quiet and unassuming man, with a kind countenance, a "rare geniality, sincerity, modesty, and lack of affectation." He eventually worked for the Orthodox United Synagogue and Union of Hebrew and Religion Classes in north London at the middle-class neighborhoods of Dalston and Stoke Newington. Some of his lessons were private and "voluntary," and for which, by tradition, he received no compensation. As one reference held, "as a teacher he produced most excellent results instilling in his pupils a rare love of their religion and its teachings." By 1917, he also received the recommendation of the chief rabbi of the United Kingdom, Rabbi Joseph Hertz, as a "learned man, honest and honorable," who was "held in the highest esteem" and enjoyed a "large circle of friends."[20]

His son, Maurice, when not in school or at his home lessons, enjoyed the life of the street. Boys played rounders (a version of American softball), football, or cricket. Girls played net ball. Kiosks and vendors lured them with a repertoire of foods: homespun items like bagels, fresh bread, raisin and prune syrups, salted fish and pickled cucumbers, or borscht with matza balls; and English fare like fish and chips, whelks and cockles (snails and mollusks), and meat pies. The streets and lanes were close-knit, yet always led outward to the avenues and boulevards of the greater city. Electric lines passed through Commercial Street at Spitalfields, near the market there, on their way to exotic places, free for the walking in the fresh air (*frishe luft*): like weekend visits to Bloomsbury Street and the British Museum, or Kensington and Hyde Park. Families made Saturday afternoon visits to Victoria Park or to the Tower of London, what the Russian Jewish exiles called the fortress (*krepost'*). Thanks to the Holiday Fund, inner-city children spent summer weeks in the country, with rural families in places like Sussex.

The family prospered over the next few years as Solomon's teaching and tutoring positions improved, enough to move out of the overcrowded ghetto, eastward and northward, following the pathways of upward mobility in Jewish London. They first moved from Stepney to the safer and richer Bethnal Green, where the "roughness" was "slowly decreasing by 1910."[21] They eventually settled at 54 Lea Bridge Road in a quieter neighborhood at Millfields, across from one of the city parks. Here they lived quite apart from their former Hasidic-centered traditionalist neighbors, closer to the more established and wealthier German and Dutch Jews of long standing. Solomon became a modern-minded Hebraist, teaching young men the Talmud, Torah, and Hebrew language.

Maurice spent his earliest years growing up in the East End, filled with tightly packed row houses and tenements, along some rather rough and tumble alleyways. But it was not all squalor. The city established a Free Public Library and Reading Room at Whitechapel Road in 1902. Bethnal Green also had a free library. The London County Council Schools reached out to the growing Jewish population with adaptive policies, like a subsystem of schools and teachers all their own; and with assimilation practices, like ritual circles of the Maypole in spring, student-staged productions of Shakespeare and Gilbert and Sullivan, or readings of Thomas Carlyle's *On Heroes*. The family arrived at a fortuitous time. Maurice was one among the rising numbers of Jewish school children, which grew in the state schools from 7,838 in 1894 to 28,224 in 1911. Son and father both rose with the rising tide of Jewish immigration.[22]

Maurice attended the Bethnal Green and later Millfields Central primary schools, redbrick buildings with strict regimes of reading, writing, and arithmetic. For most children, compulsory schooling ended after they turned fourteen. These young teenagers then joined the workforce. Maurice's poorer schoolmates, as a rule, went on to work in the sweatshops of the garment and tailoring trades, or in family-based cottage industries, making brushes or assembling matchboxes, working up to fourteen hours a day. The lucky and smart ones received scholarships to continue their studies, at either vocational or grammar (college preparatory) schools. Maurice was among them, taking a series of examinations at the age of twelve to receive one of the few available scholarships to continue his education. Among these were the London County Council Junior Scholarship and the Hickson and Starling Scholarship. In the twenty years from 1893 to 1914, only one thousand were awarded. He was one of the able and lucky ones.[23]

Maurice enrolled at the prestigious Central Foundation Boys School on Cowper Street, at Shoreditch, just outside of the Jewish ghetto. He attended for the first two years of high school, from August 1913 to December 1915. The students were known as Cowperians or the Boys. The school offered prizes for history and geography, scriptures, English, drawing, German, physics, chemistry, and mathematics. Its best graduates moved on to Cambridge University and London University, often with special distinctions in the sciences and mathematics.

The Central Foundation School enjoyed a vibrant community of learning and doing. A school parliament offered students the chance to act out the roles of prime minister and cabinet posts, something like the later Model United Nations. There was a wide variety of sports: cricket and rounders, swimming, shooting, and gymnastics. Sports Day saw races and long jumps, tugs of war, and three-legged races. There was a range of dramatic productions: *Macbeth* and *Hamlet*, *The Pirates of Penzance*, and *The Mikado*. The chess club fared well in the London Secondary Schools' Chess League, often winning against the Strand School and Whitechapel Foundation School. The school's successful graduates, the Old Boys, sent in their reminiscences and travelogues about the British Empire, from exotic places like India, British Columbia, and South Australia.[24] Within this masculine culture of sport and empire, Zucrow probably first learned to box as a young man, either at the school itself or one of the nearby "lads clubs."[25]

The campus had been a place steeped in Anglican spirituality, though this was less pronounced in Zucrow's years. In 1900, 90 percent of the students were Christian, as were all the teachers, save the German language master. The few Jewish boys left for home early on winter Fridays for the Sabbath and had major religious holidays free. By the time Zucrow attended, Jewish students were making their mark, dominating the Debate Society and winning their propositions on some rather dramatic votes. In a series of debates in winter and spring of 1914: "Rabinowich moved, 'that a very large Navy was a necessity for England,' and Epstein opposed." Or "Rabinowich moved, 'that War, under any circumstances, is unjustifiable,' and Maccoby opposed."[26] This one debater, Simon Maccoby (sixth form), had already published a piece in the student magazine celebrating the French Revolution, which he described as an explosion of democracy and class struggle borne out of the "intense misery of the people, and the crushing weight of the tyranny of the nobles." He went on to the PhD, becoming a talented historian of English radicalism, a career track Zucrow himself thought worthy, as we explore later.[27]

As an East Ender, Maurice was also something of a Cockney. Given the kinds of neighborhoods in which the family lived over most of his fourteen years, and the schools Maurice attended, he broke out of the strict isolation of the Yiddish enclaves and interacted with all kinds of people, young and old. They were native-born English Protestants, or Irish and Italian Catholics, or Polish and Russian Jews. Each came from their own religious and ethnic communities, identifying themselves and identified by others as such, usually based on name, or accent, or dress, or religious observance. Sometimes these differences became tense or violent. Local Anglican children might display airs of superiority over Catholics. Irish dockhand workers sang ditties against the Jews. But these communities were also united by a shared place and struggle against poverty and ignorance. There were often common grounds for good humor, mutual respect, and tolerance.[28]

Though he did not graduate from Central with a degree, Zucrow still shared the honor of attending with several Jewish graduates. Selig Brodetsky, a 1908 Old Boy, went on to become a noted mathematician, biographer of Isaac Newton, and one of the founders of Tel Aviv University. Jacob Bronowski, a 1927 Old Boy, soon with a PhD in mathematics from Cambridge, was perhaps most famous for his work as an ethnographer, and his masterful book and television series *The Ascent of Man* (1974).[29] Like Zucrow, both rose from East End poverty and made remarkable careers in scholarship and civic service. Each went their own successful way: Brodetsky the scholar to the new state of Israel, Bronowski the social scientist to the radio halls of the BBC, Zucrow the engineer eventually to the aerospace industries of Southern California. The main office at the Central Foundation School did not share such high hopes for the young Zucrow. When he dropped out after two years, setting off for Boston, the school administrator filled in the career path box on his transcripts with these simple words: "commercial clerk."[30]

## BOSTON'S SOUTH SIDE AND THE WAR

The outbreak of the First World War in late summer of 1914 changed Maurice's life in a dramatic way. By Christmas, when the expected Allied victory had not come, Solomon sent the whole family to the US. They traveled in December from Liverpool to Boston. If moving to London was Solomon's great risk in life, moving to Boston was Dova's, crossing the Atlantic just a few months before the German Navy sank the *Lusitania*. She and the children moved in with relatives at Roxbury, Massachusetts, the war and their future unknown. The passenger manifest listed the "Zukroffs" as mother Daba, daughters Pesia (Bess) and Nehame (Nancy), son Moses (Maurice), and youngest daughter Sprane Lea (Lillian). The move was quite rare for the time, given the family's successes in London. Solomon expected the family to return after the war. Instead, they cajoled him to come to America and join them.

Solomon was wise. The family made it just in time. Within a month, by January, the German Reich began its aerial bombing campaign of London, which pummeled the city from zeppelins and giant bombers armed with aerial torpedoes, often falling on the East End. The war emptied London of young men, and in these raids, police whistles called women and children to the Tube for shelter.[31] Over the course of the next year, the Central Foundation School gave up its own sacrifices. By Christmas of 1915, several groups of students had "narrowly escaped destruction" in the raids. Youngsters offered concerts at the Bethnal Green Military Hospital for the wounded. Letters from the front and obituaries of fallen Cowperians, the Old Boys, filled the school magazine. The Roll of Honour counted the enlisted, wounded, and killed in action, many of them rank-and-file riflemen and privates, or minor officers. It held nearly five hundred names and covered nine pages of the magazine. Editorials encouraged students to be patriotic and hopeful but warned them to be prepared for a long war, one fraught with "very great anxiety and uncertainty," with "great sadness and trouble."[32]

By the time he arrived in Boston, Zucrow was an immigrant twice removed, eventually settling near the south side, in the working-class neighborhood of Dorchester, at 39 Wilcock Street. The London East Ender now became something of a Boston Southie. His was a life of accents, it turns out, as his later Purdue students were fond of remembering. He spoke with a little bit of a London accent, in his attention to the King's English, and with an obsession for proper spoken and written forms. He spoke with something of a Boston twang. From his parents and childhood neighborhoods, he was also fond of Yiddish expressions, as in his overuse of the "z" sound, or his reliance on the occasional maxim, accompanied by a shrug or a wave of the hand. Among his favorites were "It couldn't hurt" and "This too shall pass."

Zucrow enrolled at the Boston English High School, graduating in June of 1917. It was then located at Warren Avenue, right next to Boston Latin High School, and very much the poorer cousin of the two, though both were counted among the best in Boston. The two schools were old neighbors and old rivals. They shared the same building on Bedford Street between 1844 and 1881, then shared the same location in two separate buildings to 1921. The rivalry taught Zucrow some of his first lessons in the class system. Boston Latin was the oldest public school in the US, founded in 1635, where John Hancock and Samuel Adams both graduated. It became the college preparatory school for Harvard. Its classical curriculum required intense study of the humanities and sciences, with five years of obligatory Latin. It also demanded excellence in writing and public speaking, as in the declamations, recitals in class graded by style and delivery.[33]

Boston English, on the other hand, was founded first as a trade school to prepare young men for business, the advanced technical fields, and basic engineering. It had some of its own illustrious graduates. They included Samuel Pierpont Langley, director of the Smithsonian Institution at the turn of the century and a leading aeronautical pioneer; Frank Gilbreth, the time and motion engineer, later a guest lecturer at Purdue University, where his wife Lillian taught; and Louis Sullivan, the Chicago architect and pioneer of the steel-frame skyscraper. Zucrow remembered the mathematics and science courses as especially foundational. He also fortified his language and speaking skills: the basics of penmanship, spelling, and grammar. These were the very high standards that Zucrow later brought to his own graduate students at the Rocket Lab, much to their vexation, as he was a strict and demanding editor. Boston English also taught Zucrow to love football. Beginning in 1887, English and Latin played a great rivalry football game at Thanksgiving each year.[34]

The year after graduation was a difficult one for the young man, the US now preparing to fully join the European war against Germany. He could not afford college, spending the time instead at odd jobs, until he finally enlisted in the US Army in June of 1918 and was directed to the Student Army Training Corps (SATC) across town from his home, at Tufts College in Medford. SATC was a turning point for Maurice Zucrow, and for American higher education. The War Department established the corps under the authority of the Selective Service Act of May 18, 1917. The Committee on Education and Special Training organized it, which ultimately included 140,000 men at 525 colleges. The military's objective was to accelerate the preparation of young men into college life for wartime vocational training and college education, this for the preparation of specialists and officers in the fields of the sciences, medicine, dentistry, engineering, and administration. For their commitment to serve, enlistees enjoyed a number of rewards. Enlistment was voluntary, an act of patriotism and service in its own right. They could choose a preferred specialty and course of training. They could also get part of a college education for free, live near home, and delay being sent over to the actual fighting at the front.[35]

Local civic and educational leaders in the Boston area were keen on providing support for SATC. One meeting of the Boston City Club with local high school principals, including the principal of Boston English, represented some 27,040 eligible students. "There is an imperative demand," wrote city leaders, "for a multitude of young men that can 'do things,'" and only the high schools and colleges could muster them to training and action. The objective of the early SATC was practical. The prime beneficiaries, the colleges, had a patriotic interest, but a practical one too. The wartime draft had expanded to include the ages between eighteen and forty-five. This drastically reduced undergraduate enrollments and "meant the doom of the universities and colleges." As the young men went off to war, SATC saved some young men for war training, offering financial and institutional recompense to the colleges, even if by militarizing the campuses. From Zucrow's future school, Purdue University, President Edward Elliott advised the national College Section of SATC; and its dean of engineering, Andrey Potter, worked on the Vocational Section. Eleven hundred students enrolled there. Fraternity and rooming houses closed. The Army built barracks on campus. Life at Purdue was rendered into "a state of coma" and "took on a military air." Normalcy did not return until the next fall, after telescoped fall and spring semesters.[36]

SATC was a dramatic intervention by the federal state, through the US military, in the life of the independent university. Built into the program were elements of respect for college traditions and policies. SATC officers were to "observe the general usages" and to recognize the college as the educational authority. Their operating principle was innocuous: "to utilize effectively the plant, equipment and organization of the colleges for selecting and training officer candidates and technical experts for service in the existing emergency." This included the whole college curriculum: from the liberal arts to engineering, agriculture to pharmacy, business administration to law. Yet the commanding officers had a duty, at all times and places on campus, to apply military law and "enforce military discipline." Their SATC units were armed, engaging in a range of marches, drills, calisthenics, and firing practice. Drills comprised eleven hours every week. Coursework was set at forty-two hours, calibrated to the ages of the recruits and to a new quarter (rather than semester) system. The National Committee also had the authority to balance the hours of instruction for each week of training, and to suggest various allied subjects, like foreign languages and sciences. It also required the War Issues Course, part political science and part indoctrination, on why America was fighting the war.[37]

The Office of the Chief of Engineers at the War Department made some remarkable interventions, planning for a long and arduous war. It worked with engineering schools like Tufts to survey the numbers of undergraduates in each class, to canvass the academic policies and standards of the colleges, and to complete rankings of potential candidates (by age, academic, and character criteria) before they were allowed into the Enlisted Engineers Reserve.[38] SATC taught both the US government and university administrations to work together as partners.

For Zucrow, along with the Central Foundation School and Boston English, SATC was a dramatic new horizon, an even wider circle of experiences. There were units posted throughout the greater Boston area: at Newton Technical High School and Northeastern College, at the Franklin Union School and Harvard University. The enrollment lists included young men of Anglo-Saxon, German, Irish, Jewish, Armenian, and Italian descent. African American units were segregated at this time. Zucrow was assigned to Tufts College, Division 21. He was one of 1,014 enlisted there in the fall of 1918. He was one small part of a vast national cause of some 4 million troops. It was an experience both minimizing and magnifying. Zucrow joined recruits across the country in a powerful symbolic moment. They assembled at noon, all at their different posts, on 1 October 1918, for the simultaneous raising of the American flag, national anthem, and pledge of allegiance. They wore their new uniforms with a service hat and olive drab cord, along with a collar insignia of a bronze disk stamped with "US." They heard a statement from President Woodrow Wilson, that they had joined "the entire manhood of the country," as "comrades in the common cause of making the world a better place to live in." They were about to contribute to winning a new kind of war, "fought with all the devices of science and with the power of machines." [39]

He did not know it at the time, but SATC was also Zucrow's first education in the triumphs and travails of the US "Military Establishment," as the records called it at the time. It was an education that he received again in the Second World War, and once more in the Cold War. SATC was one of the first bridges between the government, military, industry, and academia. The initial planning was a collaboration of the War Department, US Army, Westinghouse Electric and Manufacturing Company, American Telephone and Telegraph, Carnegie Foundation, and leading universities like MIT, Case School of Applied Science, and the University of Pittsburgh. The SATC experience had its pitfalls, at least according to the administrators who managed the program. Some disorganization was inevitable. The War Department's priority was in preparing and sending troops and material across the ocean to fight the war in Europe. SATC came second, but with accelerated timetables all its own, as in organizing the system of training within a few weeks, then actually commencing the training of recruits within a matter of months. There was a lot to do: approve draft exemptions for the instructors, create curriculums and entrance exams, train officers to run the program, negotiate with the colleges, build all the facilities. [40]

Financial issues were especially complex and contentious. The SATC representatives and university administrators had to agree on purchase orders for equipment. They had to buy internal combustion engines to train mechanics, or electrical components to train wireless operators. They had to agree on contracts for labor to expand kitchens, dining rooms, barracks, and sanitation for thousands of men. There were detailed procedures on the installation and use, care and storage, and damage and return of government property. Most contentious were the per diem rates for room and board, which averaged between $1.00 and $1.50 per soldier but were open to all kinds of exceptions and experiments, as well as disagreements and rescindments, depending on what the men were eating or in what fields they were training. [41] At Tufts, the per diem settled at $1.60 a day, of which 60 cents was for tuition and the rest for housing and subsistence. Administrators there, as at the 684 other colleges and universities that contracted with the US Army, felt that they were being cheated. They had simply never had to break down unit and daily costs of a college education before. On one point they agreed. SATC was not at all liberal. The per diems were too low. [42]

Herman C. Bumpus, president of Tufts, could not restrain his criticisms and frustrations with SATC. "The Government has undertaken to do a perfectly stupendous piece of work with confusion in practically every administrative office," he wrote. Notwithstanding the best characters and intentions of the people involved, mismanagement and chaos were the rule. Tufts "is endeavoring to do all that she can to help win the war, but hardly a day has passed during the last two months when we have not received orders in contradiction of earlier orders, and regulations in contradiction of earlier regulations."[43] The war ended too soon for SATC to ever prove or redeem itself. After two brief months of its existence, it never had a chance to correct its inefficiencies or recover from them. Andrey Potter, Zucrow's future mentor and colleague at Purdue University, participated in the national organization and management of SATC while still at Kansas State and appreciated its model. On a tour of SATC facilities, he was amazed at how the vocational sections had succeeded at "teaching the man not the subject," and doing so with speed and efficiency. SATC offered a powerful means of instructing skill and building confidence. He also saw this practical dimension to learning as key to renewing the engineering curriculums in the universities.[44] This was a lesson he took to Purdue.

Others saw SATC as an exercise in nation building, one more in a series of foundational events going back to the great land-grant legislation of 1861 and the era of transcontinental railroads afterward. As Charles Franklin Thwing, president of Western Reserve University, put it, "The colleges became, like the railroads, essentially government institutions." This was all for the good, to invigorate college life by way of the "democracy of war and the democracy of education." The college undergraduate, like the soldier, would grow in pride of "his manlier bearing, his fuller chest, his larger and harder muscles, his clearer eye, his greater robustness."[45] For Zucrow, SATC was more prosaic. It offered him his first real pathway to higher education and career success. He also graduated as a private with an honorable discharge from the US Army.

By this time, Zucrow's father, Solomon, had already emigrated to the US. He arrived on December 14, 1917, the day before Maurice's eighteenth birthday, at a time when his son had graduated from high school but still had not started SATC and his university studies. Meanwhile, Solomon slowly made his own pathway for career success as one of the founders of the Hebrew Teachers College (Boston) and one of its first professors of Talmud and rabbinic literature. It opened in 1921 at a former synagogue at Fourteen Crawford Street, Roxbury, sponsored by the local community and the American Jewish Society. The college also turned some of its offerings into a Hebrew High School (*Prozdor*), this to enable children from the Talmud Torah schools to prepare for college, while also offering its own stand-alone and supplemental college degrees.[46] Here Solomon befriended and advised some of America's best scholars of Jewish religious thought: George Foot Moore, Harry A. Wolfson, and Nathan Isaacs at Harvard, and M. M. Kaplan of New York Theological Seminary. He was also the head teacher at Mishkan Tefila Hebrew School. As in London, so now in Boston, he gave freely of his time and attention, often tutoring his students on Saturdays or during summer vacations without pay, in honor of Jewish tradition. Only in the most difficult of financial straits did he accept payments, not for the gift of teaching, but according to one precept, for the time spent.

With most instruction in Hebrew, the college bridged tradition and innovation. There were innovative courses in history and linguistics, literature and social sciences, with public lectures by visiting Zionist and socialist thinkers. One alumnus called it a "well-rounded humanistic Hebrew education" centered on "aesthetic nationalism." Solomon was joining a faculty of well-traveled, notable scholars. The dean of the college was Dr. Nissan Touroff, a former imperial superintendent of Russian Jewish schools and a founder of schools and teacher colleges in Palestine (1907–1919). Among the teachers were Jacob Newman, an expert in Jewish history, and Israel Pollock, Hebrew linguist and author of several language textbooks.[47] It was a small faculty of Jewish intellectuals gathered from around the world.

Solomon and Dova opened their home to Talmud classes every week, centered on the companionship (*havruta*) pedagogy, a dialectical style that paired students to analyze and debate a reading, under the guidance of the mentor, thereby engaging and questioning and teaching each other. This was always followed by refreshments and camaraderie. His students remembered that Solomon's "patience bordered on the Saintly. One could see the joy on the face of this precious soul," a man with "a total certainty of his mission and values."[48] These were impressions shared by others at his death in 1932 at age fifty-seven. Hundreds came to his funeral to pay respects to this "true '*tsaddik*'" (righteous leader) and "saintly scholar." He was "beloved as scholar, teacher, author, friend," a "typical English gentleman," said one close friend. He was that "rare blending of the bloom of the finer flowers of Judaism and his surrounding civilization."[49]

This humility and kindness were expressions of deeper values, honed over a career that spanned the shtetls of Ukraine and metropolises of Kyiv, London, and Boston. In London, where he built his career over sixteen years, he learned how to be a man of the people. He mentored those who, in such crowded and impoverished conditions, saw their family lives and marriages dissolve, and desperately needed advice on how to adapt and survive given the very strict complex of Jewish customs and laws. In Boston, he published these teachings in a series of articles in the *Jewish Advocate*, later republished in book form, and which the editors praised as a "course in rabbinics," one that offered truth and justice above the "mystical and fanatic," a "learned and highly instructive discussion on the fundamental virtue" of the human person.[50]

Solomon admitted that he was not the distant rabbi of the synagogue and service, the stern judge of marriage and family law. He was "only a layman," a teacher and counselor. That was his charisma. His scholarly writings expressed values of tolerance and flexibility in the application of the Jewish law, if from an orthodox or conservative perspective. By this he meant authority balanced with compassion, teaching the "divine law" within the "new ideas and altered conditions" of the modern world. We are living in "a revolutionary period," he wrote, filled with "radical changes in public conceptions of right and wrong." The faith needed adjustment to these changes, not callous disregard. He drew in parallels from Greek, Roman, and English law. He cited the great rabbis of the Talmud on how "life and religion might always be harmonized." This was true for diet and behavior, for family life and financial dealings. It was especially true for the treatment of women, to protect them from "a life of misery and wretchedness," for example to allow them to remarry in order to fully provide for their children.[51]

Solomon raised the image of the woman and wife as the "helpmate and companion" of their husband. He advocated for the progressive values of mercy and generosity for the poor and downtrodden. His second book included an essay, "The Dignity of Man," on the human being as the "temple" of the whole universe, a person deserving of respect, not shame. For this he cited the Talmud and the Roman Catholic tradition, freely quoting Hillel the Elder and the poet Novalis.[52] Maurice, the son, paid to have his father's two books published, the second one during the Great Depression, when Maurice was in dire straits and out of work.[53]

One of Solomon's Boston students stands out above the rest. Solomon taught the Torah and Talmud to a young Leonard Bernstein, the future conductor. One family story even held that Bernstein originally dedicated his popular *Jeremiah* symphony to his beloved teacher, Solomon: "This piece is dedicated to the teacher who taught me Jeremiah."[54] The story made sense, given the composer's estrangement from his own father at the time. The end of the story, if true, rings even better: the reconciliation and final dedication of the symphony from the son to his real father, Samuel. This was a denouement with which Solomon Zucrow would have heartily approved.

## HARVARD ENGINEERING

SATC was a short assignment, what with the outbreak of the Spanish Flu by that fall of 1918 and the Armistice in November. Yet for Zucrow it was crucial. His courses at Tufts in the fall and spring semesters, encompassing higher mathematics (algebra, trigonometry, calculus), chemistry, and physics, all during his whole freshman year of 1918–1919, clinched his decision to get a college degree. He paid his way through Tufts by "working the night shift in the baggage office at the busy South Station, Boston," just at the edge of downtown, for twenty-five cents an hour. It was one of the largest railroad terminals in the world, with tens of thousands of passengers every day. At Tufts, he received good enough grades for a scholarship to either MIT or Harvard. Though most of his high school friends were already at MIT in engineering, he wanted more of a university campus atmosphere. So Zucrow chose Harvard instead.[55]

Zucrow was one of two Saltonstall scholarship recipients in the new engineering school, founded in 1918 after a brief experiment between Harvard and MIT on the coordination of engineering education. Harvard now instituted a broadly liberal curriculum, meant to educate "good engineers" as "good citizens." Some of his first courses were shared with Harvard College, the core undergraduate program of the university. Students took trigonometry, geometry, calculus, chemistry, and physics, along with humanities courses. With his head start from Tufts, Zucrow was able to enroll early in more advanced engineering courses at Harvard, like mechanics and thermodynamics. The centerpiece of the engineering complex, Pierce Hall, housed offices, classrooms, and a laboratory. The new Gordan McKay Engineering Laboratory offered top-notch laboratories and workshops and a machine shop to craft apparatuses. Here Zucrow learned the principles and practices that occupied the rest of his career, even far into his Rocket Lab years. He learned heat engineering as applied to steam boilers and engines, as well as to internal combustion engines. He studied and used the instrumentation essential to analyze and measure heat combustion. He learned hydraulics, the "flow of water through orifices, nozzles . . . reciprocating and centrifugal pipes, turbines." He learned the properties of materials and even cryogenic engineering, "the production of very low temperatures, liquefaction of air and other gases."[56]

The curriculum included a significant cooperative dimension, a program modeled after the University of Cincinnati's pathbreaking "Cincinnati Plan" (1906). At Harvard it required at least six months of supervised work and actual experience in an industrial setting, usually during the third year and over the course of two summers. This alternated "two months of class-room and laboratory work and two months of industrial training and engineering practice." The objective was to educate the student in scientific principles and how to apply them "intelligently to the solution of entirely different kinds of problems." Harvard wanted to inculcate a "sympathy with human problems," to "teach men to think rather than to cram their heads with facts." They needed to become "real engineers . . . not handbook engineers." The logic here was that "industrial plants, transportation systems and public utilities" had all "become increasingly complex," both in the sophistication of their technical and design aspects and in their labor and management issues. The program sent students into industrial enterprises to learn the whole scope of their operations: work, manufacturing, testing, and management. With research questions in mind, they worked in the enterprises, wrote detailed technical reports (with sketches and photos), then submitted them to the class for analysis and debate.[57]

Zucrow participated in the cooperative option, which sent him to the mills and factories of Lowell and Lawrence, Massachusetts. He was an apprentice in the machine shop at General Electric in Lynn and also learned shop work techniques at Cambridge Technical High School. Over the summer of 1920, he was in classes and shops all day, writing detailed reports at night. They became the models for the very reports he later wrote at the Aerojet Corporation and the Purdue Rocket Lab. In the summer of 1921, he studied the resistance of materials and water power engineering, with visits to some of the very sites designed by his teacher, H. M. (Howard Moore) Turner, a hydraulics engineer at the Turners Falls Power and Electric Company. As Zucrow remembered, with some modesty, and perhaps still with a glimmer of amazement at the power of the industrial revolution: "I learned a great deal about the design, behavior and capabilities of machines, and certain processes. I also learned about the behavior of different materials under a variety of manufacturing conditions." By his senior year, and then during a one-year master of science program (1923), he studied internal combustion engines under Lionel S. Marks, professor of mechanical engineering, a fellow Englishman whom he remembered as "a first-rate lecturer." Zucrow graduated in 1922 with a bachelor of science in mechanical engineering and with a specialty in automotive engineering, and a year later with a master of science in heat engineering and internal combustion engines.[58]

These were exciting if trying times for Harvard. Under President A. Lawrence Lowell, the administration experimented with a new trend toward democratic and assimilatory values. Lowell was a fierce advocate for freedom of speech, even for the radicals and socialists with whom he disagreed. These were the years of the Red Scare and rising nativism (against Jews, Catholics, and African Americans). The Harvard Liberal Club, with its weekly luncheons and public lectures on socioeconomic and political justice issues, was branded as Bolshevist and anarchist in the local papers. It held a free-speech banquet in the winter of 1920 to protest the Palmer Raids of the US Department of Justice. The campus also hosted human engineering talks by the likes of Frank Gilbreth and lectures on labor activism by Sidney Hillman, along with the occasional speech by a left-wing "flaming crusader" advocating prison reform, or revolution.[59]

Zucrow enrolled at a time of peaking Jewish enrollments, rising from 7 percent in 1900 to 21 percent in 1922, a trend that frightened some college administrators, as if Harvard might be swallowed up by an immigrant wave. Elite and small liberal arts colleges around the country also leveraged against the trend with quotas, restricted admissions, and varieties of anti-Semitism. Lowell also sought to limit Jewish enrollments, fearing Jews would not assimilate but isolate.[60] It is interesting to note how wrong he was, at least in one case. At Harvard, Zucrow fell headfirst into the American experience, remembering his own naturalization ceremony, on November 28, 1921, as a second birthday of a kind. He also served as a witness when his father was naturalized, just after he graduated with his master's degree.

Zucrow actually entered Harvard thinking he would study history, which to him meant to earn a PhD in order to teach at the college level. "I started out in college with the idea of becoming an historian." But the prospect of so many more years in school frightened him somewhat, and so he took what he thought was the easier course in engineering. "My reason was financial; it was the only profession that one could enter with only four years of college education."[61] He had fallen in love with history at Boston English. From childhood, he also recalled readings of Jewish history and folklore, as from the Pentateuch and other religious books. His father often told stories from the Mishnah (the first part of the Talmud and its oral traditions) about rabbis and wise men or the great Hebrew kings, and of the travails and triumphs of the Jewish people. At Harvard, with some of his friends in the humanities, Zucrow sat in on a number of their courses to offer some respite from the tough basic engineering studies. Over forty years later in his retirement interview he still remembered the courses taught by leading Harvard professors. They included William McDougall, professor of psychology, a fellow Brit and advocate of an activist emergent evolution; George Lyman Kittredge, professor of English literature and editor of the works of Shakespeare; historian Robert Howard Lord, who taught European and Russian history, authored major studies on Russian imperial foreign policy, and was a former member of the Paris Peace Conference; and Arthur Eli Monroe, who taught the courses Economic Thought and Institutions in History.[62] Zucrow heard them all.

Had he stayed in history, he might have taken a course of study under Roger Bigelow Merriman, who was already publishing his four-volume work *The Rise of the Spanish Empire* (1918–1934), noted for its narrative vistas upon the European conquest of the Americas. Or with Frederick Jackson Turner, whose monograph version of *The Frontier in American History* appeared in 1920. Turner was famous for defining the western frontier as the source of American energy and dynamism, a frontier now complete and closed. For Turner, America needed new intellectual horizons. He was the talk of campus in these years and may have inspired Zucrow's later choice of Purdue University and graduate work in engineering. To the Midwest Zucrow went, where "lies the strength of America," according to one Harvard appeal. Or as he later told his own students, "Our new frontiers are no longer government lands, but the frontiers of science and invention."[63]

Zucrow never lost his sense of history, or his love of civics, as illustrated by one of his favorite books. It may have been a bit outdated, but it was one he treasured, by his fellow Harvard alum George S. Morison, *The New Epoch as Developed by the Manufacture of Power* (1903). Drawing from a nineteenth-century optimism in the industrial revolution, Morison defined engineering "as the art of directing the great sources of power in nature for the use and convenience of man." Zucrow similarly asked his students, "What is the professional engineer?" His answer: "An applied scientist, one who applies scientific knowledge to the uses of mankind." He was squarely within the progressive engineering tradition, one dedicated to "personal integrity, independence, and social responsibility," as Edwin Layton has written. "As stewards of technology," engineers saw themselves as "agents of social progress."[64] This issue of application, by adapting ideas and designs to actual processes and products, was by no means simple. It meant bridging theory to practice in real time. It meant the weighty responsibility to make things safe and useful for society at large, often in times of war and crisis.

KYIV, LONDON, BOSTON. THIS WAS QUITE A SERIES OF MOVES WITHIN TWENTY years. In each case, the driving force behind the migrations was father Solomon, ambitious for education and advancement, for himself and his family. The support of Dova's clan, the Smushkins, also mattered. This included her father, the patriarch Shalom, who financed the moves to Kyiv and London in the 1890s, and the various Smushkin relatives in Boston who welcomed Dova in 1915. The Smushkins were prolific. Zucrow's maternal grandfather, Shalom, did not have a large family for the time, but their descendants, through daughters Dova and Chava, and son Iakov, migrated far and wide within Russia and across the seas. Chava (Zucrow's aunt) married Zalman Feiman, had eight children, and lived in Zvenyhorodka, Ukraine; her son Solomon Faiman (wife Ida) migrated to Baltimore in 1921, where their son-in-law established a metal plating business, the Almag Corporation, with several important Cold War contracts. Shalom's son, Iakov (Zucrow's uncle), managed a flour mill and was a business agent for the local Polish magnate. During the battles of the Russian Civil War, when the White armies occupied Stavyshche in 1919, Iakov was one of a committee who paid a ransom to save its Jews, though bandits eventually killed some one hundred in the nightmares of later assaults. Iakov's son Leib migrated to Boston in the later 1920s, changing his name to Louis Smullin, and whose children included sons Ben and Sam Smullin. A distant cousin of this wing was Dr. Louis Smullin, a renowned MIT electrical engineer and Zucrow contemporary.[65]

These kinds of pathways were multiplied by thousands of lives at the turn of the century. They included the Jewish engineers and scientists who came to or were educated in the US, the very ones Zucrow met through his engineering career. People like Purdue dean of engineering Andrey Potter and head of the physics department Karl Lark-Horovitz; or the brilliant Caltech aerodynamicists and founder of Aerojet, Theodore von Kármán; or leading engineers at the National Advisory Committee for Aeronautics (NACA), Ben Pinkel and Abe Silverstein. Like Maurice, these were Jewish migrants, or sons thereof, who were devoted to acculturation and its highest forms, scientific learning and civic duty, and who made significant contributions to aerospace engineering in the US.[66] They were also *research* scientists and engineers, a term that perfectly describes their temperaments and careers: to search and traverse, discover and contribute to the modern world, with some inspired old-world values in mind.

This takes us back to the notions of the aerial man (*luftmensch*) and muscular Judaism. Maurice never addressed them in his writings. Still, he matched the classic profile of the modern, "muscle Jew," dedicated to physical fitness and strength, self-advancement, and social consciousness. He lived it, as a boxer and railway worker, as army recruit and practical engineer. He worked it through his love of engines. Engines for cars, airplanes, and rockets. His was a passion for speed and airy freedom. This was part of his family milieu, as among his own sisters, who were a dynamic group. Nancy and Lillian were activists, writers for Boston's paper the *Jewish Advocate* on social and literary causes. Lillian was also a graduate of Radcliffe (in the liberal arts, with distinction) and of Simmons College (in the sciences). Bess had perhaps the strongest spirit of adventure, traveling to Jerusalem in 1920, where she worked in the Palestine Secretariat of the Jewish Agency, to defend and promote the land of Israel. She later wrote Maurice from Addis Ababa, where she was living with her husband, Nathan Marein, then serving as attorney general and president of the Supreme Court of Ethiopia: "We are both very proud of little 'Israel.' It sure is showing the world that Jews are not cringing animals like the anti-semites would have the world believe but virile men and women who lay down their lives for their country and flag."[67] Bess and Maurice were close. She helped him to prepare and type the two volumes of his *Aircraft and Missile Propulsion* (1958) for publication.

# 2

# PRIME MOVERS

## Purdue University and Its Power Plants

PURDUE UNIVERSITY, LOCATED AT THE CITY OF WEST LAFAYETTE IN WEST central Indiana, enjoys a unique role in the conquest of two of America's great frontiers: the Midwestern prairies and outer space. They may seem completely unrelated. But they were both vast new frontiers, at first appearing empty, soon to be occupied by settlers and their machines. Purdue helped master the technology of the steam locomotives that traversed the forests and prairies of the west. The university's engineers and graduates also helped to create the technologies that conquered outer space, on the way to the Moon landings.[1] Most of Purdue's contributions to locomotive or rockets were not very dramatic. They were more routine affairs, the enterprising research and teaching that prepared engineering students, and the nation, for a new age of machines.

As early frontier institutions, both the state of Indiana and Purdue University were part of a unique experiment in republican self-government in the Ordinance of 1787, which established the Northwest Territories on paths to statehood, apportioning parts of their public lands for educational use. Federal legislation later marked these territorial and intellectual frontiers for development. Amid the struggles of the Civil War, an independent Department of Agriculture began to actively promote modern farming. The Homestead Act (1862) encouraged internal improvements, farm ownership, and forty acres to any family willing and able to cultivate it. The related Morrill Act (1862) authorized and funded the new land-grant colleges, devoted to agriculture and the mechanical arts. Purdue was rather unique among the land-grants, as it bridged both public and private endowments, first founded by the philanthropy of John Purdue, a local farmer and railroad promoter in Lafayette.

The conquest of the North American frontier claimed its high social and political costs: displacing Native Americans from their habitats, extending African Americans in slavery in the South, and fostering such corruptions as land speculation and pollution. Purdue's first historians, William Murray Hepburn and Louis Martin Sears, still valued the foundation myth of an "idealism acting on a warm-hearted race of pioneers," dedicated to the cultivation and manufacture, the building and progress of the settled frontier. They identified Purdue as part of an American "enlightenment," a pragmatic "People's University" that "opened doors of opportunity to the plainer type of citizen," by way of the "agricultural betterment" and the "liberalizing of the mechanic arts."[2]

## THE BOILERMAKERS OF PURDUE

As a land-grant institution, Purdue shared more with the railroads than we often realize. The railroads were land-grants too. Through the nineteenth century, the federal government apportioned "full and final title to more than 130,000,000 acres of public land" to the railroads, as historian John Stover reminds us. "Federal land grants aided in the building of 18,000 miles of track in 26 different states." The government realized its return on these investments when the number of individual farms tripled between 1860 and 1900, opening a whole new era of commercial farming.[3] Purdue's fortunes rose with these new tracks and farms. By the first decades of the twentieth century, it was one of the country's premier technical universities, in its own words, best known for its studies of the steam engine, the world's most efficient prime mover. This is not a phrase we use much anymore. But back then, the power plant as prime mover expressed the wonders of the industrial age, the machine as our driving force. Seemingly invisible, within factory walls, or behind locomotive cabins, or under car hoods, these engines kept America on the move.

Prime movers sometimes referred to people. Purdue people, that is, also known as Boilermakers. The term began as an insult, one that Purdue adopted as its own, even as a point of pride. The newspaper from nearby Wabash College first coined the term in print to bemoan a series of bruising football defeats between 1889 and 1891. It was hurled along with "pumpkin shuckers," "hayseeds" and "rail splitters." Purdue students accepted "boilermakers" as their own, a mark of honor for the engineering school it was becoming.[4] The term made sense, named after the craftsmen who made boilers and furnaces for industry. Purdue was, like them, gritty and hard-working, a university with something to prove.

Professor William Freeman Myrick (W. F. M.) Goss staked his reputation on this Boilermaker tradition. With a PhD from the Massachusetts Institute of Technology (MIT), he was a Purdue instructor in practical mechanics (1879), then the first dean of the schools of engineering (1890). Goss rededicated Purdue to the new "engineering sciences," a fortified undergraduate education balanced with research.[5] He founded Purdue's Locomotive Testing Laboratory in 1891, housing the test locomotive Schenectady No. 1. It measured the test variables that were otherwise impossible to gage on a train moving along normal tracks. Thus, "Purdue was the first institution in the world to have a locomotive on wheels on a stand with brakes, and it covered tens of thousands of miles without moving a fraction of an inch."[6]

The colorful Goss framed the test locomotive's arrival from the Schenectady Locomotive Works in New York State as a campus origin story. It arrived at a railroad switch nearby, only one and a half miles from its future home, at the central Heavilon Hall, itself built in the style of a nineteenth-century railroad terminal. But wheat fields and clover pastures obstructed the way. And the locomotive weighed eighty-five thousand pounds. So teams of workers had to build temporary skids to move it along, over those fields and a few gravel roads, pulled by three pairs of horses. The workers moved the skids every few feet, a gauntlet of push and pull. At the four right-angle turns, they used "cross-blocking" under the tracks. As the locomotive reached the threshold of campus at the State Street main gates, the administration canceled classes, allowing the student body to pitch in, as if dragging Purdue into the twentieth century. And then, as Goss wrote, in biblical terms, "on the eighth working day after the start, the engine arrived at the laboratory."[7]

As an amateur historian, Goss framed the Locomotive Lab as a new chapter in world history. He gave generous recognition to Alexander Borodin, chief engineer of the Russian Southwestern Railway, who later visited Purdue, for establishing the first laboratory tests on locomotive engines. Goss also recognized Purdue's Locomotive Lab as a turning point in American history: between the closing of the American land frontier in the west and the opening of new frontier in engineering science. The Schenectady, like most locomotives that conquered the west, was a 4-4-0, or "four leading wheels on two axles." It was a versatile machine, with its "ability to run over rough track, cheapness, reliability, and ease of maintenance." Now that the land frontiers were fulfilled, Goss and his Purdue teams turned to improving or replacing the 4-4-0, opening a new era in advanced locomotive engineering, especially in the fields of "high power and fuel efficiency."[8] The US needed to modernize its railroads.

The first Locomotive Lab burned down, along with Heavilon Hall around it, on the night of 23 January 1894. It happened only four days after the lab was formally dedicated. The fire started, of all places, in the boiler room.[9] But Purdue rebuilt Heavilon Hall "one brick higher," as the new slogan went, bell tower included. Mechanical Engineering eventually brought in three more Schenectadys as experimental locomotives, along with a variety of other models.

In terms of engineering science, Goss refined his own pathbreaking studies in heat transfer and gas dynamics, establishing a legacy of studies leading all the way to Zucrow's future Rocket Lab. His tests on superheated engines, beginning in 1906, added heat to the steam, increasing its temperature above the boiling point in order to improve engine efficiencies. As one historian explained, "Goss tested a Cole-equipped engine, in which superheat temperatures were about 150 degrees above saturation, at various boiler pressures." This "engine gave a fuel saving of 13 to 17 percent compared to the same locomotive running saturated as originally built. . . . The 13 percent was obtained at 200 psi" (pounds per square inch). Although not immediately applicable to actual practice, Goss's results were widely disseminated and influential.[10]

Most of all, Goss had primed Purdue for research. He even used a wind tunnel and mathematical models to study the aerodynamics of the locomotive in action. Halvor C. Solberg, who earned his BS and MS degrees in the summers of 1895 and 1896, conducted most of the research. The Wright brothers had not yet even begun their own aerodynamic experiments with wind tunnels and airplanes. Yet Goss and Solberg had already applied wind tunnels to study the motion and drag of trains. Designers later applied their very principles to create the streamlined "aerial" locomotives of the 1930s.[11]

The Purdue Locomotive Laboratory, devoted to these advances in "experimental engineering," was a national pioneer. The American Railway Master Mechanics' Association called its work the "most important" in the country. President James H. Smart of Purdue even leveraged its successes to move the university into the extension work of research partnerships with industry. He wanted to build a series of national "mechanical institutes" to educate the young, teach them to engineer the next generation of shipyards, bridges, ocean liners, and factories. In effect, to rebuild the world.[12] Dean John Burgess of Columbia berated such optimism. Purdue and the other land-grants were mere "vocational" schools, he claimed, not "true" universities like Columbia, Harvard, and Yale. Yet Purdue's research initiatives were part of a national trend, as when MIT built its famous partnership with the new General Electric Laboratory in 1900; or when Western Electric, DuPont, and Eastman Kodak soon followed with research labs of their own; or when nonprofit research institutes like the Mellon in Pittsburgh and the Battelle in Columbus, Ohio, formed industrial and educational alliances as well.[13]

Purdue's foundational strength, in agriculture, was not detached from engineering but intimately connected with it, thanks to the provisions of the Morrill Act. There was no prosperity for the farmer without mechanization of the farm itself, and of the transport systems around it. Mechanization, at least at first, meant power by steam engines.[14] The Department of Agriculture also taught Purdue its first lessons about educational outreach and research, thanks to the Hatch Act (1887) and the Adams Act (1906), which established the Agricultural Experiment Stations in the land-grant universities, centered on the research "project." The government was creating and financing a network of expert academic specialists, called "investigators," as well as county agents, for farm assistance and innovation. Professor Alfred C. True, dean of the Graduate School of Agriculture at Ohio State University and one of the drafters of the Adams Act, defined the "project" as research that "presupposes a definite aim and a definite problem to be solved, a specific end to be attained rather than a mere accumulation of data."[15]

The project was pragmatic, applied research. The accent was on competence rather than competition, buying the country's best advice rather than the cheapest.

The Smith-Lever Act of 1914, though scarcely remembered nowadays, expanded these initiatives with the revolutionary project system. The emphasis was on research for "the development of practical applications of research knowledge," as well as for "instruction and practical demonstrations of existing or improved practices or technologies in agriculture." The project system created verification mechanisms to ensure that government monies were being spent wisely and efficiently. Smith-Lever redefined projects as "plans for work," delivered from the university to the Department of Agriculture before any budgets were approved. Projects became obligatory sets of discrete "tasks," experiments, and reports. They had to set forth a "definite effort," one both "scientifically sound and technically fit." They had to be both flexible and equitable, a mark of the "sympathy of purpose" between the government as patron and the university as recipient. A project was not a "desultory, rambling, vacillating effort" but had "a clear purpose" and was "restricted in scope, and systematic in its plan." It was "designed to be constructive, and to advance on the basis of what is already known."[16] The government was now in the business of sponsored research.

Universities around the country went project crazy by the fall of 1914. The Department of Agriculture sponsored some 2,300 of them. There were normal projects, special projects, sleeping projects for those on pause, projects for farm women, and even projects for projects, as when an agricultural extension agent might need to finance one to help manage the others. The results were promising. T. L. Haecker of Minnesota, an old-school researcher in dairy husbandry, did not even know what the term "project" meant a year before. But now he was following the rules and giving his work focus, scope, and continuity. William Frear, a professor of agricultural chemistry at Pennsylvania State College, agreed: "We like the project system." It was orderly and systematic, avoiding duplication and leaving a solid record for communication and cooperation. The "project basis" meant that "whoever is in charge of it shall present some well-matured plan of action that will be purposeful and lead to some definite and progressive end." By definition, projects were temporary, meant to begin and end, and to adapt to new circumstances and needs, what the system called "shifts of emphasis."[17]

Indiana was one of the wealthiest states in terms of the Agricultural Experiment Stations and the project system, taking in sales and revenues of over $385,000 during 1915. It was followed closely by Ohio and Minnesota. Here was a windfall of new funding, and a spiritual mission for applied science. "The ideal investigator was to be a crusader," said one administrator, "who enlisted with almost religious fervor under the banner of truth." In answer to Dean Burgess of Columbia and the elitism of the Ivy League universities, this was not "pure science," nor "science for science's sake," nor "a thing apart and for a special class." It was science in a "necessary connection with the world of concrete things."[18] Purdue joined in the crusade, focused on "getting research information to the farm populace." Smith-Lever filled the campus with research specialists connected to a wide network of county agents, overseen by John Harrison Skinner, Purdue's dean of agriculture. In all, as historian Fred Whitford has summarized, the nationwide program amounted to a "watershed" of $45 million, in "one of the largest expansions of information exchange and technology transfer" in American history.[19]

Engineers soon formed their own Engineering Experiment Stations on the agricultural model. This made perfect sense, given the terms of the Morrill Act and the state of the country at the time, transitioning from a rural to an urban economy, in the midst of an industrial revolution. There were engineering tasks to be done on and near the farms: as in mechanization and electrification, or irrigation and water purification.[20] The University of Illinois and Iowa State College established the country's first Engineering Experiment Stations by 1904, followed by dozens of others at places like Penn State and Kansas State, Wisconsin, Texas A&M, and Purdue. Industry funded most of these contracts for engineering science and research projects, in smallish budgets of about $10,000, with studies of highway improvements, hot water furnaces, and metal fatigue.[21]

The First World War also set a new pace for academic research, what with the establishment of the National Research Council (NRC) in 1916, to mobilize and manage national scientific and technological resources in wartime. The presumption was that "competency for defense against military aggression requires highly developed organized scientific preparation. Without it, the most civilized nation will be as helpless as the Aztecs were against Cortez." The US joined the military race for scientific achievement.[22] Vernon Kellogg, secretary of the NRC, challenged industrial research laboratories to take up the task of research and development for product improvements and profit. The universities ought to focus instead on the "development and training of new research workers" for "scientific research and discovery." Here was an imperative born out of the war, to organize "research and training for research," to quickly "mobilize" science for the national defense, to attack "large scientific problems" with teams of well-educated scientists.[23]

This became Purdue's mandate. It turned into Zucrow's windfall. His choice of Purdue in 1923 may seem odd. Why did he not stay in Boston or on the East Coast? He did well at Harvard, one of the first two magna cum laude graduates of the new engineering program. He finished his MS in engineering there within a year of his BS. Yet Harvard did not have a full graduate program in engineering until 1934. Purdue offered him continuing graduate studies. It also gave him a job: as a research assistant (1923–1926) then associate (1926–1929) in the Engineering Experiment Station. He found a vibrant Jewish community on campus, as in the Sigma Alpha Mu fraternity, where he was an adviser. He was also an instructor in Mechanical Engineering. Networking mattered. Zucrow's mentor, Lionel Marks, sent his name and reference along to G. A. Young, head of the School of Mechanical Engineering (1912–1941), an old Harvard classmate. He was the one who did the hiring. Known as "Power Plant" Young, he was also an engineer in the Goss tradition, a patron of engineering shops and hands-on learning. True to his nickname, he also lobbied industries to send their boilers and engines to Purdue for study and improvement.[24]

Into the 1920s Purdue was still a rural campus. Its main entrance on State Street at Oval Drive was framed by vast fields, hedges of large cedar bushes, and rows of grand elm trees. Not all of the sites were that elegant. Campus buildings rose starkly from the fields without any landscaping. The steam pipes were above ground, lacing the campus with an unattractive grid of metal parts. The vast majority of the student body lived in fraternity houses or private homes.[25] According to one administrator, the Purdue student body was not in much better shape: "The most outstanding characteristic of the institution is a lack of discipline." The student body was disorganized and disparate. "The lack of any real traditions and the insular apathy with regard to the creation of any is serious." Students were often late for class, ignored their appointments, and disobeyed campus rules (like no smoking). As the administrator concluded, "There is entirely too much soft-headed and wet-eyed small town 'Christian inspiration' and not enough salesmanship of Purdue and guiding of manners."[26]

Purdue was a largely Protestant, Anglo-Saxon, middle-class community. This alone, according to several leading sociological theories of the day, set the university up for success. The mainline Protestant traditions, so these elitist theories held, predisposed the campus to scientific values, thanks to their associated tendencies toward naturalism and secularism, rationalism and empiricism, utilitarianism and individualism. Like other Midwestern and Southern schools Purdue was reputed to enjoy the benefits of the Protestant work ethic and its "sobriety" and "thrift," as well as the tolerance and hospitality of small-town democracy. These were places with a "high degree of social mobility and minimum class consciousness and traditionalism" and welcoming of Jewish and Catholic migrants. They were places where science meant "advancement in prestige and economic status." Purdue was part of an "inward turning" of the physical frontier to the frontiers of science and the mind, an "intellectualizing of frontier values," on the model of Frederick Jackson Turner's famous thesis.[27] These, at least, were some of the aspirational values of the university community.

When Zucrow arrived at Lafayette in 1923, the country was just beginning a boom in college enrollments. They doubled at campuses like Purdue, the University of Illinois, and De Pauw, seeing steady growth in others. America entered an "era of mass higher education," filled with new opportunities, if still mostly for the upper and middle classes.[28] Purdue's new president, Edward C. Elliott (1922–1945) took full advantage of the trends. He was man from the West, raised in Nebraska, later working as a science teacher in Colorado. After he received his doctorate at Columbia, he became a college administrator in Wisconsin and Montana. One of his favorite books was Turner's *The Frontier in American History*. Thanks to his leadership, Purdue became a transformative campus, a university constantly on the move. Elliott created a spirit of democratic cooperation at Purdue. As he wrote, "In a very literal sense the vital essence of the University is the leadership of all.... A leadership by which one acts for the many, and the many act as one."[29]

Elliott became a national reform leader in higher education, not in any grand manifestos but in a host of rather simple and practical studies of the mechanics of college administration. These included a guide to the everyday functions and purposes of a board of trustees, and a pathbreaking study of the "unit costs" of a university based on the actual statistics of revenues and costs, including such arcane issues as overhead and depreciation.[30] Elliott was also patron of a new unit on campus within his own executive office: the Division of Educational Reference (1925), which conducted opinion polls and research studies of student attitudes, behaviors, aptitudes, and learning. Soon under director Hermann Remmers, professor of psychology and education, the division published these studies in its nationally acclaimed *Studies in Higher Education* (1926–1968).

At Purdue, Elliott assembled a powerful team of administrators, the very people who mentored Zucrow through his graduate studies in the 1920s and worked with him as his colleagues after the 1940s. This included Robert B. Stewart, the university's first chief business officer (1922–1963). Stewart, also known as R. B., was a person of sharp words, always demanding and often impatient. He created a model financial system for Elliott based on exact unit costs. He also invented a unique system of "open-end" bond financing for dormitories, called the Stewart Plan, treating them as if they were public utilities, based on their guarantee of steady income and an alumni base. With State of Indiana approvals, he applied a $50,000 donation, with $100,000 in bonds, to build the impressive new Cary Hall. It was more than a dormitory. By R. B.'s own philosophy of education, it was a residence hall, a place for student living and learning, supported by a wider infrastructure of dining, recreational, and medical facilities, and centered on the new Memorial Union, a model for the country at large. For students who could not afford the residence halls, he created a system of housing cooperatives (Purdue Student Housing Corporation), for which they contributed time and labor in exchange for cheaper rents and fees. He framed all this in an educational philosophy of the wholistic university, an organic community of giving and receiving.[31]

Andrey Potter arrived to become dean of engineering at Purdue in 1920 by way of Kansas State Agricultural College, where he had helped to create its Engineering Experiment Station and served as dean of engineering (1905–1920). A Jewish immigrant from the Russian empire, born at Vilnius, Lithuania, he remade himself in the US, eventually becoming a Presbyterian, Rotarian, Republican, and Mason of the thirty-third degree. Potter was a renaissance thinker, enamored of American democracy and free enterprise. His heroes were Benjamin Franklin and Thomas Edison. He was also a perfect fit at Purdue. After graduating from MIT (BSME 1903), he had worked at the General Electric Research Laboratories at Schenectady in the new steam turbine department and later at the engineering labs at Lynn, Massachusetts, becoming a national expert in steam power and agricultural machinery. Ever the historian, when dean W. F. M. Goss invited Potter to give his job talk in March of 1920, he did not ask for an engineering lecture. He asked for one on the history of the Russian revolution.[32]

Purdue's Engineering Experiment Station (EES) was already in place when Potter arrived, established in 1917 on the guiding principle that the faculty alone were to choose "the fields of productive research," so long as they were "of sufficient dignity and scientific character."[33] Potter transformed it into the country's largest and richest. He increased sponsored industrial research from $8,750 in fiscal year 1922 to $275,000 in 1929, "more than half of the outside research funds at all stations combined," according to historian Bruce Seely. His closest rival, at least in terms of funding, was his alma mater, MIT, whose Division of Industrial Cooperation and Research made $290,000 in 1929.[34] Potter soon surpassed it. Between 1922 and 1932, his EES brought in a total of $1.4 million in "contributions by industries, railroads, and public utilities." This money funded "cooperative projects" in contract with outside companies and agencies. The largest contributor by far was the American Railway Association, at $950,000, and which even financed a small building on campus.[35] By 1933, 90 percent of EES earnings came from outside the university. These financed a staff of associates and assistants, as well as graduate student fellows. They paid for EES publication runs: *Bulletins* filled with "original work"; *Circulars* with "compilations" of data for ease of public use; "speed bulletins" rushed into print for high-demand topics. The EES also held myriad public conferences, short courses, special classes, and lectures. Most were for a fee. Potter was priming Purdue for the business of research.

Potter also created small "station projects" based on faculty interests, some of them to fund undergraduate and graduate theses, as long as they proposed a definite timetable, budget, methodology, and expected results. He paid for them with state monies in the budget, or from EES earnings. He simply took a portion of EES profits, or part of the set charge for university overhead (a charge for the general costs of operating the campus laboratory), and created a fees fund for internal use.[36] The overarching philosophy of the Purdue EES was to apply cooperative research to educate engineers for industry. Potter's research projects prepared future engineers in the free "atmosphere" of the university laboratory, along "new paths" of knowledge. The very engineers who would "extend the frontiers of knowledge" and "develop new processes for the conservation of our resources and for the elimination of waste." It was all a radical experiment to transform Purdue into a laboratory for the state and nation, something like a local

Bureau of Standards or even a Mellon Institute. After all, wrote Potter, "Engineering is a profession of progress."[37]

In 1924, the year after Zucrow enrolled, Purdue also built its new Heating and Power Plant under Potter's guidance. He appointed his former student and current Purdue instructor, Harry L. Solberg (MSME 1923), to build it. The new power plant was his master's thesis and PhD dissertation rolled into one. Behind the decorative façade were a 75,000-gallon water tank, a three-foot switchboard and generator, and "four batteries of two boilers each," with eight stoker-fired boilers in all. Underneath the plant were four wells that pumped 750 gallons of water into the plant every minute. A railroad cut through campus to transport coal to the plant. By the 1950s, it was a 44-ton diesel electric locomotive, carrying a ton of coal a day, deposited in a great pit behind the plant, ten feet deep and holding up to 8,000 tons. An automated system transported the coal, crushed it, and sent it to the stokers for burning.[38] All of this noise and dust was at the heart of a busy university campus.

The Power Plant became a symbol of the university's engineering prowess, another apt confirmation of the campus moniker, Boilermakers. Potter and Solberg built it in a rectangular ziggurat style, with ornamental brickwork and long arched windows, punctuated by a smokestack that was 15 feet wide by 250 feet high, half the height of the Washington monument. The University of Chicago, Washington University in St. Louis, and Ohio State University all built similar power plant monuments through the 1920s. As one professional journal extolled, "Chimneys to give draft must reach far upward, like church spires or memorial towers."[39] Purdue's "powerhouse" or "smokestack," as students called it, towered over campus, soon becoming its geographic and symbolic center. It "may leave something to be desired" as far as beauty, said one nostalgic alumnus, but it was a familiar and welcoming sight. The Power Plant was originally at the edge of campus, behind the engineering complex at Heavilon Hall and the engineering workshops. President Elliott built a new Purdue around it, as a campus of factory-like brick buildings, in the collegiate Gothic and Romanesque styles.[40]

The Power Plant was also a working laboratory for undergraduate instruction and graduate research, a key to Purdue's dedication to hands-on learning. The chimney or stack had sixteen inspection doors with elevators for student observation and tests for "flue gas velocity, gas temperatures, and gas analysis." Once a summer or semester, undergraduates spent a few days within the actual plant, climbing its ladders, observing the inner workings of pumps and valves, gages and instruments.[41] Potter and Solberg also turned the plant into their own laboratory, publishing a series of twelve research studies, joining engineering science and public policy on issues like heating loss, solid fuels, coal gasification, coal composition, and costs.[42] In 1929, joined by graduate student George Hawkins, with funds and expertise from the Babcock and Wilcox Company and Bailey Meter Company, they took up Goss's earlier challenge to build the High Pressure Steam Generator Laboratory. It generated 6,000 pounds of steam per hour, at pressures of up to 3,500 psi. This was three times the maximum pressure then in normal use.[43] The team researched higher temperatures for steam power plants, up to 1,500°F, well above the 850°F then common. They investigated new metal alloys and chemicals to fight the heat and corrosion.[44] These techniques were visionary, confirming the viability of the "superpower designs" to come.[45]

Zucrow joined these vanguard experiments with high-pressure and super-heated steam, partnering with Potter to research ways to store it for later use. He also got to play the role of historian, spotlighting Purdue's research with a sweeping historical review of thermodynamics and steam power. Higher pressures, if still somewhat risky and unreliable, promised efficiencies and economies. Zucrow was confident that engineers would perfect the new metals and processes for the boilers to withstand the higher temperatures, stresses, and corrosive effects. It was a judgment that he later took to rocket engines as his unique expertise.[46]

Zucrow's relations with Potter went far beyond steam engines. They became lifelong friends, though in writing they always referred to each other as Dean Potter and Dr. Zucrow. They had much in common. Both emigrated to the US as teenagers. Both finished growing up on the same streets of Boston. Purdue engineering labs were also close-knit communities, located in and around Heavilon Hall near Potter's main office. Potter often visited to get to know the students and review their research projects. Someone was taking an interest: "a comforting and inspiring thought," remembered Zucrow.[47] Here were models of mentorship and teaching that Zucrow also made his own. He dedicated the first volume of his *Aircraft and Missile Propulsion* (1958) to Potter, his "inspiring teacher, wise counselor, and true friend."

## DOCTOR OF PHILOSOPHY

When he first arrived at Purdue, Zucrow actually started out in the School of Civil Engineering, dragooned there to fill a need and based on his Harvard work experiences. On the staff of the Purdue EES for two years, he taught and researched hydraulics under F. W. Greves. He also published his first scholarly articles, hardly glamorous, dealing with sewage sprinkler nozzles and the measurement of flows through pipes. But they were his first, and were prominent.[48]

Zucrow lived across the Wabash River, in the city of Lafayette at 105 Adams, just beyond the Ninth Street Hill, in a neatly packed middle-class neighborhood. There he befriended a local Jewish family, the Zovods on nearby Kossuth Street, who introduced him to a cousin, "a local girl," as he remembered. He married Lillian Feinstein on 2 August 1925 at the Sons of Abraham Synagogue in Lafayette. Their daughter, Barbara, an only child, was born on 7 February 1927. The Zovods owned Zovod's Poultry Company at 103 South Second Street; the Feinsteins raised and processed the chickens in Chalmers, a small town north of Lafayette. The families had emigrated in 1906 from Heniches'k, in the Kherson district just north of Crimea, part of the Russian Empire. Zucrow also became a fast friend with Harry Freedman, who owned a clothing store in Crawfordsville, just southwest of Lafayette, and who had married Lillian's sister, Sophie. Zucrow thus became something of a local boy, a new man of the Middle West. He loved driving his "muscle car," a Franklin Roadster, as fast as he could along Indiana's rural highways. He once took it all the way to Champaign–Urbana in record time, and on only twelve cents of gas. Sometimes Purdue was an exciting place to be.

By 1925, Zucrow quickly reestablished himself in a relatively new machine: the internal combustion engine, as applied to both the automobile and airplane. He joined the Internal Combustion Engine Laboratory (known as ICEL) under Claude S. Kegerreis and Harry A. Huebotter. Mechanical Engineering had been one of the first in the nation to set up an Otto Cycle gas engine for research, in 1892, followed by labs for gasified coal research and automobile steam engine testing. When Zucrow arrived, the lab had nine internal combustion engines and one coal–gas engine for study, along with the country's first university carburetor testing apparatus.[49] The carburetors were on loan from the Zenith-Detroit Corporation, Stewart Company, Ball and Ball, and the Carter Rayfield Companies. Zucrow turned the lab data and findings, supplemented with his own experiments, into readable articles in the *Bulletin of the Purdue Engineering Experimental Station*. They were "the 'Bibles' for the internal combustion and carburetor engineers of the 1920s," he recalled.[50]

Purdue's carburetor studies were renowned. Lionel Marks applied them in his pathbreaking work *The Airplane Engine* (New York: McGraw-Hill, 1922). Major universities, including the Big Ten, New York University, and MIT, taught them. They were featured in *American Automotive Digest*, *Automotive Industries*, and *Scientific American*. They meant close working relations between Purdue and the US Army Air Service at nearby McCook Field in Dayton, Ohio. The Air Service hired Opie Chenoweth from Purdue (BSME 1921) for its "National Army Experimental Aviation" tests at McCook. "You would be surprised how well the research at Purdue is considered here," he wrote back to his teachers. The highest praise came from Dr. H. C. Dickerson, director of the Bureau of Standards. Purdue carburetor studies, he said, "in all phases was the most fundamental and far-reaching in its theoretical and practical applications that had been carried on by any university." If applied to manufacturing, he even predicted "a saving to gas users" of tens of millions of dollars nationwide.[51]

These pulls toward industry took their toll on the university and its educational function. Kegerreis once complained that too many companies were calling him and the EES about doing consulting work. He refused most offers. "The University is the place to conduct fundamental research and not detail[ed] development work," he warned.[52] Kegerreis's stand here was part of the new trend toward engineering science, what Harry Solberg later called the "gradual shift in emphasis from so much hardware to an emphasis on fundamental principles." Traditional engineering was about how to refine and finish products, centered on memorization and rote practices. It was weighed down by handbooks, T squares and triangles, slide rules, logbooks, and enthalpy diagrams.[53] Engineering science was about mathematical and mechanical first principles, empowering the engineer as creative designer. Harry Huebotter (BSME 1912, MSME 1923), Purdue's associate professor of gas engineering, embodied the new trend. He had just published a cutting-edge study of the internal combustion engine, applying differential and integral calculus to teach "the ability to apply basic principles to concrete problems in a practical manner." With chapters on piston assembly, the crankshaft, and water and air cooling, Huebotter's focus was on the "actual proportioning of parts," so as to "eliminate unnecessary weight and at the same time give adequate strength."[54] This was an important lesson for Zucrow, as when later in life when he confronted the all-important thrust-to-weight and heat transfer factors in rocketry. Huebotter was Zucrow's most important mentor. A bachelor who was totally dedicated to his work, a "real brain" as Zucrow put it, he enjoyed that "unusual combination of an excellent theoretician and a real pragmatist," as well as a "prodigious worker and skillful experimenter."[55] He signed his name with the simple but venerable, M.E., Mechanical Engineer, a practice Zucrow made his own.

Once he completed his Purdue master of science degree in mechanical engineering (MSME) after two years of coursework, Zucrow became the test case for Purdue's brand-new PhD program. Harry Solberg remembered the graduate program in 1923, when Zucrow started, as little more than advanced fourth-year undergraduate work, overloaded with teaching and laboratory duties. It had "no formal organization." Doctorates were "rare" in engineering. So Purdue needed to do it right, "wanted to be sure that the doctorate was awarded only to a man with a well-rounded education in addition to a specialization."[56] That man was Maurice Zucrow. Purdue made its first formal mention of the doctorate in the official catalog of 1924–1925 for fall semester 1925, just when Zucrow began it. For the next three years, he was Purdue's single doctoral student. His coursework and examinations became the precedents. His major fields were automotive engineering, internal combustion engines, kinematics of machinery, mathematical physics, and power engineering. His minor fields were hydraulic engineering and industrial engineering. He also read a series of classic texts under his favorite physics and mathematics professors, among them A. G. Webster, *The Dynamics of Particles and of Rigid, Elastic, and Fluid Bodies: Being Lectures on Mathematical Physics* (1922); J. H. Jeans, *Electricity and Magnetism* (1925); W. F. Osgood, *Advanced Calculus*, 1925; and E. N. da Costa Andrade, *The Structure of the Atom* (1927). It was the beginning of a lifelong love of reading.[57]

Zucrow then spent six months preparing for a three-stage examination sequence, a gauntlet meant to prove both his and Purdue's fitness for the PhD. First came the preliminary oral examination, held at Dean Potter's office for about five hours with nearly forty professors attending. This was partially a review of his Harvard humanities courses, centered on a recent book given him by Potter, probably Alfred North Whitehead's *Science and the Modern World* (1925), which Zucrow said he "must have read three times before I understood it." President Elliott posed the first question, based on Zucrow's reading of Andrade's *The Structure of the Atom*. "On the basis of atomic structure," he asked, "give me a theory on the fatigue of metals." Zucrow said he "improvised" and was "inaccurate" but spoke with unflappable confidence, enough to please Elliott. A philosophy professor asked about Whitehead, and Zucrow made the answer what seemed like forty-five minutes long, with "nods of approval from Elliott," though he left "pretty much exhausted." One of his professors had to drive him home. Second came an ordeal of three eight-hour days of written examinations on a dozen scientific and engineering questions. Third came the final oral examination, open to all faculty, on 22 May 1928 in the Trustee's Room of Eliza Fowler Hall. Two hundred people were there for a late afternoon exam that lasted two hours. The young engineer faced the seven-person examination committee and the audience, answering questions about his "major subjects" and his dissertation. "If you are a little nervous to begin with, and I was, that many onlookers can make you awfully uncomfortable." It became something of an interrogation. Zucrow remembered a few professors in the crowd who tried to stump him, out to "impress" President Elliott with a knockout question to make him fail. But mostly he remembered a few kind faces, and "what seemed like thunderous applause" at one point, when Elliott cautioned one professor not to interrupt the young scholar again.[58]

The final oral examination was a highlight of the 1927–1928 academic year. It was also the first in a series of dramatic public performances that Zucrow gave in his career. This one happened in a rather small venue to a local audience. But it was still science as public drama, displaying his knowledge and character, his ability to explain complex subjects with precision, integrity, and honor.[59] Twenty years later, before a major conference of the American Society of Mechanical Engineers (ASME), he gave another command performance on jet propulsion; and twenty years after that, another series of talks on spaceflight before the Panel on Science and Astronautics of the US Congress.

Zucrow received his degree on 12 June 1928. He is sometimes remembered as Purdue's first PhD, sometimes as its first in engineering. Neither are exactly true. The first PhD degree at Purdue was Daniel T. MacDougal in agriculture, with his dissertation titled "The Curvature of Roots" (9 June 1897). It was awarded based on his work at the University of Leipzig, and well after he had left campus. MacDougal soon became one of the country's leading botanists and plant physiologists, a welcome alumnus. To honor him and avoid controversy, Andrey Potter and Maurice Zucrow both agreed to the title "the first earned PhD granted by Purdue University."[60] From then on, in university publications he was known simply as M. J. Zucrow (PhD 1928).[61]

The centerpiece of Zucrow's doctoral defense was his dissertation. Titled "Discharge Characteristics of Submerged Jets," it was immediately published in the *Bulletin of the Engineering Experiment Station*. Zucrow researched it using precise lab equipment that he designed and built himself. It was a capstone to his graduate work in hydraulics and combustion studies. In a marriage of engineering science and industrial application, Zucrow studied the reasons for the improper distribution of fuel mixtures in the internal combustion engine. He conducted a scientific survey of the distribution of gases, as well as the fuel condensation, heat, and vaporization, plotting the geometries and mathematics of carburetor jet flow. His solution was for an engine redesign, namely in the induction system, to distribute the fuels more effectively. He turned his findings into a practical guide for engineers to improve engine performance.[62]

Zucrow wrote the dissertation, in the tradition of Potter's EES, as research for development, engineering for progress. Yet it was also much more. This dissertation stands, if at the very early margins of the history, as one of the first applications of the new science of jet propulsion. Although hydraulics and maritime engineers had used the term most often, aeronautical engineers also began to apply it after the First World War to describe a variety of possible propulsion systems: ramjet, turbojet, and even rocket craft.[63] The pathbreaking work of Ludwig Prandtl and his collaborators also gave engineers the analytical tools to study external aerodynamics, the flow of air on the surfaces of aircraft and wings, as well as internal aerodynamics, the flow of fluids within power plants.[64]

The dissertation was not yet a study of the turbojet. But Zucrow did apply the best of his studies at Harvard and Purdue in thermodynamics and fluid mechanics, hydraulics, and gas engineering, along with the latest in aerodynamics. This included several textbooks, several reports out of the National Advisory Committee on Aeronautics (NACA), and some of the work of his former Purdue colleague, Opie Chenoweth. The two had worked together in 1922 under Huebotter on the airplane supercharger. This was an air compressor that increased the density of air at the intake of an airplane engine in order to increase its oxygen supply and energy capacity. Chenoweth soon became one of the country's leading experts on internal aerodynamics as applied to airplane supercharging, working for the Army Air Service at McCook Field.[65] Like Chenoweth, Zucrow's innovation was to adapt the study of the external aerodynamics of the airplane to the internal aerodynamics of the automobile and airplane engine, or what he termed "the motions of liquid through jets."[66] He applied dimensional analysis, extrapolated from the flow of air over the wings and structures of the airplane, to study the fluid flow of carburetor jets within the internal combustion engine.[67]

Chenoweth and Zucrow had all the critical pieces of the jet propulsion puzzle in place. They were part of a cadre of research engineers poised to discover the turbojet engine. But they did not. Chenoweth later berated himself for not thinking more creatively. He had failed at engineering "the turbosupercharger . . . into what should have been the turbojet. We should have seen that but we were all too stupid."[68] What held them back? Traditional thinking, he said. In service to the military and industry, both he and Zucrow served practical developmental needs over visionary research. They were both also very glad to have jobs in the difficult economic downturn of the early 1920s. Potter's EES, within which they both thrived, was in structure and function organized for development, for business, for the cooperative research it contracted with industrial interests. The EES had shaped them, but for existing power plants like the internal combustion engine. Their work was a false start toward the turbojet. It was nonetheless a start and set them all up for later success in the transition to turbojet propulsion and rocketry.

Zucrow stayed on at Purdue for the 1928–1929 academic year, teaching Thermodynamics, Fluid Mechanics, and Vibration Theory, each according to the new model of engineering science. The Vibration Theory course had an interesting backstory. In these years, Zucrow worked closely with H. M. Jacklin, associate professor of automotive engineering, who replaced Huebotter and who expanded ICEL into an automotive engineering laboratory. He and Zucrow studied ways to improve engine performance and fuels, applying precise measurements by way of micrometer screws, sonoscopes, and stethoscopes to measure the compression ratio of "the incipient detonation" of the fuels. They also collaborated on a visionary project to study the human "psychology, physiology, and engineering" of riding in automobiles, this "to establish for human beings a curve representing sensitivity to shocks and vibrations." They reached out to industrial sponsors around the country, receiving the praise of the ASME and its Riding Qualities Subcommittee. Here was another Purdue innovation: the country's first chassis testing equipment to compare the relative riding qualities of different makes of automobiles. For this they invented the Purdue accelerometer. It was a set of precise instruments "used to determine graphical records of vertical vibrations, acceleration and deacceleration, horizontal side vibrations in automobiles, street cars, day coaches, pullman cars and airplanes." Zucrow did not know it yet, but here was another foundation skill for his later work with rockets, which by their explosive thrust also caused intense vibrations and oscillations.[69]

All of these interventions with the ASME and Society of Automotive Engineers gave Zucrow a national reputation. He was a regular reader and discussant for their technical articles. His commentaries reveal a mastery of the literature, both old and new, in English, French, and German. He balanced scientific accuracy with practical application. In a critique of one sophisticated article about axle ratios and transmissions as they applied to engine performance and the acceleration of the automobile, Zucrow offered the simple reminder that "part of the energy goes into accelerating the various parts of the engine" too. About another article filled with elegant references to calculus, he reminded readers that "a function that satisfies a differential equation is a solution no matter how obscure its origin, and one that does not satisfy it is not a solution, no matter how illustrious its pedigree may seem to have been." Mathematics only really mattered if it applied to real-life solutions.[70]

Zucrow spent a total of six years at Purdue, two more than at Harvard. He married and started a family while at Purdue. The university gave him his first career job and built a graduate program around him, awarding him the doctorate. He made lasting friendships at Purdue, taught there, and returned to teach again for the rest of his career after the Second World War. This all raises the question of alma mater. Was it still Harvard or now Purdue? It's not really a fair question. The term is so packed with emotion, with connection to one's intellectual birthplace. Zucrow was loyal to both. He remained a faithful alumnus of Harvard over the years, a member of the Class of 1922. He even rooted for the Crimsons when they played the Boilermakers: "I swallowed the 19 to 0 football defeat Purdue gave Harvard" (1927), or at least so he claimed.[71]

## PURDUE IN THE AERIAL AGE

As Zucrow left campus in 1929 for what became nearly twenty years of work in industry, the campus began to change again in dramatic ways. Unknown at the time, these were the very ways that enabled him to return and flourish after the war. Purdue took a leap into the aerial. It came rather late relative to national leaders and nearby institutions. Purdue was not an academic juggernaut like MIT. No school could compete with its influence. MIT was the first to establish an aeronautical engineering degree program, in 1914. Pioneering programs soon followed at the University of Michigan, Caltech, New York University, the University of Washington, the University of Minnesota, the University of Detroit, Rensselaer Polytechnic Institute, Stanford University, and the Worcester Polytechnic University.[72]

Purdue established an aeronautical engineering elective in 1919. These were four to six optional technical courses (nine to twelve credits) that students took in their junior or senior years. Campus interest peaked with Charles Lindbergh's New York to Paris flight in 1927, with rising enrollments to follow. Purdue also shaped an aerial identity. Weldon Worth (BSME 1931) became Purdue's own celebrated "Lindy." He had begun flying at Arsenal Technical High School in Indianapolis, then went on to spend five years piloting for a geological survey in California and barnstorming in Indiana. He was most famous for landing his two-person Curtiss Jenny biplane, with a fellow student on board, in a hayfield next to campus at the university farm south of State Street. They flew in to register for classes.[73]

Purdue was ready for success in aeronautics, thanks in part to the university's strengths in internal combustion engines. Cars needed them, but so did airplanes. Yet aeronautical engineering curriculums required money, funds that Purdue did not have and was at a loss to receive. Like other state universities, Purdue did not benefit from outfits like the Carnegie Corporation and Rockefeller Foundation, which gave generously to privileged schools, mostly for basic research. Harvard, Yale, Chicago, and Columbia received one-quarter of all the private funds awarded to universities.[74] Caltech and Stanford, Purdue's main rivals for engineering prominence on the national stage, were close behind and well funded too.[75]

There was another crucial source for monies: the Daniel Guggenheim Fund for the Promotion of Aeronautics (1926), to sustain developmental research in the universities and link them with the fledgling aviation industries. New York University's School of Aeronautics received $500,000, MIT $264,000, Georgia School of Technology $300,000, Caltech $350,000, the University of Washington $290,000, Stanford University $195,000, the University of Michigan $78,000, and Syracuse $60,000. Purdue was not among the winners, even though it was the nation's second largest engineering school (MIT was first); even though by 1929 Purdue had the "largest number of students enrolled in aeronautics of any university in the United States." Purdue graduates filled the aviation industries. Yet the Guggenheim Foundation rejected Dean Potter's appeals on the pretext that Purdue was most famous for "locomotive engineering," not aeronautics. Purdue did trains. It was not ready for planes.[76]

The National Advisory Committee for Aeronautics (NACA), established in 1915 to promote the new science, often supplemented the Guggenheim grants with its own funds. After the First World War, NACA adopted the project system, inherited from the Agriculture and Engineering Experiment Stations. NACA division chiefs assigned projects to their own laboratories for efficient planning, supervised by section heads, serviced by project engineers. They also contracted external projects, as sponsored research, in order to avoid the burdens of competitive public bidding and to choose the best project contracts for experimental research and development (R&D). NACA sacrificed cost and competition for quality and results. In 1939, for example, Stanford's aeronautical engineering program went bankrupt when its Guggenheim funding ran out after twelve years. NACA stepped in, with lucrative funding to revive the program. Stanford was too big to fail. Some universities were luckier than others.[77]

Rebuffed by the Guggenheim and NACA, Purdue found strength in the nearby resources of the US Army Air Corps (so renamed in 1923), headquartered at McCook and Wright Fields in Dayton. These were leading national centers for research, having designed and improved the famous Liberty engine and prepared and tested the Fokker T-2 for the first nonstop transcontinental flight (1923); and having launched the Douglas World Cruisers, the first airplanes to reach around the globe (1924), supervised by the logistical genius of Purdue alumnus Frederick Martin (BSME 1908). By late fall of 1926, the Army Air Corps established the new Materiel Division at Dayton, dedicated to R&D. The command's Experimental Engine Division, Airplane Engineering Department, and Air Corps Engineering School soon amounted to a $160 million physical plant. It tested superb wind tunnels. It researched magnesium alloys for lighter-weight aircraft. It studied engine superchargers, in cooperation with Purdue, for added power.[78]

Beginning in 1926, Purdue hired Army Air Corps veterans from Dayton to direct and teach its new aeronautical engineering courses: Maj. William A. Bevan (1926–1929) and Lt. George W. Haskins (1929–1937). The partnership paid off. Bevan, for example, a former instructor at MIT, took his aeronautical engineering seniors on field trips to McCook and Wright Fields. He brought back some of America's latest military planes and engines for research and study. Most of the labs, in the days before the Purdue airport, were in Heavilon Hall, the very location for earlier research in locomotives. Airplanes now took their place.[79] In exchange, Purdue sent some of its best students to Wright Field. Besides Opie Chenoweth, the high-flying Weldon Worth went to work at the lab, there inventing excellent new methods of propulsion cooling and lubrication, later directing some of the country's first projects in ramjets and liquid-propelled rockets.[80] All this meant that Purdue's aeronautical engineering deserved its "place in the sun." A campus editorial playfully pictured the engineering deans perched on an airplane flying into this brighter future.[81]

Purdue also found purpose in its own culture of aspiration, crafting initiatives in the field of university research. In January of 1926, Dean Potter partnered with university benefactor David Ross, vice president and general manager of the Ross Gear and Tool Company in Lafayette and a self-made millionaire and leading member of the Purdue Board of Trustees. Together they made a sweeping industrial and academic tour of the East Coast, mostly by train. Ross first visited with Vernon Kellogg at the National Research Council in Washington, DC. Potter joined him to consult at the research labs of American Telephone and Telegraph and the General Electric Company's Schenectady Research Laboratory. They visited the major eastern universities, including Columbia, Yale, and MIT. While in Boston, they took Sunday off and visited with Potter's mother. They moved on to Pittsburgh, with meetings at Westinghouse Electric and Manufacturing, as well as at the Mellon Institute and University of Pittsburgh, models for private and public research cooperation. The trip ended in Detroit, with visits to the Ford Motor plants, General Motors, and the University of Michigan, Purdue's benchmark peer university.

This was a talking tour, a way to gather information on the hiring and research needs of industry and on the status of research at universities. It was also a business trip to advertise Purdue. The discussions were mostly fulfilling. As Potter and Ross reported, the "industrialists" wanted better employees and more cooperative research out of the universities. One of them was so impressed by Purdue's high standards that he "agreed personally to investigate the record of every Purdue graduate in the employ of his company and to arrange for the advancement, transfer or dismissal as the individual case might warrant." Another executive immediately ordered the transfer of "one outstanding Purdue student," who was working in a "lonely desert pumping station" (simply out of convenience, and his "reliability") to a higher position fulfilling "his proper destiny in life."[82]

On their return, Potter and Ross organized the first industrial conference at Purdue, joining the governor, and leading industrialists and manufacturers, with the aim of selling campus research to industry, mostly to aid in the economic development of the state. As a premier technical university, Purdue needed to become even more of a campus of research laboratories, centered on undergraduate and graduate education, to inspire young people to "scientific attainment" and rewarding careers. Potter led tours of the campus laboratories for agriculture, railway locomotives and brake systems, limestone quarry development, telephone buried cable, and carburetor research. Purdue, he declared, was contributing to the "wonders of our age," beyond the light bulb and telephone, extending the US as a country of invented things. The initiative continued into 1928 with the new Department of Research Relations with Industry, this at the peak of the economic boom of the 1920s. President Elliott hired G. Stanley Meikle as its first director in May of 1928 to pursue a unique "project in progressive education." The department was charged to support the new doctoral program, baptized by Zucrow, for the "intensive education of productive creative manpower" by way of contract research with industry.[83]

It's worth spending a moment to introduce Meikle, in that he helped institutionalize the culture of research under which Zucrow later thrived. Meikle had wide experience. He had been an electrical engineering student at Worcester Polytechnic Institute for three years, then graduated in 1913 with a BS and MS in civil engineering from Union College at the same graduation ceremony. Along the way, he served as an "assistant to the manager of one of the largest districts of the United States Steel Corporation," helping to direct "several large development engineering projects." Between 1912 and 1918, Meikle worked as a scientist in one of the country's premier laboratories, the General Electric Schenectady complex, there assisting director Willis R. Whitney, Nobel Prize in Chemistry laureate Irving Langmuir, physicists William D. Coolidge and Albert W. Hull, and the famed mathematician and electrical engineer Charles Steinmetz. They were "great kindly men of endless scientific curiosity and boundless enthusiasm," he remembered. At Schenectady, Meikle researched a Purdue specialty: heat transfer in furnaces of new designs. He also invented the famous tungar rectifier, a bulb with an ionized gas that acted as a more efficient principal current carrier, saving time and energy in the recharging of batteries for automobiles, railway signals, fire alarms, and telephones. In the last months of the First World War, he was director and officer in charge of the US Army Research and Development Unit, directing over 160 scientists and engineers in designing battlefield weapons and gear. Meikle also helped perfect an ultra-sensitive microphone for underwater detection of German U-boats and designed the country's first true gas mask, both flexible and fitted, for American troops already fighting in Europe. This was all quite a series of calling cards.[84]

Meikle's new department at Purdue faced immense obstacles. The Board of Trustees stalled. Business hiring scouts rejected it. The complex and constraining rules of finance and taxation held Meikle back, as did the ensuing Great Depression. In response, Elliott and Ross created the breakthrough Purdue Research Foundation (PRF) on 30 December 1930 with seed money of $50,000 each from David Ross and Josiah Lilly. Meikle was its first director of research. The brilliant Ross kept its foundation charter treasured in a velvet embossed edition, hoping the PRF would become an adjoining campus all its own, a real community of science, built along the lines of Francis Bacon's "New Atlantis." Instead, it simply became an educational "corporation" established for "encouraging, fostering and conducting scientific investigations and industrial research" as well as "developing" the students and faculty to serve them. Purdue basically crafted the PRF on the model of Potter's EES, as an administrative machinery to mobilize the "research project." By a Board of Trustees directive, "all sponsored cooperative research, except in a rare case of expediency, must perform an educational function."[85]

This directive was more than just words. Meikle built it into the actual practice of the PRF. It was his constant refrain and standard. Zucrow later made these procedures his own. Business interests would pay for research, but the university would jealously guard its academic freedoms. In fact, Meikle insisted, "the individual shall be free to select the subject matter of his graduate research, free to dictate the direction it shall follow, and free for concentration and conscientious work prerequisite to the development of creative thinkers." Meikle even hoped this new partnership would tame "unbridled capitalism" and revive American democracy.[86]

Amid these high ideals, the PRF became in practice a "designee of Purdue University—an independent contracting agency—to promote bequests, donations, gifts, grants, pledges, and funds from any source for the development of basic scientific research as an educational function at the graduate level." As a private, nonprofit foundation, separate from the university, the PRF was able to negotiate contracts and overhead, assign research projects, and attract and manage wealth and property in the university's interests. R. B. Stewart also created lucrative special trusts and tax-exempt annuities for the PRF and its university donors, enabling attractive stock purchases and buybacks for them. As he once admitted, all this avoided "the IRS looking over your shoulder."[87]

There were a few other organizations like it. The Research Corporation for Science Advancement, established in 1912, helped several universities fund their applied research, including Yale and the University of Southern California, the Rensselaer Polytechnic Institute, and George Washington University. MIT's Division of Industrial Cooperation and Research morphed into the Office of the Vice President for Research Administration after 1932, which managed patent rights, copyrights, contracts, and overhead for sponsored research.[88] The University of Michigan created several university administrative offices for engineering research and research administration after 1920 for the same purposes, taking a percentage of the contract price for its industry–faculty projects. The University of Wisconsin created its Alumni Research Foundation in 1925 to protect faculty patent profits.[89] Alone among these enterprises, the PRF was a pioneer for the large land-grant public schools around it, offering a "separately incorporated research organization" for sponsored research. Ohio State University, the University of Illinois, Iowa State, Indiana University, Auburn, and Rutgers quickly adopted the Purdue model.[90]

The rise of the PRF was not easy. Meikle's initial relations with faculty were strained. He had little university administrative experience. Faculty did not trust the new "business" of research, fearing the university would profit at their expense. They were worried about losing sovereignty over their own "scientific investigations."[91] Meikle eventually found a balance, assisting faculty with their patent protections and focusing his efforts on chemistry and physics. He worked closely with Richard B. Moore, the dean of the School of Science (1926–1931) and head of the Chemistry Department. Moore was the inspiration for a second major Purdue appeal for Guggenheim funds, this time for a dirigible institute. He had discovered a new method for the industrial production of helium, one of the best of the lifting gases. But the Guggenheim rejected Purdue once again, in favor of the University of Akron. Perhaps this was all for the best. Akron's dirigible program collapsed in 1938 with the Hindenburg disaster. Purdue was spared.[92] Moore had more success on campus, attracting one of the world's leading physicists, Karl Horovitz, to head the Physics Department. This was part of a Purdue plan to compete with the rising star of Lawrence Livermore at the University of California. Horovitz had received his PhD at the University of Vienna's Physics Institute in 1919. He also married Betty Lark in 1926. She was an artist who eventually taught psychology at Purdue. Deeply in love, and an advocate for women's causes, he changed his name to Lark-Horovitz to honor her. Arriving on campus in 1928, he taught Contemporary Physics Theory, and such specialized courses as Atomic Structures and The Electron Theory of Matter. Lark-Horovitz raised the profile of the department as its head until 1958, in a style most describe as autocratic.[93]

The PRF was also instrumental in Elliott's and Ross's designs to build the Purdue University Airport, the first such campus facility in the US. Ross donated the land, 122 acres at the former Edgewood Farm, in June of 1930. Meikle oversaw its building and development. R. B. Stewart secured the finances from the New Deal, with extra funds to build nearby dormitories, instructional halls, and the golf course as well.[94] This was all a curious amalgam of development, research, science, and public relations. Amelia Earhart personified it. Elliott promoted her counseling work with women students on campus. He and the PRF also financed the experimental Lockheed Electra airplane for her dramatic round-the-world flight (with navigator Fred Noonan), an earnest drive to raise the university's national profile, ending in tragedy in 1937.[95]

Purdue also revived its aeronautical engineering program in the fall of 1937, with the hiring of professors K. D. Wood and Joseph Liston, who quickly raised the program as a national trendsetter. They also recognized formidable challenges. Purdue was an underdog, "handicapped" without the Guggenheim funds, even though its work in aerodynamics and engines was as good as any in the country. Their solution: a dual program in aviation technology and aeronautical engineering to train "outstanding men" in "the multiplicity of technical problems involved in the development of new airplanes" and engines; as well as an aeronautical laboratory "to place Purdue in number one position among the colleges giving work in aeronautical engineering" and the new trend of research and development. Aviation had made great strides in the last twenty years but had reached foreboding altitude and speed ceilings. To move beyond them, Purdue needed to study streamlining to reduce parasite drag, to test new alloys and fabrication of steel and aluminum for lightness and stress, and to design improved flight instruments. Wood and Liston made their strongest case for engines. They proposed a vanguard test stand and laboratory for "engine performance studies" and to investigate "the effects of different variables on increased power output." This included fuel and lubricant studies for "improved fuel characteristics." Most dramatically, since even these improvements were short-term, given the inadequacy of the internal combustion engine, they made a pitch for jet propulsion. To reach higher and faster in the air, the world needed new and "different types of prime movers." It needed "not only radical designs of gasoline engines, but also rockets."[96]

This futuristic appeal to rockets actually made sense at the time. Europe was already leading the momentum for high-altitude and long-range rocketry, based on the work of Hermann Oberth and the German Society for Space Travel (1927–1933). The German military shifted much of their work to the secret installations at Kummersdorf and Peenemünde, under the direction of Wernher von Braun, along the way to building the A-4 (V-2), the world's first liquid-propelled ballistic missile. In the US, Robert H. Goddard was a media sensation through the 1920s and 1930s who published some of his key achievements and findings in 1936. The American Rocket Society (ARS) had also begun to build serious liquid-propellant rockets in the 1930s under pioneers James Wyld, Lawrence Manning, and Alfred Africano. Several of them joined together to form the East Coast's first great rocket company, Reaction Motors Inc., in 1938.[97]

Caltech joined the trend in March of 1935 when graduate student William Bollay gave a seminar paper on the potentials for a rocket plane. His mentor, Theodore von Kármán, the world's leading expert on aerodynamics and director of the Guggenheim Aeronautical Laboratory at Caltech (GALCIT), gathered a motley crew of experimenters in solid- and liquid-propellant rockets. It was a dynamic group, including several talented technicians: Jack Parsons and Edward Forman, and Caltech graduate students Frank Malina, A. M. O. Smith, Hsue-Shen Tsien (aka Xueshen Qian), and Weld Arnold. Martin Summerfield joined them in 1940. Von Kármán, an émigré who had participated in Germany's early debates over liquid-propellant rocketry, understood the significance of the new field, in the very terms Wood and Liston had described. Aeronautical engineering needed to overcome the limits of the internal combustion engine with the radically new ways of internal aerodynamics and jet propulsion.[98]

Von Kármán educated the country's first generation of MS and PhD engineers upon these principles. They were Zucrow's future colleagues and patrons. Some were civilians, like Bollay, who went on to work for the US Navy, Harvard University, and North American Aviation; James E. Lipp, soon to work at the RAND Corporation; and others in Joseph Charyk, Francis and Milton Clauser, Frank Wattendorf, Homer Stewart, and Harrison Storms. Still others were military, one admiral and two rear admirals to be; along with a small cadre of officers from the Army Air Corps out of Wright Field, including Donald Putt and Charles Terhune, a Purdue alumnus.[99] Caltech was educating a new generation of experts for jet propulsion. And Purdue wanted to be a part of it.

Although Wood and Liston failed to adapt Purdue to turbojets and rockets in 1938, they eventually got their aeronautical engineering lab, funded by the Works Progress Administration. The university also built upon this momentum to make an ambitious stand just a few years later, an attempt to contract for a major NACA national laboratory. In 1940, NACA received $8 million to design and build a new aircraft engine research laboratory. Its Site Committee called for fair and competitive bids from cities and installations around the country. The top competitors were Los Angeles, Chicago, New York, Detroit, Dayton, and Indianapolis. Cleveland won the prize, given its leading airport and large manufacturing base for aircraft parts.[100] Purdue's administrators knew the university had a minutely small chance of winning. But they submitted a proposal anyhow, losing in the earliest round. The payoff, according to the PRF, was to help "focus attention of national leaders in aviation upon Purdue as a possible center for aeronautical research and education." Moreover, "the Foundation proposal was the only non-political, non-industrial proposal of the 63 presented," and "the repeat and frequent contacts with the committee members created the favorable impressions intended." Here was a small consolation prize for one of the losers, again.[101]

Zucrow could have been a part of all this. He was scheduled to teach at Purdue in the 1929–1930 academic year. Mechanical Engineering needed his teaching and research talents. But he asked for a short leave to gain "industrial experience," a normal request for the time. He took a risk and entered the job market in September of 1929, an extraordinary moment in US history. Never before had so many firms been so invested in research labs, and so many universities invested in the production of PhDs in the sciences.[102] It was an auspicious start at an inauspicious time. The stock market collapsed just two months after Zucrow left Purdue, followed by a decade of the Great Depression. He was constantly employed in some way, at least on paper. In reality he was intermittently unemployed for several years, and in these times remembered the sting of anti-Semitism in the ranks of American business.[103] Still, he was in his element, the breach of new enclaves and new horizons, and nearly twenty years in private industry.

His first job was in Chicago as chief engineer and eventually vice president in charge of engineering (1929–1934) at the Paragon Vaporizer Corporation, a "manufacturer of low-grade fuel burning equipment for industrial engines" like tractors, air compressors, and "large inner-city motor coaches." These were cheaper and smarter alternatives to diesel engines, without their initial costs and maintenance complications. Zucrow joined "laboratory and field" work to improve "the exhaust-gas-heated vaporizer-types" of these "heavy-fuel carburation-systems." He basically created new kinds of carburetors. The results paid off. In his tests on a fleet of five buses traveling daily between Chicago and St. Louis, covering hundreds of thousands of miles, his "heavy-fuel carburation system" saved money on the fuel bills. It met the expected performance in "fuel consumption, dependability, oil dilutions, power output and flexibility." He submitted at least two formal patents for his inventions and published several related scholarly pieces, a mark of his desire to return to the academy.[104]

But "the bust caught Paragon," he remembered. "Our customers were going bankrupt and so did we." Zucrow was unemployed at least for a year, let go as the most recent hire. The family trekked back to Lafayette to live with relatives. Ever the scholar, Zucrow spent his days in the Purdue library writing two textbooks: one on fluid dynamics, another on the "balance of reciprocating engines." They were models of his teaching and textbooks to come. Scholarship always seemed to call him back. For a time in Chicago while unemployed he also finished the coursework for a law degree, but he never took the bar examination.[105]

Zucrow found work again in Chicago, in the enclave of a Jewish family, the father and son team of Jacob Spitzglass (1869–1933) and Albert F. Spitzglass "Sperry" (1900–1962). The elder Spitzglass had emigrated to the US in 1904, received a BS and MS from the Armour Institute, and was a prolific inventor. His son Albert, who Americanized their family name during the Second World War, was born in Odesa in 1900 and was a BSEE graduate of Rensselaer Polytechnic Institute (1922).[106] Zucrow joined their firm, Republic Flow Meters Company, as director of research (1934–1937), specializing in the design and manufacture of monitoring instruments and control apparatuses for industrial power plants and public utility companies. The work offered a stable income and a series of important scholarly publications in national forums. He turned these pieces into classroom lectures, teaching power plant engineers about the basics of boilers and steam generators, all in order to make them more efficient and turn a profit.[107] But even this was too "serene," he later recalled. "The ants got in my pants." He and Sperry established two offshoot consulting firms in downtown Chicago at 415 N. LaSalle Street. One was Spitzglass and Zucrow, the other was the Hubbard Engineering Company (1937–1940). Zucrow conducted valuations for public utilities in Iowa, serving as an adviser for the Moscow Dam case, a contentious legal dispute about a "gigantic waterpower generating plant." A Hubbard team prepared the plans for the fifty-thousand-kilowatt plant at a cost of $14.6 million. The local newspaper described how "Dr. Zucrow, hydraulics expert," in two days of testimony, "amazed the court with the long array of degree letters after his name and with his recounted professional experiences." The experience turned Zucrow into a public policy expert, an advocate for the progressive public utility agendas of the New Deal.[108]

Zucrow and Spitzglass then formed the Ring Balance Instrument Company, where Zucrow worked briefly as vice president and general manager (1940–1941), directing engineering and research, designing and manufacturing "industrial instruments and automatic control devices." The company's premier product was the ingenious ring balance meter, what Zucrow later claimed was his own invention. It precisely and efficiently measured steam, water, gas, and air "at static pressures up to 1,000 pounds per square inch" and came in several adjustable varieties, with calibrated weights and protective steel casings.[109] Though all of this, Zucrow did not forsake science. In a published lecture for the American Association for the Advancement of Science, he devised a thoroughly original method to solve problems in "industrial temperature measurement and control." One top editor affirmed that it was "a practical guide to the solution of practical problems," easy for both the nonmathematical engineer and technician. It offered "a method that will reduce days of calculations to perhaps not more than half an hour."[110]

Zucrow did not forget his achievements in Chicago. When he built the Purdue Rocket Lab in 1948, he modeled its systems on the instrumentation and controls for the Hubbard Engineering Company's steam power plants.[111] Throughout his Chicago years, as chair of the Translation Committee of the Industrial Instruments and Regulators Division (ASME), he also "did all of the translations" of technical literature from French and German. They were also key to his later success. He threaded his scholarly articles with the best of European learning, assembled valuable engineering bibliographies, and mastered the classic German works by Aurel Stodola and Gustav Flügel, key among the foundational texts on gas turbines that Zucrow later used for his own textbook on jet propulsion.[112]

## ENGINEERING THE HUMANITIES

As Zucrow immersed himself in the scrappy business life of industrial Chicago, Purdue's administrators continued their transformative trend back on campus, preparing the way for his eventual return. Between 1922 and 1931, Potter raised engineering enrollments from 1,790 to 3,032. Purdue also finally surpassed MIT as the country's largest engineering university. Potter did so by raising standards, not lowering them. There were now "minimum essentials" in all courses. The required undergraduate courses rose from 52 to 59. He created 120 new courses, including a true graduate program built around Zucrow's PhD. He established new instruction for engineers in finance, accounting, and management. He did so "with no increase in the teaching staff." The faculty were, quite simply, overworked. They were also underpaid. In terms of the "salary budget per engineering student," Purdue was the absolute lowest of its peer institutions, at $105. Michigan was at $281, Michigan State at $233, Illinois at $225, Wisconsin at $218, Ohio State at $208, and Iowa State at $155. Purdue was actually quite proud of its "low cost," based on a "greater scrutiny of expenditures and low overhead," along with its "lower salary scale" and "excessive load."[113]

Potter and his staff were also at work bridging technology and society, engineering and the humanities. He was a dedicated patron of a humanities education for his engineering students. This was not so unusual for any large American technical university. MIT and Caltech, like the programs at Columbia and Tufts, were also integrating engineering with the sciences and liberal arts. The trend was driven by higher enrollments and market demands for higher professional standards, as well as for management and labor expertise. Engineers needed to be accomplished speakers and writers, as well as upstanding citizens.[114] Potter became a national leader with his "rule of thirds." This meant that one-third of an engineering degree was devoted to the sciences and mathematics; one-third to engineering courses; and one-third to general studies in languages, the social sciences, and the humanities. More than any of its peer institutions, he "liberalized" the Purdue curriculum with mandatory courses in English composition and speech communication and with mandatory humanities electives (then housed in the School of Science). He hired nationally renowned professors for the Department of History, where Louis Martin Sears and Victor Albjerg taught popular courses in US diplomatic and European history. Potter advised Stanford on how best to organize its new School of Engineering, which soon applied his "principle that the undergraduate curriculum should be broad and liberal."[115]

Potter built his rule of thirds beyond the curriculum into the very life of the university. For his engineering faculty, he sponsored a book-of-the month club to promote intelligent discussion and professional camaraderie. Potter's personal papers are filled with reading lists: books he had read or that he wanted to, books for his faculty, books for his engineering students. One of them was Oswald Spengler's weighty two volumes *The Decline of the West* (1926–1928). For students, he sponsored the annual Hanley Speech Contest: five minutes "on any subject." First prize was a seventeen-jewel Swiss watch, second prize a Shaffer desk pen set, and third a year's subscription to *Fortune* magazine.[116]

By 1940, Purdue's All-University Literary Contest and literary banquet was also in its thirteenth year. Robert Frost spoke at that year's banquet, with formal prizes for best short play, poetry, short story, informal essay, historical essay, and "best essay on a scientific subject." The student literary magazine, the *Scrivener*, featured the winners. Undergraduates wrote essays about the majesty of the railroads, about the reward of "research in pure science," about the "chemistry of vinyl resins," or about the "development of atomic physics." They offered fine pieces of local Indiana history, covering the New Harmony commune, the historic Beck's Mill, and the Battle of Tippecanoe. They shared their love of flight. A young pilot, Janette Morris, beautifully described the earth from the air and from the controls of her plane, "aware of a feeling of infinite possession along with a feeling of infinite insignificance." Another student wrote a marvelous piece on the propeller, in all its physical and mechanical dimensions, as "nothing more than a wing that is made to revolve in a circle."[117] Potter was also patron of the *Purdue Engineering Review*, one of the country's largest and most vibrant of campus magazines. Students ran it and wrote for it.

Potter also sent his own engineering professors into the liberal arts, most notably Warren Howland, who joined the civil engineering faculty in 1926 and taught popular courses on the ethics, art, and economics of engineering. Always ready with a quick quote from Herodotus or Goethe, Howland married the liberal arts to Purdue engineering. He enlarged the offerings in "humanistic courses," added a comfortable reading room to the main library, and advanced attendance at local churches, dances, and cultural events. This was Howland's answer to President Elliott's 1933 call to strengthen the "four pivots" of a university education: "intellect, ideals, individuality, and independence." Howland was also fulfilling Potter's guiding principle that "engineering is the modern type of a liberal education acquainting the student, as it does, with the processes, devices, and methods which make our civilization distinctive." But Howland's purposes were also rehabilitative and pragmatic. With so many of its students studying the growing specializations in engineering, and located in a small Midwestern city, far from the larger and more vibrant urban centers, Purdue needed culture. Its student body, overwhelmingly male and still consumed with competitive intramural sports and fraternity hijinks, needed cultural "redemption."[118]

These initiatives encompassed a unique university spirit. R. B. Stewart, Purdue's chief business officer, was also an advocate for the liberal arts, what he considered essential to protect individual rights and freedoms, and the proper democratic balance between job "training" and a "liberal education." Every American deserved both, regardless of gender or ethnicity, because "training is not always useful and education is not always useless," he wrote. "There is a very real long term need for philosophy with science rather than for either alone." Or as one student editorialized, in less equitable terms, "We are among those who boast that a technical man is a manly man, but we are also aware that technical training alone does not fit one to live effectively. Cultural education is necessary for the fullest development of character."[119]

Ever the humanist, Potter was also the sponsor of the Goss Library. Edna, the widow of W. F. M. Goss, donated his library of some nine hundred volumes in 1928, supplemented by two hundred volumes from Professor Michal Golden, and an endowment for more. It grew to thirty-five hundred volumes and included the textbooks and further donations of Purdue's own professors. Potter placed it prominently at the ground floor library of the Engineering Administration Building in a beautiful set of rooms with wooden bookcases, handsome tables and desks, and leather-upholstered chairs. It was to serve as an engineering reference library and inspiration for faculty and students. Most of the books were about transportation, especially locomotives and early automobiles and aircraft, but also covered natural philosophy (electricity and metallurgy, for example). Potter made a point about buying history books. The collection was the Goss Library of Engineering History. He wanted coverage of the many engineering "successes and failures of the past." These were essential lessons for the aspiring engineer. "Learning to look backward with reverence," said Potter, "should enable the engineer to build a better today and tomorrow."[120]

One of Purdue's most famous engineering alumni, C. C. Furnas (Chemical Engineering 1922), threaded this tenet into his popular history of the future, *The Next Hundred Years* (1936), a selection of the Book of the Month Club. He was already well-known as a track runner in the Olympic Games in 1920. Spurred by the continuing Great Depression in the US and the rise of Nazism and Fascism in Europe, Furnas published a set of sober and balanced predictions. He offered hope in machinery to save humanity from exhausting labor, in new medicinal and hormonal and dietary treatments, and in a new chemistry of synthetic and plastics. He expressed faith in the human ability to "*improve and apply*" measured progress. But he also warned against the "dangers of overconfidence," blindness to our own hazards and limitations.[121] A painting that once graced the third floor hallway of the Purdue Memorial Union relayed a similar message about history, if less optimistic. *Frankenstein* (1926), by the Hoosier artist George Aldrich Ames, portrayed a frightful burst of fire and smoke in one of south Chicago's steel mills, framed by the small workers and their homes below. Ames, who usually painted tranquil rural scenes, was bridging romanticism and gothic horror. A gift of President Edward Elliott and David Ross, the painting's message to students was that "for the good of mankind, the social sciences and humanities must keep pace with technological progress." Their motto was from Ovid, "Education defines character" (*Abeunt studia in mores*). Human beings needed to take care of each other as much as of their machines and industries. The alternative was monstrous: Frankenstein mutated into all manner of ecological and social disasters.[122]

These initiatives created a most creative culture at Purdue, one both liberal and practical. Jobs depended on it. In his first year as dean, Potter established the new Personnel Placement Service thanks to a $3,000 grant from the Indiana Manufacturing Association and comprising an initial network of six hundred companies. The office collected the "personnel ratings of engineering students," meant to help them "in the improvement of their personalities." These were a series of personnel "cards" that freshmen and juniors completed: five from classmates, five from faculty, and five from other persons. The standards were high: "Every Purdue University engineering student is encouraged and assisted to develop a good address, agreeable manners, cheerful attitude, cooperative ability, pleasing disposition, untiring industry, sane judgement, well balanced initiative, creative leadership, love of fair play, and the spirit of service." Potter was appealing to the business caste, future employers, with his promise "to train engineers for responsible and effective citizenship. This means obedience to law, interest in the common good and tolerance toward the lawful rights of others. They are cautioned not to allow sentiment of prejudice to color their views and are impressed with the interdependence of human rights and property rights, the evils of mob control, and the dangers lurking in aroused class consciousness."[123]

At the start, the system had its failings. Faculty neglected to complete the cards. Students exaggerated the character traits of their classmates. No one seemed to care. These were the early complaints of the Personnel Placement Service director, William Steinway "Bill" von Bernuth, a man of intense dedication and high energy. A veteran of the First World War, and one of the heirs to the Steinway piano fortune, he was also the wrestling coach and assistant freshman football coach. His motto was "Work Fast, Give Them Service, Sell Purdue." In time, his office established a routine, expanding into "counseling delinquent freshmen students," career counseling, summer work for juniors, and job placement for seniors and alumni. By 1950, the Personnel Service had amassed some thirty thousand records. Three hundred fifty seniors used it that year, matching their credentials with fifteen hundred employers, firms like International Business Machines and International Harvester Company, Westinghouse and Ingersoll Rand, and American Sugar Refining. Potter boasted that Purdue beat out all the other engineering colleges in the US by recruitment visits, including MIT and Caltech. More telling: two-thirds of Purdue engineering alumni were already in the "top management positions" in business and industry.[124] Campus culture mattered. Here was a high standard that Maurice Zucrow was soon to keep, as well.

BY THE TIME MAURICE ZUCROW RECEIVED HIS PHD FROM PURDUE IN 1928, THE campus was in a second genesis. These were transformative and sometimes awkward years for Purdue. Its administrators forged innovative ties with private industry and state government through Dean Potter's EES. They secured New Deal funding and development through the pioneering PRF. They partnered with the US Army Air Corps at Wright Field on the campus airport and aeronautical engineering program. Purdue fulfilled its own mandate for R&D, with power plants to conquer the land and air: locomotives and automobiles, airplanes, and even plans for rockets.

True, Potter's devotion to the EES and cooperative research, including profitable project contracts with business and the military, meant that he relegated vanguard and risky research, research without immediate development, to second place. Purdue might have pioneered the new science of jet propulsion. But it did not. Only Caltech took those risks. Yet each of Purdue's initiatives established the foundations for Zucrow's later successes with the Rocket Lab. Purdue bequeathed him some of the world's best teaching and research in "heat transfer and transportation," its signal specialties.[125] Potter and Solberg, joined by Hawkins, formed the team that confirmed the university's authority in thermodynamics, high pressure, and heat transfer as applied to steam engines. They were the team that brought Zucrow back to Purdue in 1946. Their insights, and his wide-ranging studies at Purdue in hydraulics and combustion, high pressure and instrumentation, previsioned his work in jet propulsion, preparing him and his students to become the "boilermakers" of rocket engines.

Trains and rockets may seem distant, unrelated technologies. But ever since once of the first English locomotives, nicknamed the "Rocket," publicists attached this metaphor of explosive speed and aerodynamic ease to locomotives, as if leaving their grounded tracks to race through the air. It remained a favorite advertising draw into the 1950s, before commercial air passenger travel. The Rock Island Company out of Chicago famously inaugurated a "rocket line" in its "fleet of stainless steel, streamlined, Diesel powered passenger trains." The Twin Star Rocket, between Des Moines and Minneapolis, was first; followed by the Peoria Rocket and the Kansas City–Denver Rocket, among others; and eventually the Golden Rocket from Chicago to Los Angeles. They represented reliable and "streamlined beauty." Most of all, they were fast. Like the "beauties who dance in Radio City's Music Hall" (the Rockettes) and "modern propulsion flying methods."[126]

Rockets were a thing on Purdue's campus, an emblem of modernity. Students had fun with them in campus publications as the stuff of science fiction, ships to escape Earth and travel to Mars. Eventually they considered them as serious science, celebrating Robert Goddard's experiments that literally promised the "threshold of adventures" by actually traveling into "interplanetary space." Engineers need only deliberately study the concept of "step rockets" in order to achieve "higher energy value per pound."[127]

Purdue students, like undergraduates everywhere, tended more to the utopian. Some of them imagined boldly. Take one of the first proposals for a university mascot, made by Zucrow's brother in the Jewish fraternity Sigma Alpha Mu. Israel Selkowitz (BS physics 1941) put forward what he called a "mechanical robot." It was to be the "symbolic representation of the engineer who builds, the agriculturalist who feeds, the scientist who plans." It was also "to symbolize the invincibility, the immortality, and the ingenuity which is the spirit of Purdue." He offered a thoughtful design, with a body made out of a boiler, fashioned in the campus shops, and the arms made up of "agricultural instruments," to be eventually "operated by a system of remote control." The robot was too avant-garde for university leaders. Instead, they soon chose the Boilermaker Special as the university's official mascot, built by the Studebaker Corporation and the Baldwin Locomotive Works. It was a train.[128]

# 3

# RESEARCH AND DEVELOPMENT

## American Rockets for War

RESEARCH AND DEVELOPMENT (R&D) BECAME A CATCHWORD PHRASE IN the US through the 1920s and 1930s. Industry coined it during the First World War. The Lynite Laboratories of the Aluminum Castings Company, for example, established a "special research and development laboratory" in the summer of 1915, concentrating some one hundred researchers in a twenty-two-thousand-square-foot building all its own. They did research in the physics and chemistry of metals for the development of new molds and castings for aluminum alloy automobile and aircraft engines. Aviation, in this sense, had one the first named R&D laboratories. From these humble beginnings, industry and the military disseminated the new phrase, defined as a new union of basic and applied research, a category all its own.[1]

The R&D model filtered into the universities during the Second World War, thanks in large part to the successes of the Office of Scientific Research and Development (OSRD).[2] It directed the country's intense race for new weapons, radical innovations like radar, the atomic bomb, the proximity fuse, and battlefield rockets. These were some of the US advances during the war. It fell behind in other areas, like turbojets, rocket planes, and ballistic missiles. Nazi Germany's Messerschmitt 262 turbojet fighter, first deployed in July of 1944, and Messerschmitt 163 rocket plane, first deployed in May of 1944, impressed American scientists and engineers as foundational achievements, though they were both only ever used for short-term tactical surprise in the air. Germany deployed its first Vengeance V-1 missile (with a pulsejet engine) in June of 1944 and V-2 missile (with a liquid-propellant rocket engine) in September. Both were vengeance weapons, used to target Western European cities with surprise terror bombings, this to break civilian morale. Neither of them was decisive in the war.

Frank Wattendorf of Caltech, one of Theodore von Kármán's proteges at the Guggenheim Aeronautical Laboratory (GALCIT), speculated that the Germans were "an order of magnitude ahead" of the US in terms of these varieties of jet propulsion, along with their higher altitudes, speeds, and capabilities.[3] He was one to know, part of America's premier rocket laboratory for jet propulsion devices and guided missiles at the new GALCIT-Aerojet complex, soon charged to catch up with Germany's new weapons. This chapter covers the story of these leaps and gaps, laced with the contributions of several Purdue personalities, including Frederick Hovde on rockets, R. B. Stewart on finance and management, Andrey Potter on education, and Maurice Zucrow on propulsion. These were not the glamorous achievements of the German kind, but simpler ones. They helped win the war on the ground, at sea, and in the air. Not by terrorizing civilians, but by winning battles.

## ROCKETS OF THE OSRD

There is a curious fact built into the history of the Second World War. A lesson about R&D. Claude Monson, vice president and general manager of Northrop Aircraft, pointed it out. "All the airplanes used in World War II were models developed before 7 December 1941. Yet all the rockets used from them by air, or with them by land and sea, were not developed until after December 8."[4] The country's massive production of traditional piston-engine, propeller-driven airplanes helped win the war. Witness the strategic bombing campaigns of the B-17 bombers and P-51 fighters, or the air and ground attacks of the F4U Corsair. Their armaments, in new solid-fueled rockets, helped win the war with offensive, annihilating effect.

The story of America's battlefield rockets begins in June of 1940 with the establishment of the National Defense Research Committee (NDRC) at the direction of Vannevar Bush. As president of the Carnegie Institution of Washington, DC, and chair of the National Advisory Committee for Aeronautics (NACA), he was inspired to action by Nazi Germany's aeronautical advances, so reported by Charles Lindbergh after his recent visit there. America needed to compete. Bush was joined by James Conant, president of Harvard; Karl Compton, president of MIT; and Frank Jewett, president of Bell Telephone Labs. These "four horsemen," as the media dubbed them, were about to wield apocalypse upon the enemies of the US, in the form of American science and engineering, to assist the UK in the war and accelerate the R&D of new weapons.[5]

Bush was the catalyst. He was a consummate diplomat and organizer of people and things. Among his wartime posts, he was a member of the board of NACA, director of the NDRC and OSRD, chair of the Military Policy Committee and adviser to its Manhattan District (for the atomic bomb), and chair of the New Weapons and Equipment Board of the Joint Chiefs of Staff. He was one of the most powerful figures in Washington, DC, enjoying FDR's confidence and support. Associates remembered that he had more authority than most generals.[6] His easygoing nature defined his style. One interviewer described him as an eccentric country schoolteacher: "tall, gaunt and angular, with a humorous twinkle in his small blue eyes and an obstinate lock of hair shooting forward from his slightly bulging forehead . . . his leisurely manner, his pithy speech, bespoke the classroom rather than the battlefield."[7]

America's new romance with the rocket actually began with a much-publicized mission to the UK in February of 1941, across the Atlantic by way of Bermuda and Portugal. The mission initiated a "secret program of collaboration" over "secret vital war research projects." The small group was headed by Conant and Compton from the NDRC and included a young Frederick Hovde, former Rhodes scholar, who had a network of contacts in England and was set to remain and build the wartime partnership in weapons R&D. The future president of Purdue University, like his predecessor Edward Elliott, Hovde was a man of the West. He grew up in Devils Lake, North Dakota, graduating from the University of Minnesota with a BS in chemical engineering (1929), where he was also quarterback of the football team. In 1940, Hovde became the "representative of American science" to the British leaders and scientists, the fellow who "built up the office to a state of excellence" in the chaotic conditions of actual war.[8]

Hovde was at work in London from February of 1941 to June of 1942, about a year and a half, one of two NDRC scientific liaison officers there. His official titles were resident secretary and head of the London office (Mission) of the NDRC and head technical aide and special adviser. He oversaw the transmission of personnel back and forth between the US and UK on topics like radar and battlefield rocketry. He accommodated specialists from both countries, giving technical experts the room to educate each other. He cleared them for all the right permissions between the US State Department and British Foreign Office and all the various military and intelligence services. He moved information too, reports and technical data, which demanded strict attention to scientific accuracy, secrecy codes, and above all speed. Between March and August of 1941, the office prepared 138 memorandums, 121 diaries of NDRC specialists, 151 telegrams, and most importantly, 947 official reports. The safe became too small to protect the growing library of secret reports, so a Marine guard took on the job.

Hovde also wrote his own reports, based on the conferences he held and the knowledge he gathered, focused on "urgent policy and general trends." These were reports for Bush and Conant back home, often placed as a secret addendum to technical documents, explaining Hovde's priorities for necessary action by the NDRC. They encompassed quite a range of topics, such as arming bombers with improved gun turrets, radar, electronics issues of many varieties, tank designs, toxic agents, uranium research, rocket propellant charges, and the Frank Whittle turbojet engine. Hovde called these "problems of policy and future research." He was especially interested in air-to-air and air-to-ground rockets, thanks to his close work with Sir Henry Tizard, their patron. The British Air Ministry was already invested in battlefield rockets. The Russians were using them against Nazi tanks. "Such development should also be pushed in the States," he advised.[9]

Hovde also became America's "atomic" man. In October of 1941, he flew to London to deliver a secret note from Franklin Delano Roosevelt to Prime Minister Winston Churchill, proposing cooperation between the US and UK on the atomic bomb project. The note was hidden in a belt fastened under his clothing. More than a courier, though, Hovde was also an NDRC scientific liaison and informal diplomat for this effort. Roosevelt called him the "head of the London office of our scientific organization," a plenipotentiary to delve into the atomic issues more carefully and advise the British on the available US resources. Only four people had access to American and British technical secrets at the time: Bush, Conant, Compton, and Hovde. The British delayed a response for several months, seeking to gain an upper hand over the Americans. Hovde actually did very little, by design. The delay perfectly served James Conant's ulterior purposes, as he was one of Hovde's mentors and the NDRC manager in charge of the bomb project, and was keen on an American monopoly in the atomic field. By spring and summer of 1942, the US was forging ahead, mostly on its own.[10]

Hovde experienced dramatic events during his tenure in London: the continuing Blitz and Battle for Britain; Lend-Lease in the spring of 1941; Pearl Harbor by the end of the year. He made several hazardous trips across the Atlantic, once in the belly of an Army Air Corps bomber. But it was not all hardship. There was the camaraderie of friends, the drinking of many beers, and weekend trips to the English countryside. There was the prestige of working at the US Embassy at Grosvenor House. Hovde was fond of one elegant supper there, on 23 July 1941, to celebrate the Anglo–American partnership: *Saumon Fumé, Crème Faubonne, Caneton Rôti à l'Anglaise, Petits Pois Fins*, and *Framboises Melba*. He saved the menu.[11]

While Hovde was in London, Bush retooled his approach to wartime R&D, establishing the OSRD in July of 1941, a new executive agency to supplement and encompass the NDRC. Both were within the Office of Emergency Management. Bush had organized the NDRC by way of the "committee form," borrowed partly from NACA, which was essentially an advisory committee for aeronautics. The NDRC was a committee of committees, including subcommittees, special committees, and ad hoc committees, and a whole series of accompanying divisions, sections, and panels. Civilian scientists were in the lead, in liaison with military officers, directing and engaging R&D in a maze of secretive units, alphabetized by the last names of their directors. It was not an advisory body but an executive agency. Most significantly, "rather than establish any large organization of its own," Bush "decided to use its available appropriations through contracts with universities and industrial firms." In its first year, NDRC negotiated 270 contracts with 47 universities, and 53 contracts with 39 industrial firms, with several thousand scientists already in the ranks, like physicists from Caltech and Bell Telephone Labs, chemists from the University of Illinois and Standard Oil.

Building from aviation's "revolution in transportation," its eradication of space and time, the NDRC mobilized science to help realize "the potentialities that lie hidden in the future."[12]

The OSRD redoubled the NDRC, joining it with a Committee on Medical Research, empowering its administration to coordinate scientific research in the War Department, Navy Department, and federal government. The crucial factor in all its deliberations was speed: to work fast, and to look only so far ahead as to what was most achievable and lethal. It was research *for* development, the making of battlefield weapons to win the war. Costs mattered, though less so, in order to balance scarce resources with quick results. What made it work? Bush called it his "pyramidal unitary organization," one that was hierarchical but also "associational," animated by committee democracy and peer review. By early 1943, the OSRD was managing two thousand contracts with two hundred industrial labs and one hundred university labs, comprising some six thousand people.[13]

Amid these dramatic events, Hovde returned home in June of 1942, appointed by Bush to serve James Conant, the chair of the NDRC, as his executive assistant, a post he held until September of 1943. Bush wrote Hovde a motivational note at this time, playing on one of his favorite aviation metaphors, that the war effort now had "lift," encouraging him to push others to keep working hard and smart for the long-haul flight. Hovde needed the boost. He not only served as Conant's assistant, but also as head scientist of the NDRC in a technical and administrative capacity. He was closely involved with radar work and at one point was considered for chair of the Radar Development Planning Committee. He coordinated several NDRC divisions and sections for scientific warfare. He was the primary liaison with the War and Navy Departments. He also served as acting chairman of the NDRC when Conant was absent, taking on the duties "in his stead." Hovde was responsible for "watching all research projects to see that there is no duplication of effort." He reported "on the status of research projects and the development of new projects," as well as on the "utilization of accumulated data in the perfection of these projects." This all happened at a crucial turning point in the war, as America was just beginning to actually engage the enemy in battle.[14]

Conant needed the help. He was hardly ever at the NDRC between the fall of 1942 and the fall of 1943. Instead, he was busy working out of the Dumbarton Oaks mansion, or traveling to the West Coast, transitioning the atomic bomb project from the OSRD to the US Army and its Manhattan District. This was his "$2 billion bailiwick," as biographer James Hershberg phrased it. As a result, Conant left the NDRC in downtown DC "under Hovde's supervision" and "with much to do."[15] The organizational charts and official histories never revealed it, and Conant never relinquished his title, but Hovde ran the NDRC on a daily basis for about a year, with business trips of his own to places like Pittsburgh and Dayton, Los Angeles and Los Alamos. He was also acting NDRC chair for the crucial reorganization in fall 1942, when it transitioned into nineteen divisions, ranging from War Metallurgy to Sub-Surface Warfare to Radar and Radio Communication. Hovde reduced their administrative functions, giving the scientists who ran them more independence and authority for creative and cooperative work.[16] Following the lead of Bush, he inspired the new teams to devise smarter strategies, work better with their military liaisons, telescope their work from ten years down to two, and get their products into mass production for immediate battle use.[17]

Hovde's longest tenure in the OSRD was to serve, at Bush's appointment, as chief of Rocket Ordnance (Division 3), formerly known as Special Projectiles, between September of 1943 and the summer of 1946. It was a logical fit. In London he had worked under Charles Lauritsen for US investments in aircraft, antiaircraft, barrage rockets, and jet-assisted take-off (JATO) rockets. Hovde advocated an American version of "lightning" technical war, meaning rocket weapons as "mass in motion," weapons to "absorb the thrust and rebound to the attack."[18] He was no stranger to Division 3. Bush had already appointed Hovde there as a liaison and staff aide in late 1942. He eventually ran the division from his own offices at Twenty-First Street, on the campus of George Washington University. The division was initially in disarray, given its rapid and vigorous expansion to two coasts: Section H, under Clarence Hickman and Ralph Gibson, at the Aberdeen Ballistics Laboratory and Proving Ground, and the Allegheny Ballistics Laboratory; and Section L, under Charles Lauritsen, at Caltech and the Naval Ordnance Test Station. Bush needed a disciplined leader. That leader was Hovde, whom Bush later congratulated for his "energetic handling" and "great skill" in the new post.[19]

Hovde's task was to consolidate all rocket work under his authority, in agreement with the military services.[20] The work was complex. Neither the Navy nor the Army had any experience with the wide array of battlefield rockets then in planning. There were no engineering texts or guidelines. Rocket research meant balancing a host of shifting variables: shapes, sizes, weights, and velocities; the design of the presses for powdered charges; the design of heads, fuses, and launchers; the motor design for liquid propellants. To these Hovde added several other factors under his own purview: "our own program, the British program, and the enemy program." Rocket R&D was not easy.

Hovde took charge on 30 September 1943 at a meeting between Division 3 and War Department Ordnance officers. The transcript reveals his diplomatic but authoritative style, showing respect for the military and civilian teams, but calling for a series of essential reforms. He was also blunt. Division 3 and Army Ordnance officers had not been cooperating. "We haven't got in hand at the moment all the things we should be doing," he said. The military was not communicating its needs to the R&D teams: improvements like "larger payload higher velocity rockets," and "accuracy" and "designs using rotation," and "faster burning powder." Although focused on the weapons development program, Hovde defended and promoted "pure research." Division 3 needed to study "the internal ballistics of the motor," as well as "the external ballistics, sense of flight, dispersion, yaw, drag, and so on." These were not academic concerns, so he lectured the ordnance officers. They were essential. "It is absolutely necessary that we do everything in our power to learn more about what goes on inside a rocket motor." Hovde called for "a complete theoretical study" of solid- and liquid-propellant rockets, including long-range rockets to come.[21] Division 3's achievements covered rocket ordnance for sea, land, and air use. They were new and daunting weapons of offensive war, so useful because they offered recoilless action and speed. They included the Navy's *Mousetrap* rocket, launched from its ships against submarines; the High Velocity Aircraft Rocket (HVAR or *Holy Moses*), launched by the P-47 fighter against tanks; and the famous *Bazooka*, a shoulder-launched battlefield rocket. Hovde also oversaw Caltech's and Aerojet's new work with JATOs and their liquid-propellant rocket engines, and he occasionally consulted with Maurice Zucrow at Aerojet on their status and George Hawkins at Purdue on heat transfer issues.[22]

In all these capacities, Hovde was one of Bush's trusted "bird dogs," aides who navigated the peaks and valleys of the OSRD pyramid. Its official history named Hovde disparagingly as but "one of the members of the staff at one time or another."[23] Yet from start to finish, no one on the NDRC or OSRD staff wielded the kind of coverage that Hovde did. He enjoyed across-the-board horizontal and vertical power. In London, at the NDRC, and in Division 3, he was Bush's ultimate fixer. Besides these posts, he also served in the following capacities within the OSRD: lead scientific officer of Division 3; member of Division 4 (Ordnance Accessories); member and consultant (and lead director) at Division 5 (New Missiles); member of Division 8 (Explosives); and member of Division 19 (Miscellaneous Weapons).[24]

Hovde's wide coverage brings us to a third institution that Vannevar Bush helped create in Washington, DC. First was the NDRC in 1940. Second the OSRD in 1941. Third the Joint Committee on New Weapons and Equipment (JNWE) in 1942, part of the of the Joint Chiefs of Staff. Bush was its chair. Hovde was its NDRC representative, a member of its Committee on Guided Missiles, and chair of its Subcommittee on Rockets. These posts gave Hovde wide access to the US Navy's Bureau of Ordnance and Bureau of Aeronautics, to Army Ordnance and the Army Air Forces, and to NACA. Although little known, the JNWE was perhaps Bush's greatest legacy, in that it had lasting power, becoming the Joint New Weapons Committee (JNW) in 1943 and then the Joint Research and Development Board in 1946.

Bush and Hovde leveraged their power on the JNWE to weigh against the building of an American V-2 missile. By summer of 1943, British aerial surveillance had discovered the German V-2 missile sites at Peenemünde on the Baltic coast, and even retrieved and reverse-engineered a stray V-2 that had crash-landed in Sweden. US Army Ordnance and the Army Air Forces investigated the option of building an American version. Hovde became chief of Division 3 at this moment, in September of 1943, just after the very first intelligence reports had reached the US. They painted a clear technical image of the dramatic new weapon. Opinions in the OSRD were mixed. Ralph Gibson, a close confidant of Hovde's and leading rocket specialist in Division 3, rose in favor of researching and developing an American V-2. Maj. Gen. Gladeon M. Barnes concurred. His opinion mattered. As chief of research and engineering and chief of the Research and Development Service, Office of the Chief of Ordnance (1938–1946), his was the voice of the Army in Division 3. Barnes appreciated the potential of an American long-range missile to bomb Berlin and "demoralize" the enemy. Compared to the fleets of B-17 bombers already doing this, he argued that missiles would "be lots cheaper" for the same results. Hovde also expressed his support of "rockets for the projection of large payloads, big bombs." But by this he only meant long-range rockets of about seventy-five miles, a project that he was about to delegate to the GALCIT program in Pasadena. Hovde was equivocal about longer-range rockets of the V-2 type, with well over twice the range: "I am not at all sure in my own mind that this problem warrants the diversion of manpower it would require."[25] This was the Vannevar Bush line, applied by Hovde with both strength and tact. America would make no V-2.

Hovde was also working from the facts as he knew them. Most of his Section H experts advised against the American development of a V-2. Based on the knowledge at hand, they had made "very close predictions" about its design and performance. They had also judged it a close approximation to Robert Goddard's rockets, and quite similar in their fuel injection pumps, gyroscopic controls, and regenerative cooling. Above all, Hovde and Division 3 judged the V-2 a weapon without a "military purpose," whose investment of monies and time (up to two years) was simply not "warranted." The US Army Air Forces applied the same logic to the accelerated development and production of American turbojet fighters. Better instead to mobilize mass production of traditional aircraft already in play against Germany's declining armies, cities, and industry. In this way, as aviation historian Walter Boyne has said, "a huge quantity of good technology overwhelmed a much smaller quantity of very advanced technology."[26]

None of this came easy for Vannevar Bush. As Allied knowledge about the V-1 pulsejet and V-2 rocket became widely known even before their first use in June and September of 1944, the military services lobbied for review of OSRD's decision against an American V-2. Col. Leslie Skinner, recently back from consultations with British analysts on the V-2, asked Bush to advance research and development of American "large long-range artillery rocket propulsion." In response, Bush established a special OSRD Committee on Jet Propulsion Systems in February and March of 1944, combining Army, Navy, NACA, and OSRD representatives charged to visit laboratories around the country to survey the "present state of development of long-burning jet propulsion units." With Bush's prodding, they found it satisfactory, advising no accelerated development of an American V-2, only new coordination in a set of JNWE technical panels.[27]

Bush expressed his approach to such new weapons in a letter to Hovde, cautioning him to "settle into the collar for the long pull," to stay vigilant against too much innovation: "In the shifting scene of a technical war it is impossible to tell what may be of greatest importance. The value of one thing that comes through, offsets many an effort that does not, and yet the whole effort is necessary in order to be ready as the scene may shift."[28] The decision not to pursue an American-style V-2 as directed by Bush and Hovde fell within this category of "shift." Or in this case, no shift at all. Bush depended on Hovde as his point man in the OSRD's Division 5 (New Missiles) to keep the country's guided-missile efforts tempered and slow. He congratulated Hovde after the war, in appreciation for this work, for "directing the over-all Divisional Program," and for his "strong liaison" between the "Rocket Program and the Guided Missile Program." As Bush said, through Hovde's "advice and counsel, the Division refrained from fostering a program of guided rockets which would have been costly and probably ineffective in the war." Hugh Spencer, the nominal chief of Division 5, also thanked Hovde for how his "incisive thinking guided the Division at its meetings along fruitful channels and prevented erratic wanderings."[29]

Such words were little consolation for Hovde after the war. Once the US atomic bomb and German ballistic missiles were publicly known, politicians and journalists criticized the OSRD for having fallen behind with ballistic missiles. Hovde recognized his own "failures, difficulties, shortcomings, and errors of judgement" in Division 3. But he did not relent on the V-2. The OSRD was right not to invest in one, nor in the massive and expensive Peenemünde Nordhausen complexes and laboratories to build it. Instead, the US built conventional rockets for actual war. Workers in our laboratories "provided a greater variety of useful rocket weapons to our fighting men than any other nation." They got them out quickly and effectively to apply lethal force. Hovde called this an "economy of force," an "imperfect" but "proper balance" between current and future weapons. He also appreciated the OSRD's flexibility, its easy adaptation to the American free market and free university, the "independence" of the project system.[30] As he also recalled, the "atomic bomb program was a better bet within the time available and that program needed the available scientists and engineers." The US needed to "spend the nation's money to win the war, not to win the next war."[31]

## WARFARE AND HIGHER EDUCATION

The OSRD was an all-American institution. Not only in its use of the committee form, but also in its application of the project system, inherited from the Department of Agriculture, the Agricultural and Engineering Experiment Stations, and NACA. As Bush instructed, the OSRD would "write contracts with universities, research institutes and industrial laboratories."[32] It initiated projects and wrote contracts with institutions, projects to engage their teams of research scientists and engineers in the creation of discrete weapons and systems. The emphasis was on creative enterprise to define the problem, investigate it, confirm it, and create a final product or process within an achievable timeline. It was "a thoroughly pragmatic set of arrangements," as Don Price put it. "The government contract became a new kind of federalism." The OSRD, with the military, industries, and universities, imitated the relationships between the federal government and the states in a new separation and balance of powers.[33]

There had been project contracts before, but nothing like this: just over 2,000 new ones in some 440 organizations. The War Powers Act waived the requirement to publicly advertise for contracts to universities. The OSRD also spent a lot of money: $453,656,657 between 1940 and 1946, about half to industry and half to academia. The top industries were Western Electric at $17 million, followed by DuPont, RCA, Eastman Kodak, and General Electric. Among universities, MIT received most, 75 contracts for $117 million (about 23 percent of the total). Caltech was next, with over $83 million (16 percent of the total). Then came Johns Hopkins with over $10 million and Princeton with over $3.5 million. The Universities of Chicago, Michigan, Illinois, and Iowa received lesser millions. On the coasts, Harvard, Columbia, George Washington, and UCLA did too.[34] This was all a remarkable upsurge in government financing of R&D, especially in higher education. Keeping track of it all was the Administrative Division, under the Office of the Executive Secretary (headed by Irvin Stewart, who answered directly to Bush). Its sections for contracts, project control, central records, and procedures were staffed and managed mostly by women, a small army of administrators, typists, clerks, archivists, and secretaries who created the documents and kept them standardized, organized, copied, and moving.[35]

This was all very new. Before the war, the limited disbursement of federal funds to universities (about $140 million a year) meant that few rules applied. Industry was eligible for 100 percent claims for overhead costs, those to cover the normal business of the firm beyond the exact services provided in the contract. Industry's accounting practices and bookkeeping made this easy enough. Academia did not have such a system in place and by practice was due to receive a lesser percentage of the contract for overhead. So Bush's OSRD policy paid universities with a "no-profit and no-loss" provision and a normal overhead rate of 50 percent, initially tied to the labor payroll. This was still a prize, a substantial set of earnings for universities, along with helping to save the US and the free world from Nazism and Japanese imperialism.[36]

Purdue was not among the top twenty-five academic contractors for the OSRD during the war. Yet it had a role to play, first and foremost in the contributions of R. B. Stewart (Robert Bruce, better known as R. B.), Purdue's chief business officer and vice president of Purdue Research Foundation (PRF). Bush appointed R. B. as a special adviser to review the OSRD's elaborate projects and contracts. Several universities had lodged complaints about overpayments to their more privileged rivals. Bush sent him around the country to study "accountability for the use of the funds," especially with regard to university overhead charges for general costs. Was he "just giving money away or was he getting something?" In response to the abuses he found, R. B. devised a new contract system to finance OSRD research on college campuses. "I became the head of a team to mastermind these contracts," he remembered.[37] He applied accounting standards to accommodate all universities, both large and small. These included a sliding scale of overhead based on a percentage of reimbursable salaries, wages, and a variety of other costs. The basic rule was 50 percent overhead for small contracts up to $150,000 and 30 percent for those over $2 million, though in both cases he offered leeway for special cases and local conditions. He effectively restrained overhead gouging by the more privileged schools.[38]

Beginning in the summer of 1942, R. B. adapted this system for the US military. He became chair of the Joint Army–Navy College Contract Board and coordinator of the College Training Programs for the Army and Navy Departments, directly advising secretary of war Robert Patterson and secretary of the Navy James Forrestal. R. B.'s OSRD work was his calling card. The Navy was also impressed with Purdue. As it was one of the first universities to start Navy wartime training on its campus, R. B. gave the Navy a "set price" for housing and food and services, this thanks to his and President Edward Elliott's earlier work on the first true "unit costs" in higher education. The Navy valued the "thoroughness" with which R. B. prepared his cost figures and the "fairness" with which he "approached the problem of negotiation." Now it and the Army needed R. B. to "codify all contracts" under stricter "cost accounting" principles.[39] The practices were a mess elsewhere, the result of a culture clash between the military and university. Some naval officers, for example, expected "T-bone steaks four times a week for dinner" from the privileged ivory towers of academia, said R. B. "In every institution across the country you had civil war between the institution and the naval or army unit on the campus." People stopped speaking to each other.[40]

R. B. enjoyed a unique authority on the College Contract Board. In the interests of democracy and equity, he brought in more university representatives from around the country. He was also the "swing man," the deciding vote in disagreements between the military services and universities. He could "tell the president of a university or a staff member to go to hell to get something done." Universities did not "love" R. B., but they did "respect" him, he admitted.[41] His guiding principles were fairness, uniformity, and flexibility: to establish "contract procedures to protect the colleges against loss and the Government against profit." As to the military, it "should leave the colleges autonomous in the administration of their own institutions in order that free education in America shall not be lost during the national emergency." As to universities, he created standard rates of payment, uniform contract provisions, and a standard training unit contract. He applied strict accounting procedures and precise charges. Colleges had to charge for costs based on their own documentary records, to charge for what things actually cost.[42]

R. B.'s contributions to the OSRD and US military were not just about fairness. They were also about self-interest. Purdue may not have risen very high in the ranks of OSRD project financing, but he put the new federal monies to good use on campus. As R. B. bragged, "The real boost to PRF came during the War," with contract overheads "which we impounded and used to build capital for the foundation." He rewrote the rules, but also used them to Purdue's benefit.[43]

Through the war, Purdue administrators were otherwise focused on manpower and education: on people, not things. Between 1942 and 1943, President Elliott served on the War Manpower Commission, spearheading its committees to organize higher education for the war. He was chief of its Division of Professional and Technical Training. He helped establish the National Committee on Education and Defense. He served on the US Committee on War-Time Requirements for Specialized Personnel. He directed the National Roster of Scientific and Specialized Personnel, the Procurement and Assignment Service, and the Army Specialized Training Program. These offices may have lacked glamour, but they offered crucial wartime support by mobilizing doctors, dentists, lawyers, professors, scientists, and engineers around the country. His principles, like R. B.'s, were for fairness and flexibility, especially in representing all kinds of universities from all over the country, previously forgotten places like Oregon State College, the University of Minnesota, Texas A&M, and West Virginia University.[44]

Elliott was also initially responsible for the Engineering, Science, and Management War Training program (ESMWT), though he delegated most of that responsibility to dean Andrey Potter. Or as Potter later remembered, "I planned it." Potter was chair of the National Advisory Committee and served as its expert consultant and liaison officer to the universities for about one week of every month between 1940 and 1944.[45] He drafted the basic principles by which the ESMWT operated. They reflected his philosophical, pedagogical, and procedural approach to administration. Central among these principles was that the ESMWT's purpose was merely to assist. The universities ruled. This was a "program of the institutions" and defined by "regional cooperation" and local control. Potter chose the national administrative staff from Kansas State College, the Missouri School of Mines and Metallurgy, South Dakota State College, Antioch College, the University of North Carolina, and Worcester Polytechnic Institute. He divided the US into twenty-two regions, each self-governing with its own chair and committee.[46]

The people who ran the ESMWT credited its success to Potter's "organization and execution," his "time, energy, spark and leadership." He also made it "unique among government agencies for the simplicity of its regulations and the freedom from annoying restrictions and red tape."[47] Training in the ESMWT was diverse. It ranged from the education of engineers in the new field of radar at MIT, to simple refresher courses in mathematics at Ohio State University, to Maurice Zucrow's pioneering course in jet propulsion at UCLA. Seventy-five percent of the enrollments were in engineering (mostly electrical); 21 percent were in management. Of the 1,795,716 students who enrolled, 282,255 were women (one-sixth of the total). Only 25,158 were African American. The University of California system had the largest enrollment, at 150,621, followed by Pennsylvania State College at 141,569 and Purdue at 64,251. ESMWT negotiated contracts with more than 227 institutions, all based on Potter's strict rule for charges, "costs to be paid but no profit," including university overhead, which he kept very low. Here was yet another contribution from the frugal Boilermakers. Congress budgeted nearly $60 million for the ESMWT. Of that, $28.5 million went unspent and was sent back to the US Treasury, largely thanks to Potter.[48] That might have been a first, and last, in American history.

Back at Purdue, as director of the Engineering Experimental Station, Dean Potter supervised the wartime research projects, eager to keep the best faculty and staff on campus and deferred from the draft. He and Harry Solberg established several aeronautical projects with the Air Materiel Command at nearby Wright Field. George Hawkins worked with Army Ordnance on Purdue's signal strength in heat transfer: how to keep machine guns and other weapons from overheating.[49] The PRF managed a series of secret agreements between the OSRD and the Chemistry and Physics Departments dealing with the atomic bomb, explosives, chemical weapons, and radar. Most of this work was conducted in national laboratories off campus, although the Locomotive Testing Laboratory did become the site of atomic energy testing in 1944, and Purdue received a release for its graduate students to conduct "extra-hazardous work" as part of the war effort. Both departments made significant contributions to the Manhattan Project. On the whole, the Second World War was also a financial boon to the university. Of the $5 million that the PRF brought in between 1930 and 1948, 88 percent came after 1941.[50]

Significant though Purdue's wartime contributions were, they were not as prestigious as the OSRD programs at the big three universities. MIT led with its radar work in the Microwave Committee and the Radiation Laboratory. The University of California at Los Angeles and Berkeley followed with their atomic R&D. Caltech was actually a double winner. Beyond its OSRD battlefield rockets, von Kármán's Guggenheim Aeronautical Laboratory (GALCIT) continued its rocket propulsion research in solid and liquid propellants, with wide-ranging funds from the military services and the National Academy of Sciences. Its small Rocket Research Group, also known as the Experiment Station of the Air Corps Jet Propulsion Research Project (1941), helped to invent American rocket science, confronting the challenges of propellants, high pressures, heat transfer, motor design, and fortified materials, culminating in GALCIT Project No. 1 on the "development of a Liquid-Propellant Jet Motor." In collaboration with Robert Truax and the Engineering Experiment Station at the US Naval Academy (Annapolis, Maryland), GALCIT decided upon its propellants of choice: red fuming nitric acid (RFNA) and aniline, the best of the hypergolic "self-igniting" varieties.[51]

GALCIT succeeded in flying a series of JATO devices in 1941 and 1942, designed to assist traditional piston-engine propeller airplanes to shorten their takeoff times, from 13.1 to 7.5 seconds; to shorten their runway distances, from 580 to 300 feet; and to extend their level flight speed by 45 mph. The US military valued these gains, especially for the Pacific theater of war, where its airplanes carried heavy loads and needed the extra power for the short runways on aircraft carriers and islands. JATOs for the US Army Air Forces and US Navy meant contracts, and contracts meant mass production. Mass production meant the need for an actual company: the Aerojet Engineering Corporation, founded by von Kármán and his group in March of 1942. It was a radical departure from traditional practice. No Engineering Experiment Station or university laboratory had yet transformed into a corporation. The Air Corps allowed this anomaly given the novel technology and the imperative of war. GALCIT needed to convert rocket science into rocket engineering, to turn research discoveries into operational products. As one of its new directors argued, "We must adapt scientific principles to working units—units which are sturdy, dependable, safe, and workable." This demanded not research scientists but a variety of "mechanical engineers, thermodynamicists, hydrodynamicists, aerodynamicists, stress analysts, designers, and practical production men."[52] The stage was set for the entrance of Maurice Zucrow.

## THE AEROJET EXPERIENCE

Aerojet Corporation was first housed in a converted car dealership at 285 West Colorado, Pasadena, filled with testing and engineering labs and production facilities. With its early successes, the company built a forty-eight-acre test site at Azusa (the Propellant Plant Proving Grounds), about twenty-five miles east of Los Angeles. Aerojet was an interesting place to work. Andrew Haley, a gregarious and hard-drinking DC lawyer, replaced von Kármán as director. His personal secretary, Frances Christesen, called him the heart and soul of the company, a person with "dynamic drive and enthusiasm."[53] Then there was John Parsons, a follower of occult leader Aleister Crowley, and his *Ordo Templi Orientis*. One of the founders of the Rocket Research Group, Parsons worked on rockets by day and engaged in magic and esoteric rituals by night. With his smart-looking goatee, he even sometimes sang the pagan ode to Pan as the group prepared its rocket tests. The proud Roman Catholic Haley loved to hear it, as

Parsons chanted, "I am gold, I am god . . . And I rave, and I rape and I rip and I rend."[54] These were some of the colorful colleagues Zucrow joined in the middle of the Second World War.

Zucrow had a personal stake in the war. In the Blitz, Nazi bombers, and eventually V-1 and V-2 missiles, destroyed large parts of his old neighborhoods in the East End of London. In Operation Barbarossa, the invasion of the USSR, Nazi armies annihilated the Jewish populations of Ukraine. Maurice's birthplace, Kyiv, became the site of one of the largest and most concentrated killing fields of the Second World War: the ravines at Babyn Yar, on the outskirts of the city, where Nazi forces murdered approximately one hundred thousand persons, including some thirty-four thousand Jews, between 29 and 30 September 1941. They were mostly the infirmed, elderly, and women and children.[55] There was no escape for his first homelands from these atrocities.

As the war waged in Europe, Zucrow left Chicago to become a research and development engineer for the Elliott Company at Jeannette, Pennsylvania, near Pittsburgh. It was an act of patriotism and national service. There, in cooperation with the US Navy, he helped to design and improve one of the country's first gas turbine power plants, the Elliott-Lysholm Marine Gas Turbine plant, first operated in October 1944. The gas turbine was nothing new. Inventors and engineers had been experimenting with prototypes since the nineteenth century. The Brown Boveri Company in Switzerland had more recently created the world's first practical model. Zucrow's special skill set in thermodynamics and fluid flow, as well as power plant design, made him especially valuable in planning the "two-turbine arrangement, the high-pressure turbine driving the low-pressure compressor, and the low-pressure turbine driving the high-pressure compressor and supplying the useful output."[56]

He was not there long. For "the siren beckoned to things more exciting—jet propulsion." That siren was none other than Theodore von Kármán, who called Zucrow long-distance with an urgent invitation to come work at Aerojet. Von Kármán made his pitch, looking for "an engineer with a knowledge of heat transfer, turbines, combustion, and other things." Zucrow thought he meant turbojet engines but soon realized he was talking about rockets, the vanguard business of internal aerodynamics. Haley and von Kármán needed Zucrow for his mastery of engineering science, and for his experiences in product development. Thus began Zucrow's "California days" between November of 1942 and July of 1946. The "war years, rockets, jet units, excitement galore."[57]

Zucrow was hired with about sixty new employees to service the Air Corps and Navy contracts for JATOs. The company's culture drew a clear divide between the pioneer founders and the newcomers. Its cult of the five men celebrated Kármán, Malina, Parsons, Forman, and Summerfield for doing what Robert Goddard never could: design and test solid- and liquid-propellant rockets for actual use. They turned a $2,500 investment into a company worth over $1 billion. Von Kármán and Summerfield were PhD physicists. Malina was the group's only PhD engineer, though he and the others had little in the way of practical engineering experience. Parsons and Forman were brilliant technicians, without degrees.[58]

The working dynamics between GALCIT and Aerojet are not easy to untangle. It is a history yet to be written. They remained independent, at times with rivalries and resentments, if usually working in close collaboration. Each had its own codes of secrecy and loyalty. Parsons worked seamlessly for both GALCIT and Aerojet, as did Summerfield at times. Malina was embedded mostly at GALCIT, where he was chief engineer of Project No. 1, though he answered to Hovde in the OSRD on "liquid developments" for both GALCIT and Aerojet. His official title at Aerojet was engineering consultant, where he worked in the Executive Office and Engineering Division. He and Zucrow had common interests and education, and they might have worked more closely together into the future, though Malina tended to monopolize and compartmentalize his own power within GALCIT.[59]

In his memoirs, Malina often diminished Aerojet as a mere production and business arm, a passive vehicle for GALCIT "know-how." Summerfield did the same: when asked to compare GALCIT and Aerojet and their "principal influences" on American rocketry, he quickly answered that GALCIT was the true creative force. "There's no question about that," he said. "It developed propellants. It developed chemistry. It developed knowledge of injection. It developed combustion chambers. . . . It developed nozzle shapes. It developed everything." Zucrow had a different perspective. Yes, he also recognized the initial landmark research of GALCIT Project No. 1. But it was Aerojet that turned GALCIT's "hazy" and rough "art of rocket jet propulsion," its "kitchen chemistry," into a "well grounded" and "well defined science."[60]

Zucrow had varied titles and roles within Aerojet. At times he published under the titles of staff or executive, division or design engineer. Most documents refer to him rather modestly as the "engineering technical assistant" or "technical adviser" to the executive vice president and contract administrator, William Zisch. Zisch was an unlikely combination of chemist, pilot, and accountant, a recent graduate (1939) of Pasadena's Sawyer School of Business. He was ultimately responsible for project contracts, getting reliable and finished products to clients on time. Zisch assigned Zucrow to discrete tasks and project teams, and to miscellaneous or additional problems, as his enforcer on the ground floor. He trusted Zucrow as Aerojet's point person in technical negotiations and reports to the Army Air Forces and to the Navy.

In essence, Zucrow was Aerojet's roving project engineer, on call, someone able to address myriad issues along the project timeline, from initial design parameters, to the efficiencies of propellant mixtures, to cooling features or nozzle shapes, to the final stages of development and production. He was first widely known as "Doc" in these early Aerojet days. The appellation was already in use in the engineering and industrial workplace, given the few PhDs in a rather rough business. Popular culture made it even more popular, as in the Doc Savage and Doc Smith series of science fiction stories, or in the famous line "What's up, Doc?" first delivered by Bugs Bunny in the Looney Tunes cartoon short *A Wild Hare* (1940).[61] Doc was a sign of respect and familiarity, in Zucrow's case for an engineer who was also a teacher and mentor.

Zucrow had much to learn as well. He worked closely at Aerojet with the brilliant Fritz Zwicky, also a professor of astrophysics at Caltech, who shaped his grand new morphology of jet propulsion during the busiest days of the R&D groups in May of 1943. They were facing the formidable challenges of weight reduction. Dr. Zwicky "brought up the fundamental point that weight reduction will inevitably accompany improvements in efficiency," suggesting that "tables should be compiled of the ultimate free energies available from different fuels and oxidizers."[62] Like a modern-day Carl Linnaeus, Zwicky saw the need for a whole new taxonomy of jet propulsion things. He called it "power generation through the utilization of chemical reactions," meaning "the ground work for the new large-scale applications of jet propulsion as a means of motive power." Rocket science needed to explore fuel-propellant combinations ("self-contained propellants") and chamber temperatures and pressures for the best specific impulses and exhaust velocities, including nine thousand feet per second, reaching for extreme altitudes. Zwicky framed Aerojet's achievements, in one report to the US Navy, as nothing less than the foundations for a new world of amazing energies and power plants.[63] He was thinking through all the practical pathways to spaceflight.

Zucrow's other close collaborators at Aerojet included Dave Young, the company's first employee. Young had a Caltech BSME (1942) and wide industry experience, including a stint doing hydraulics at the Lockheed Aircraft Corporation. George Sutton also joined by 1943, with a recent associate's degree in engineering from Los Angeles City College, a forthcoming MSME and teaching experience from Caltech, and wide experience with internal combustion engines, wind tunnels, and the "recoil mechanisms of large caliber guns." Zucrow helped get his former student at Purdue, Bernhardt Dorman (BSME 1928), an Aerojet job. Zucrow also had one equal on the staff in terms of education and career experience: R. H. Krueger (PhD, engineering, Technische Hochschule, Darmstadt Germany, 1931), the chief engineer in the Liquid Development Division. Krueger had quite a resume: He had been a development engineer in automobile plants, Bell Telephone plants, and power plants. He had designed wind tunnels, sea walls, and transportation tunnels. He had completed naval contracts for torpedo mounts and hydraulic devices. He had designed production facilities for Rolls Royce Merlin engines and for the Packard Motor Company. He had helped to build the steel mills at Magnitogorsk, USSR, and oil refineries at Ploesti, Romania. With such a wide array of talent, Aerojet was primed for success.[64]

Aerojet was an exciting and exasperating place to work. The staff toiled day and night over the Christmas holidays of 1943 to meet their first Army Air Forces JATO contract: Work Order No. 44. It was for one hundred of the 25-ALD-1000s, the famous Aerojet liquid droppable units. To fill the order, nearly on time, they ate bacon and bread sandwiches on the go. Once the order was complete, Haley rewarded them with their belated Christmas holiday on 4 January 1944, with a party at the Pasadena Athletic Club ballroom. He delivered two hundred pounds of turkey, a dance with director Kelly Kelso and his orchestra, and two thousand sterling silver service pins with the Aerojet trademark red rocket. Haley's sister wrote the lyrics to the company's "Aerojet Song," played to the music of the popular Irish dance "Gerry Owen." The Haleys meant it to express the thrill of exploration and discovery at the frontiers of science, research, and development in jet propulsion, "expressive of fighting men and freedom." It was also the favorite ballad of Gen. George Armstrong Custer's 7th Cavalry, infamous for the last stand. Among its lines were "We are the boys who have the right / To probe the darkness of the night.... Oh, let the world your voices hear / For we will ride the stratosphere." And "To the stars, the moon the sun away / Our Aerojet will reach someday."[65]

Undercutting the victory were some deep problems. In terms of research, the company was beset by duplication of efforts, sloppy experimental and laboratory tests unfit for the scientific method, and "delays and lack of coordination." Miss Christeson, secretary to Haley and also the administrator of the library and central files, offered caustic commentary on the "great many prima donnas" who created dissension and discord. There were far "too many bosses." Many were selfish, obsessed with patent rights and "personal credit." As she wrote, "It requires a big person to direct research, to get out of others all they are capable of performing, without their being afraid they are being used to further someone else's ambitions." Two of Aerojet's engineers were such "big" achievers: Zucrow and Krueger. She also praised Malina, Zwicky, and Zucrow as equals for their ability to navigate the arenas of research and engineering development with ease. Under their leadership, Aerojet brooked no artificial divisions, no "cloistered research." Dr. Robert Swain, chair of the Chemistry Department at Caltech, agreed: Aerojet, he believed, had the "best Research and Engineering Department in the United States."[66]

The records also show myriad development challenges. Haley once gave the dictation: "It is a marvel that we have done as well as we have." As the company grew in 1943, it failed to establish routines and rules, like "inventory control, material control, quality control, all sorts of procedures." There were issues with the perchlorate in the solid-rocket cartridges, along with production and moisture problems, igniter performance troubles, and nozzle erosion. Whole batches of motors were rejected in production.[67] On discovering one failing contract with hemorrhaging losses, the front office investigated. Zucrow found several extra unused drums of furfuryl alcohol that "probably can be returned to the tune of $10,000." In another case, as the war was ending, Zucrow made a bold claim that "Aerojet is encumbered by too much paperwork, and by duplication of records," hampered by "waste and confusion." The front office proposed a series of reforms in response.[68]

What were, then, Aerojet's achievements and Zucrow's contributions to them? My observations below are on his "contributions," one of his favorite terms, on his leading role within the cooperative work of engineering teams.[69] GALCIT had achieved its first tentative, experimental successes before Zucrow came on board. He was tasked with improving the technical design of the JATO units, to turn research into development, experiment into production. For example, he studied several improvements for the solid units.[70] But most of his initial work was on the liquid-powered 25-ALD-1000 JATO, built first for the Army Air Forces, then for the US Navy. Some of his early work in 1943 was centered on finding the best new materials to contain the corrosive acid propellants and to resist their high exhaust heats. He designed, fabricated, and tested tungsten carbide–lined nozzles (as opposed to copper). He was also on a consultation team with Malina, Zwicky, Summerfield, and Apollo M. O. Smith focused on how to improve the liquid unit's injectors, materials, shape and size, and especially on how to increase the operating pressures with novel cooling jackets and internal cooling sprays. As the US Navy liaison phrased it: "The principle of Film Cooling by seepage of coolant through the inner wall of the motor chamber" was a radically "new approach." No one else in the country was doing it.[71] It became one of Zucrow's trademarks.

Zucrow also contributed to one of Aerojet's most secretive projects, appropriately named Project X, and which even had an "X File" in the company's central file archives. This was a design for what he called "a rocket power plant for a rocket interceptor plane." The company called it a "super-performance" aircraft, with "abnormally large accelerations and level flight velocities or rates of climb," for only one minute or less. This was a major leap beyond the JATO.[72] By May of 1943, Zucrow was one of the few engineers involved in this secret project, along with von Kármán, Malina, Summerfield, and select engineers (George Sutton, Chandler Ross, and Rolf Sabersky) and military liaisons.[73] Zucrow contributed as principal development engineer in the Liquid Development Division. As coordinator of Project X, Summerfield appointed him to design its "prototype jet propulsion unit," with attention to "thrust, power, propellant consumption, control, weight, cruising thrust, accessories." Quite an array of challenges. Zucrow and his team studied combustion pressures and cooling techniques (both regenerative and film cooling) with special care. They conducted extensive tests on acid-resistant materials for the fabrication of the tanks (via special metals, sprayed aluminum, and plastic coatings), including a search for better welding techniques. With his characteristic confidence, Zucrow declared the results reliable, if "not perfect." The "ultimate solution" was "a combination of metal spraying plus a plastic coating." The Air Corps liaison deemed Zucrow's heliarc approach to welding, borrowed from his contacts at the Northrop Aircraft Company, a success.[74]

One troubled design in the Project X portfolio was the Aerotojet, named for its "roto" or rotational effects. As team leader Ernest Vogt described it, "The Aerotojet consists, basically, of a combination of jet propulsion motors and pumps so arranged that a tangential reaction derived from the motors is utilized to drive the pumps which, in turn, transfer propellants to the motors."[75] Chandler Ross remembered talking to Rolf Sabersky once at lunch about Project X, what he called "the development of a rocket engine using a Hero Turbine to drive the pumps." Zucrow overheard the conversation and casually noted they were "pretty far off base trying to use a contraption" like that. It was better use simple "impulse wheels."[76] They did not listen. It did not work.

In the end, Zucrow was part of the team that contributed to the Aerotojet's replacement: the XCAL-200 rocket engine. It was the engine used for the powered tests of the Northrop Flying Wing (the XP-79B, in its secret version known as the MX-324) on 5 July 1944. At Harper Dry Lake, near Barstow, California, a P-38 fighter plane towed the Flying Wing to eight thousand feet altitude, where it broke away then fired its engine and flew to 130 mph for a ten-minute flight. Pilot Harry Crosby became a rocket man in the flight. He lay prone in the aircraft: "the combustion chamber and nozzle lay between the pilot's legs while the propellant tanks were outboard on opposite sides." He was also protected by neoprene shields. Miss Christeson took Haley's dictation: "Early this morning at Lake Harper, the first flight of a rocket powered plane went off successfully. Many months of work and anxious planning went into this XCAL-200 unit." The results were "very impressive and beautiful." Zucrow had helped to design and build the engine that propelled America's first rocket-powered plane.[77]

Another Aerojet engine represents a case study in Zucrow's many hidden contributions to the company and country: the 38-ALDW-1500 rocket motor, which later became the rocket engine for the Corporal missile. The No. 38 was an important early JATO in the Aerojet arsenal. It was originally a Navy model, designed and built by Robert Truax at the Naval Engineering Experiment Station at Annapolis. The Aerojet contract with the Navy's Bureau of Aeronautics was for redesigns and improvements. The Navy sent two lieutenants, Truax and Roy Stiff, to monitor the developments and school themselves in Aerojet ways. Zucrow obliged. He was the lead engineer on the contract team for the No. 38. With E. M. Walsh and B. L. Dorman, he conducted elaborate tests at the new Azusa plant, where Dorman was principal test engineer. The contract was for "investigation of mechanically improving liquid jet propulsion units and their accessories." Their improvements included a new design for "a uniform rate of opening for the Propellant Control Valve," along with "development work leading to the design and construction of successful propellant (fuel) cooled jet motors." They also made an "extensive study of heat transfer" and a variety of "material and design considerations," including injector designs, pressure studies, leak-proof fittings (O rings and gaskets), special tubing for acids, and corrosion- and solvent-resistant materials.[78]

On this last score, after a detailed study of metallurgical and physical properties, Zucrow chose Uniloy 19-9DL as the optimum high-tensile, corrosion-resistant stainless steel for tank construction. He assumed leadership of this earliest company initiative to construct corrosion-resistant pressure vessels. He consulted with the US Spring and Bumper Company, Northrop Aircraft Company, and Universal-Cyclops Steel Corporation and found that Uniloy's strength and lower weight was just right for "high pressure corrosion resistant tanks" for use in liquid jet propulsion units. Haley and von Kármán were anxious about this risky choice, especially "the difficulties encountered in forming 19-9DL for tanks." Yet Miss Christeson reported that Zucrow was "working hard to find the answer." It took him only a month. Between 4 May and 14 June 1944, he found the process for forming it for tanks. Zucrow's work on Uniloy 19-9DL was no small thing. It answered the Navy's concerns about Aerojet fulfilling its current and future work orders. It offered a boost of confidence.[79]

In the end, Aerojet's new No. 38 "unit met contract requirements." Zucrow and his teams reengineered Truax's early prototype into a workable and reliable production model. Truax later recalled that "of all the liquid propellant JATO developed, the 38-ALDW-1500 came the nearest to service use." Historians have called it America's "first practical, successful regeneratively-cooled acid-aniline motor." By the summer of 1945 Aerojet produced one hundred of these units, the Aerojet 38-ALDW-1500 Droppable Jet Unit for the PB2Y-3 sea plane, in use by the US Coast Guard for quick sea rescues off the California coast at San Diego. In the Pacific theater as a whole, over the final four months of the war, sixteen such JATO seaplane squadrons saved some fifteen hundred aviators and sailors by their quick response.[80]

In a classic Zucrow move, he turned this development moment into a research and teaching moment too. He and his team also tested mixed acids and monoethyl aniline (rather than RFNA and aniline) as propellants at various chamber pressures and mixture ratios for improved exhaust velocity. They studied the rate of heat transfer and coolant effects. Their extensive testing found positive results, like reduced heat loading and less fuming (for mixed acid), but also negatives, like a loss of performance with the new propellants, and greater ignition lag. Here were some of Zucrow's first signature studies in heat transfer and film cooling, and the whole set of cascade effects that he later studied at his Purdue Rocket Lab.[81] He also gave freely of his expertise to younger colleagues. Take, for example, Aerojet Report No. 58, "Some Considerations on the Heat Transfer in a Liquid Cooled Jet Motor." Zucrow played a silent, invisible role. He once spent a day with Rolf Sabersky on improving and correcting it. Summerfield and Sabersky signed their names to it. But Zucrow was the one who rewrote it for them.[82] The episode offers a clue about how engineering happens, so often outside of public recognition, so often the result of unsung cooperative effort.

Accompanying Zucrow's detailed reports was a small textbook of sorts, an impressive booklet on design and testing, neatly organized in his vintage fashion, minutely describing the fulfillment of the contract and the lessons learned on motor design calculations, heat transfer, injector design, and tank size and shape. In it was a Zucrow Aerojet teaching moment for GALCIT Project No. 1, based on his objections to its existing design for the JATO combustion chamber. He recommended a whole new motor shape in the form of a "thermos-bottle," built "so that the inner chamber is cooled by the flow of fuel passing through spiral passages surrounding it"; built to save weight, corrosion effects, erosion, and shock resistance.[83] The lessons went both ways, of course. GALCIT taught Zucrow lessons on the importance of O rings and on its own discoveries with corrosion-resistant metals.[84] Still, for evaluators in the Air Forces, Army, and OSRD, Aerojet was the trendsetter in jet propulsion, far more than GALCIT. It was America's premier "engineering and research organization," brimming with "scientific 'know-how.'" It was the leader for testing, interfacing, and quick thinking.[85]

In one of the first in-house histories of Aerojet, Zucrow celebrated the anonymous achievements of the company's three hundred engineers. Between October of 1943 and August of 1945, they created a superb "long-duration, light-weight liquid propellant rocket motor" of three-thousand-pound thrust, based on high-pressure gas feeding the propellants to the rocket motor.[86] This was, according to Martin Summerfield, Zucrow's own project-engineered XLR7-AJ-1, a pump-fed, regeneratively cooled, liquid-propellant (RFNA–aniline) engine. Aerojet teams improved upon this threshold between 1944 and 1945 by devising a "light-weight, self-feeding, turbine-driven pumping plant" of six-thousand-pound thrust. It was an early competitor, with Rocket Motor Inc.'s design, as a possible power plant for the X-1 rocket plane.[87] Although not chosen, the Air Force still called it America's "first successful chemical-gas, generator-driven, turbo-pump fed, regeneratively-cooled rocket engine." Aerojet celebrated it as "the highest thrust, continuous-duration liquid jet engine produced up to that time in the United States."[88]

Along with advances in solid propellants, Zucrow also noted Aerojet's further innovations in developing "nitromethane as a liquid monopropellant" (to avoid the weight of the dual oxidizer and fuel), as well as in creating a "coordinated organization capable of conducting all types of engineering tests ranging from hydraulics to electronics." His teams had "developed much special measuring equipment for studying the phenomena in rocket motors," along with "methods of tests" in order "to secure basic information with accuracy and dispatch."[89]

An example from the history of North American Aviation (NAA) confirms Zucrow's judgments as to these Aerojet achievements. NAA was the firm that the Air Force chose for a contract to build a V-2 style long-range ballistic missile, up to five hundred miles in range, what eventually became the MX-770 Navaho and whose liquid-propellant power plant sustained a long pedigree into the Redstone, Jupiter, Thor, and

Atlas missiles, with eventual iterations for Saturn and the space shuttle. But by the summer and fall of 1946, before they ever got hold of a V-2 engine, NAA engineers conducted their first investigations and experiments, in the company's parking lot, on a series of Aerojet engines that Zucrow had helped to redesign: the 25-ALD-1000 JATO unit and the 38-ALDW-1500 rocket motor. No doubt, North American engineers under William Bollay's direction achieved key milestones from their own reconstruction and retooling of the German V-2 by fall of 1947. That, in fact, was their mandate. But it was a year earlier, in the fall of 1946, that North American first learned how to operate and manage liquid-propellant rocket engines from the Aerojet and Zucrow models.[90]

Zucrow had yet another contribution to make through Aerojet. Von Kármán assigned him to create "a course in jet propulsion theory" specifically designed to teach "graduate engineers of the local aircraft industries." Aerojet sponsored it. Zucrow taught the course at UCLA under Andrey Potter's Engineering, Science, and Management War Training program (ESMWT). This made perfect sense. With the largest enrollment in the program, 150,000 students, UCLA had become "a humming war plant—the largest in the nation," as its president boasted. Its ESMWT courses were spread out over Los Angeles at sites in Inglewood, Burbank, and Long Beach. Most had rather innocuous titles, like Industrial Accounting II, or Personnel Counseling, or Aircraft Materials and Processes.[91] Then an announcement appeared for the rather stunning Principles of Jet Propulsion, the first such course in the US. Zucrow first taught it in the summer of 1943 at the downtown War Training Center (714 Hill Street), every Friday from 7:00 to 9:30 p.m., for sixteen weeks. Prerequisites were a BS in engineering, including courses in thermodynamics and fluid flow, and a design engineering position in the aircraft industry. He taught it two more times into the summer of 1944, to classes of sixty students each. The course was free, but students could request extra gasoline rations if they had a long drive.[92]

By the command of war censors, Zucrow offered readings and information only from the technical press, covering the basic principles and "generalized operating characteristics." Building from his Purdue education, he taught the breadth and depth of propellers, high-speed compressors, the Whittle turbojet engine, "thermal jet systems," JATOs, and rockets. Applicants needed to pass a qualifying examination of Zucrow's own making. It was nine pages long with twenty-five questions, mostly centered on equations and formulas. Here are three of the simpler questions. One asked: "If a perfect gas has a molecular weight of m, and is at a pressure of P lb. per ft. abs. and at a temperature of T° Rankine, what is its specific volume?" Another posed this challenge: "A gas is compressed rapidly so that there is no heat added to or removed from it. If its volume is halved by the compression, what is its change in entropy?" Yet another asked the student to "sketch the relationship between the left coefficient on an airfoil up to the burble point."[93]

Zucrow's ESMWT courses in the summers of 1943 and 1944 helped to create the country's first cadre of about 180 aerospace engineers. It was also the first GALCIT or Aerojet venture, as a course and textbook, to publicly apply the "Jet Propulsion" moniker. GALCIT only renamed itself as the Jet Propulsion Laboratory (JPL) in an official 20 November 1943 report, signed by Hsue-Shen Tsien. One favorite foundation myth was that JPL's founders chose the name to raise their standing in the military and scientific community by avoiding the fantasy and science fiction sound of "rocketry." But the name was already embedded in the idioms and practices of GALCIT and Aerojet. They were making jet-assisted takeoff devices, and their reports after 1941 commonly applied the term "jet propulsion." Zucrow's ESMWT course and lecture notes in summer of 1943 simply gave the term its first full public currency.[94]

In the last few years of the war, with von Kármán situated on the East Coast, Malina's energies focused on administration, and both traveling frequently, Zucrow enjoyed more independence of action at Aerojet.[95] He was busy communicating, via phone calls and business trips, with the Navy Bureau of Aeronautics in DC, the Applied Physics Laboratory at Silver Spring, Maryland (run by Johns Hopkins University), the Army Air Forces at Wright Field, and NACA Lewis in Cleveland and Ohio State University in Columbus, along with the many points in between. To illustrate one such point, he consulted for the Brooklyn Polytechnic Institute on "the metallurgical, fabrication, and design problems involved in cooling rocket and intermittent jet motors by the diffusion of fluids through porous metal combustion chamber liners." The goal was to improve the mixing and combustion of fuel and oxidizer in rocket motors.[96]

By 1944 and 1945, Zucrow was also pursuing several projects at the main line of American jet propulsion and rocketry. Although he never admitted it at the time, many years later he laid claim to having designed "much of the basic engineering" of the first US Army missiles, to which we now turn. There is, for one, the continuing saga of the No. 38 liquid-propellant unit. It was one of the power plants for the Corporal, America's first ballistic missile. By agreement of Hovde at Division 3 and Gen. Barnes at Army Ordnance, the OSRD commissioned GALCIT-JPL to develop America's "long range rockets of the V-2 type." Barnes offered a $3 million contract in January 1944, a bonanza for GALCIT-JPL, now retooled into Ordnance-CIT (ORDCIT), with initial plans for a ten-thousand-pound thrust engine to reach a seventy-five-mile range.[97] With von Kármán and Malina away, by January of 1945 Homer Stewart designed the actual Corporal missile, known in one of its versions as "without attitude control" (WAC). It faced significant challenges in structure and guidance and control. Propulsion proved easier, thanks to Stewart's choice of an Aerojet engine. The WAC-Corporal engine was a unit improved at Aerojet and GALCIT, with Zucrow's contributions, using RFNA as the oxidizer and aniline (mixed with 20 percent furfuryl alcohol) as the fuel. Regeneratively cooled, with fifteen hundred pounds of thrust for forty-five seconds, it had a minimum effective exhaust velocity of sixty-two hundred feet per second and a one-hundred-thousand-foot altitude. Ray Stiff described it as the "first liquid fueled sounding rocket to be developed with government funds in the US."[98]

Zucrow also helped to design and improve the second stage for the Nike antiaircraft missile, planned to intercept enemy aircraft flying at sixty thousand feet and 600 mph speeds. It was quite a challenge. Beginning in February 1945, Army Ordnance contracted to Douglas for the airframe and to Bell Telephone Labs for the guidance system. Aerojet's teams designed the propulsion system, including both a solid-fuel first stage and a liquid-propellant second stage. Zucrow's contributions were mainly in the development of the liquid-propellant stage, the X21AL-2600 engine, designed and developed in late 1945 and test-fired in May of 1946. Historians call it "the world's first operational surface-to-air missile." Another Zucrow-assisted first. This engine, in modified forms, also served as the propulsion system for North American Aviation's (Rocketdyne) NATIV aerodynamics and guidance and control tests, essential research building blocks for its future development of rockets and missiles. It was also the power plant for the Aerobee sounding rocket, with its high reliability record of over 90 percent, and used as the second stage of the Vanguard satellite missions.[99] Zucrow's improvements left their mark.

Zucrow also transitioned into new work on propellant combinations as a lead engineer for Aerojet's earliest tests with liquid oxygen and liquid hydrogen. In December of 1945, the Navy's Bureau of Aeronautics contracted with Aerojet to develop an oxygen–hydrogen power plant as a surface-to-air-missile and as a possible launcher for its Earth Satellite Vehicle Program. The initial product, the XLR16-AJ-2, with two thousand pounds of thrust, was a small-scale version of a preliminary engine for a rocket missile.[100] In July of 1946 Zucrow consulted for the Army Air Forces' contract with Ohio State University on the project, now titled Liquid Hydrogen Propellant for Aircraft and Rockets. The OSU Research Foundation created a rocket research laboratory in 1946 for this purpose. It was originally a cryogenics laboratory in the Department of Chemistry, run by Professor Herrick Johnston. Zucrow helped the lab "get started." A year later, it produced the first liquid hydrogen–liquid oxygen thrust chamber test in the US, with Zucrow's novel injector design for the propellants, and film cooling using his porous chamber walls.[101] His wartime resume was complete.

## SECRECY AND SPECTACLE

On 6 and 9 August of 1945, the US destroyed Hiroshima and Nagasaki with two atomic bombs. The country's most secret wartime project suddenly became its most public. At the heart of its power was a further paradox: it represented the control of the atom, if now unleashed in a bomb that destroyed whole cities. Humanity now faced losing control over its own destiny. These were paradoxes compounded by the pilotless V-1 buzz bomb and the remote-controlled V-2 missile. They foreshadowed what military leaders and publicists now called "push-button war." Atomic bombs were scary, but robotic war made it scarier still. War could start and end within minutes.[102]

Hovde participated in this odd moment of secrecy and spectacle over the summer of 1945. He maintained a busy schedule of cross-country trips, including to Alamogordo, just before and after the Trinity test of the first atomic bomb. Members of his OSRD teams had been delegated there to work on its fuse mechanism. He met with Charles Lauritsen and Robert Oppenheimer about these personnel questions when, at a 6 August meeting, they all first received word about Hiroshima, when the bomb became "death, the destroyer of worlds," in Oppenheimer's dark terms.[103] At this very same time, Purdue was secretly negotiating to hire Hovde as its next president. He formally accepted the offer by 14 August.[104] Vannevar Bush recommended him in the following terms: Hovde was a man of "great tact" and dignity, "a very unusual individual, combining scientific and technical ability with managerial skill" and diplomacy. He was "well-liked by all those around him."[105] Hovde also enjoyed a strange connection to the university. In 1928, as the Big Ten top scorer, he played quarterback for Minnesota against Purdue, winning 15–0, assisted by his all-star fullback, Bronko Nagurski. Guy "Red" Mackey (future Purdue director of Athletics) played that day. "All I saw of Hovde was heels," he said.[106]

The spring and summer of 1945 also saw a strange combination of secrecy and spectacle in the US military's treatment of the German wonder weapons: its series of turbojet and rocket planes, and of course the V-1 and V-2, formidable weapons that might have lost the Allies the war. With the war now nearly won in Europe, Gen. H. H. "Hap" Arnold, commander of the Army Air Forces, sent top-secret delegations from Caltech to investigate the German research sites and interrogate their experts. Zucrow's close colleague Fritz Zwicky wrote one of the most poignant reports centered on the "psychology of technical men in a dictatorship and in a democracy," along with the "ease" by which German professionals flocked to Nazi promises of "jobs, problems, technical equipment to solve them," and all the "buoyant activity promising a bright future." The dictatorship had mobilized German inventiveness in science and engineering, but within the "large hierarchies" of statist science, filled with "perversion" and "calcification." Better were America's free individuals or small groups of the "little fellow" in the nation's service, guided by "just one thing, a diligent and sincere effort toward its future security, progress, and social justice." He advised paying $30,000 a year to a university professor for research free from politics and commerce. Zwicky was exactly describing the future Professor Maurice Zucrow. He summarized it all by plotting global achievements in jet propulsion R&D on an x-y graph. The Germans had the advantage of a steady upward slope from 1930. But that was in the past. By 1945, the Americans were overtaking them with a steeper slope of advance. That was the beauty of the horizontal line: America had time on its side.[107]

Col. Leslie Simon, director of the Army's Ballistic Research Laboratory, was similarly impressed by his own on-site investigations in occupied Germany. He even wrote a book about it, one of Zucrow's favorites. Simon appreciated the vanguard engineering labs that he found in a defeated Germany. They were models of organization and efficiency. But like Zwicky, he castigated the centralized management of Nazi R&D for its ideological indoctrinations and its subservience to the weapons industry and war production. He instead praised the American balance between basic and applied research, clear project lines, proper respect for the "intellectual specialist," and the "free exchange of scientific information." The US wisely invested in design and development, linking the military, industry, and university in cooperative arrangements. For Zucrow, the book was formative.[108]

One of the most secret and enticing of the late wartime reports came straight from the pen of the leading Nazi rocket engineer, the "doctor professor" Wernher von Braun. In June of 1945, in a classified study for the US Army, he strategically positioned himself as the world's master of the new "art of rocketry." Von Braun cleansed the V-2 of its evil intents and effects, fashioning himself as the prophet of peaceful space exploration. He proposed a winged stratospheric rocket and offered an orbiting rocket satellite as an observation platform and station, inhabited by spacewalking rocket men. He predicted travel to the Moon and planets. These were not his own designs, but freely borrowed from Robert Goddard, Hermann Oberth, and the German Society for Space Travel. Still, von Braun made his point: rocketry had "revolutionary consequences in the scientific and military spheres."[109]

The US Army got the message, staging Operations Paperclip and Overcast to import several hundred German specialists from the V-1 and V-2 programs into the US, some to Wright Field, the core group under von Braun to Fort Bliss, Texas, and the White Sands Proving Ground. The Army Air Forces also brought over the German wonder weapons as trophies of war, secretly transported to Wright Field and associated fields in Ohio and Indiana beginning in 1943. These included the Messerschmitt 262 turbojet and Messerschmitt 163 rocket plane, the V-1 pulsejet, the Junkers 88, and Arado 234, along with over 250 tons of top secret technical documents. The Air Technical Service Command (ATSC), Technical Data Laboratory, and Engineering Division at Wright Field were also responsible for deciphering and understanding these new machineries. Zucrow's Aerojet expertise, along with his language skills, embedded him in this vast translation project. He was a consultant for the ATSC, primarily through its Technical Intelligence Directorate (Intelligence T-2) and Analysis Division. He understood the variety of power plants almost immediately on seeing them and was able to read the documents in the original German with ease.[110]

With the victory in the war against Japan, Brig. Gen. Laurence Craigie was now free to publicize these weapons, along with their promise and threat. He premiered them at the conference of the Institute of the Aeronautical Sciences in Los Angeles (August 1945), backed by officers from Wright Field and important guests like the chief engineer at Lockheed, the chief of structural design at Boeing, and Jack Northrop of Northrop Aviation. Craigie warned that the world had shrunk and a V-2 style missile war might unleash the next Pearl Harbor onto America's Main Streets. He was one to know. In October of 1942, at Muroc Field, he was the first military pilot to fly an American turbojet plane, the Bell Airacomet (XP-59).[111]

Zucrow played his own lead in these dramas during the summer of 1945. He did not join the Caltech teams that ventured into war-torn Europe. He stayed back in Los Angeles, working at Aerojet, refining his textbook. He thereby seized an opportunity to gain an even bigger professional spotlight. It was scientific/technical theater of a high order, like his PhD defense performance from years before, if on a bigger national stage. It was his speech to the Aviation War Conference (11–14 June) at UCLA, sponsored by the American Society of Mechanical Engineers (ASME), held in the auditorium of Royce Hall under its Romanesque twin towers. He gave the keynote address before twelve hundred people on the first evening session, introduced by the provost, Dr. Clarence Dykstra. The "technological advances during the war have been stupendous," Dykstra said. "The imagination is spurred to new action when we consider the peacetime potential of such developments as propulsion by jet, rocket, and gas turbine."[112]

Zucrow's paper "Jet Propulsion Principles and Rockets for Assisted Takeoff" was a characteristically cautious speech on a rather dramatic topic. He offered charts showing propulsion efficiencies and optimal speeds for the new turbojet engine. He showed a film about Aerojet's achievements with JATOs. He highlighted the scientific unknowns, the absence of "operational data" and "performance characteristics," and the technical limitations thus far, warning about high fuel consumption and extreme heats. He cautioned against any fantastic immediate applications to commercial jet travel. Still, the *New York Times* advertised Zucrow as the country's new prophet of jet propulsion. It was an academic performance in a very public setting. The journal of the ASME captured Zucrow in a photograph at the event, his small frame peering from behind the microphones and lectern, studded with the four-leaf clover logo of the group, his face somewhat pained by the size of the crowd and all the publicity.[113]

Even more significant was what happened afterward: a small barrage of his articles in the major professional journals for mechanical and aeronautical engineering. Zucrow became the voice of the revolution in jet propulsion and rocket flight, at least to America's engineers. He was also one of the very first of the country's experts to see the real possibilities of spaceflight. Zucrow canvassed a coming rocket future, the most efficient power plant yet for high-speed and high-altitude flight, for flight into outer space, to infinity.[114] His signal contribution was the article "The Rocket Power Plant," an excerpt of the final chapter from his forthcoming textbook. The article was the country's first practical "handbook" approach to the rocket engine. Zucrow offered a detailed study of its mechanical designs, thermodynamic processes, and performance data (exhaust velocities, expansion ratios, and cooling methods). He wrote as an engineer for engineers. It also stands as a monument to his practical engineering achievements at the Aerojet Corporation during the war and his coming role as America's first professor of jet propulsion. As Zucrow wrote in his unassuming style, the rocket's "thrust is almost independent of its environment" and yet "can function best in a vacuum" by carrying its own oxygen. He included a bit of history, charitably citing the achievements of Konstantin Tsiolkovsky and Fridrikh Tsander, Robert Esnault-Pelterie, Hermann Oberth and Walter Hohmann, and Max Valier and Eugen Sänger. He marked Robert Goddard as "preeminent." He recognized the German achievements in the Messerschmitt 163 and the V-2, but he did not overrate them. In his prospectus on theory and practice, he also inspired a new American confidence. Liquid-propellant rocketry was the practical basis for the country's guided missiles and human spaceflight to come. "The rocket principle," he said, was "a new source of power for attaining objectives unattainable by other methods," a means "for attaining extreme altitudes and speeds and for making space travel possible."[115]

THESE NOTIONS OF SECRECY AND SPECTACLE REVEAL A STARK TRUTH ABOUT rockets: their inevitability as both weapons of war and instruments of peace. Rocket platforms were neutral enough. Their purposes and targets, not always. The country's premier jet propulsion interest group, the American Rocket Society (ARS), understood this dilemma. Before the war, initially named the American Interplanetary Society, it was largely devoted to rockets for peace. During the Second World War, engineers who were engaged in military rocket research joined its ranks, and press attention to the V-1 and V-2 offered the society a "needed impetus to rocketry." The war inaugurated a new era of rockets, just as the First World War had heralded an era of flight. In 1943, the ARS board decided to transform from an interest group for rocketry and space travel into to a "truly national society of rocket engineers." G. Edward Pendray was one of the architects of what he called turning the ARS into a "nucleus" for a new "national engineering society" dedicated to "rockets and jet propulsion engineering" and centered in industry and the universities.[116]

The war also seeded the universities with a set of experienced administrators, primed for military and industrial cooperation. The OSRD was a school for university presidents. Two of its founders had come from university presidencies: James Conant of Harvard and Karl Compton of MIT. After the war, T. Keith Glennan, director of the Navy–OSRD Underwater Sound Laboratory, became president of Case Institute of Technology. Lee DuBridge, administrator of the OSRD Radiation Lab, president of Caltech. Clifford Furnas, NDRC administrator, chancellor of the University of Buffalo. H. Guyford Stever, OSRD liaison officer in London, president of Carnegie Mellon University. Irvin Stewart, administrator and executive assistant to Bush, president of the University of West Virginia. Lloyd Berkner, assistant to Bush in the OSRD, became president of Associated Universities Inc. Frederick Hovde became president of Purdue. They each navigated the new arena of state-sponsored research in their own individual ways, dedicated to military–civilian mutual reward and to the integrity of their universities. OSRD had taught them how.[117] They each now participated in an inherently risky venture, demobilizing from war to mobilize for a kind of peace, preparing for the complex demands of the rising Cold War. These R&D veterans were dedicated to the higher virtues of American democracy, and to calibrating the proper balance between military and civilian rule. Joining them in academia were hundreds of the country's leading scientists and engineers, those who had been present at the creation of America's first liquid-propellant rockets, including Maurice J. Zucrow.

# 4

# JET PROPULSION

## The New Science of Rocket Engines

THE SECOND WORLD WAR DEFINED THE CRUCIAL ROLE OF R&D IN WINning a war. But it was the transition to peace, and to a new kind of war, the Cold War, that institutionalized R&D in America's language and landscapes. This was not the R&D we know today. It was more the big "D" of development. One US Army circular, from May of 1947, made this quite clear. It detached development from "procurement, purchase, storage and distribution" and instead attached it as a distinct product of research, stating: "Research is a continuous process of scientific investigation prior to and during development. It has for its aim the discovery of new scientific facts, techniques, and physical laws" as applied to "new types of materiel or techniques." Moreover, "development is the application of the known scientific facts, techniques, materials, and physical laws to the creation of new or improved materiel or methods for military use." The Army was clearly interested in development, in actual things.[1]

One of the most common tropes related to R&D after the war was the scientific breakthrough or revolutionary advance, research leading to brand new technologies, real things and processes like radar, atomic energy, and jet propulsion. These technologies were driving humanity into the future, colonizing the globe. What might be next? All eyes were on the horizon, another powerful metaphor of the day. As Robert Oppenheimer wrote, basic research was "an essentially limitless universal field with expanding boundaries getting farther out as one approached them." Or as one engineer also proclaimed, as for R&D in the thermonuclear and missile age, "there is no end in sight to this process."[2] In the wartime transition, civilian and military veterans engaged in a decisive battle between two promised ideals: Vannevar Bush's "endless frontier" and Theodore von Kármán's "new horizons." This was really a battle between basic science and applied science. Meanwhile, as the country and its pundits debated these issues, the US Navy responded with its trendsetting Project Squid, a whole new way of doing R&D in practice, and the main patron for Zucrow's Rocket Lab at Purdue.

## NEW HORIZONS AND THE NAVY'S PROJECT SQUID

Bush's classic essay *Science, the Endless Frontier* was actually a series of reports to President Franklin Roosevelt, prepared between November 1944 and July 1945, to propose new federal policies on the organization of science for the national welfare and defense. Bush recommended a new kind of Office of Scientific Research and Development (OSRD), one essentially without the development factor. It was to be an independent civilian government agency, the National Research Foundation (NRF), to stand apart from politics, with significant federal funding for basic research and normal science in the universities, and to lead national science and technology policies.[3] His noble NRF was a means to collect pure research as essential capital for national security, an "intellectual bank" for any needed later withdraws.[4]

The media gave Bush's *Endless Frontier* wide play. The US Congress adapted it into legislation, the bill sponsored by Senator Warren Magnuson of Washington. Leading academic and political voices gave it support. The editors of *Mechanical Engineering* weighed in: "The war has popularized research and has generated public confidence in its usefulness." Dean Andrey Potter, acting president of Purdue (1 July to 31 December 1945), sent copies of the booklet to his faculty as required reading, committing Purdue to the new national research agenda. He also suggested a new course for undergraduates, Research Objectives and Techniques, and the preparation of a new research-centered graduate program.[5] Bush's manifesto was made all the more poignant by coincidence. It appeared with the debut of the two atomic bombs and America's first operational jet fighter, the Lockheed Shooting Star. Publicists have raised it to canonical status ever since.

Theodore von Kármán, at the behest of Gen. "Hap" Arnold, countered Bush's manifesto with one of his own, or actually thirty-three of them, in *Toward New Horizons*. It was a massive multivolume set of books, begun in September of 1944 and completed by December of 1945. Hsue-Shen Tsien was the general editor, joined by von Kármán's core Caltech teams and leading scientists and engineers from around the country. The booklets were eminently practical, dedicated to the new machinery and infrastructures of supersonic flight. They covered the full range of jet propulsion theories and practices, their apparatuses and performance levels. There were books on German and Japanese technological achievements in the war, aircraft power plants and propellants, aerial strategies and operations, guidance, explosives, radar, weather, aviation medicine, and guided missiles and pilotless aircraft. Von Kármán essentially outlined a series of engineering developmental projects for the military's attention.[6]

In terms of the competition, Bush's *Endless Frontier* did not really have a chance against von Kármán's *Toward New Horizons*. Bush's NRF became embroiled in partisan politics, under attack in Congress and from the president.[7] Bush's principled stance for basic over applied research was doomed from the start. Even his own wartime Committee on Postwar Research understood this. Its representatives from the OSRD, the National Advisory Committee for Aeronautics (NACA), the National Academy of Sciences, and the military services valued sponsored research projects the most. Projects turned the "unanticipated" into "practicable and producible products." Even Adm. Julius Furer, one of Bush's closest allies, concurred. Said Furer, "Research is not an end in itself." R&D had to "translate" new scientific discoveries into "specific devices."[8]

Bush's OSRD was also overwhelmed by events. As it demobilized, the military services and universities raced in a scramble for its resources so that they could craft their own new R&D units, divisions, and commands. Among universities, MIT and Caltech, along with Berkeley and Johns Hopkins, were the big winners in the Second World War. They simply needed to renegotiate their contracts. Frederick Terman, president of Stanford and a former student of Bush's at MIT, sought out new faculty and laboratories for special "engineering niches," negotiating with the NACA, Army Air Forces, and the Naval bureaus for massive funding, with special emphasis on electronics. His vehicle: the new Stanford Research Institute.[9] It became a model for the establishment of dozens of new university research foundations in these years, like the Armour Research Foundation (at the Illinois Institute of Technology) and the Cornell Research Foundation, pillars of Big Science and the new "University-Government-Industry Nexus" of the early Cold War.[10]

Unlike Bush at the OSRD, and von Kármán's cadres through the Army Air Forces, the US Navy wrote no grand manifestos at the end of the war. It simply got to work on R&D. It was already primed for innovation, thanks to its traditions of vast engineering enterprises, in battleships and aircraft carriers, power plants and shipyards.[11] The secretary of the Navy, James Forrestal, approved the new Office of Research and Inventions (ORI) at the end of the war, retooling it as the Office of Naval Research (ONR) a year later. Its mission was to "establish an extensive research program" with industrial, foundation, and especially university laboratories, thus to extend "long term basic and applied research" and "the bank of fundamental knowledge" in naval technologies.[12] The ONR established strong professional and personal linkages with academics around the country, and exacting standards for cooperative contract work. It took special care in respecting the freedom of university scientists to pursue research topics that were mostly of their own choosing, within their campus traditions. The Navy offered well-financed contracts and new laboratories as well as overhead costs for their institutions.[13] In its first three years alone, the ONR spent $20 million, on twelve hundred projects, with three thousand scientists and engineers and twenty-five hundred graduate students, in two hundred institutions.[14]

Purdue had a unique advantage in this whole process: Frederick Hovde. Between the summers of 1945 and 1946, he held two important posts, one as the OSRD administrator transferring its contracts to the Navy, and the other as president of Purdue, preparing to benefit from those very contracts. No one seemed to see the conflicts of interest at the time. The OSRD called the process "transfer and termination." It was literally Division 3's final project, under the supervision of a special NDRC committee, including James Conant, Hovde, and Charles Lauritsen. Under Hovde's direction, they handed over the files, sites, and functions of Division 3 to the Navy's new R&D offices, the Bureau of Aeronautics, and the ORI. In this, he enjoyed close relations with the Navy's Adm. Furer, and especially with Commander Robert D. Conrad, a technical expert with whom he had worked closely in London earlier in the war. In the transition, as chief of Division 3, Hovde advised the War Department and Navy that he was powerless to "approve projects for the postwar period." He counseled them to "continue a coordinated program of research, development, and application of thrust and jet-propulsion units," and to preserve wherever possible a "nucleus of experienced scientists" for these and related pursuits. Free them to return to their full university duties, he said, but retain their contracts and connections. Navy officers, in turn, informed him of their plans "for a program of postwar fundamental research."[15]

These relationships came in useful as Hovde prepared to take on the presidency of Purdue. He actually leveraged his OSRD position to counsel the Purdue Research Foundation (PRF) and faculty, who were actively seeking industrial and military research contracts. It was a bit of insider trading, so to speak. He confirmed the dominance of Aerojet in "rocket research and development," and was not opposed to a partnership. Yet he counseled patience, favoring Army and Navy contracts: "I believe we must give the federal agency involved first preference." Purdue professors took his advice. He was, after all, a wartime leader in "aerial-rocket research." By the end of 1945, they were keen on joining the new Project Squid, a "cooperative project with the Navy."[16]

Project Squid offers a unique case study in the mobilization of American academia, for what the Navy called the "fundamental technical problems" of rockets and pulsating jets, essentially the German V-2 and V-1. The Secretary of the Navy approved the plan of action for the project on 4 February 1946, just five months after the end of the Second World War. Project Squid, sponsored by the Navy's Bureau of Aeronautics and ONR, was based on the Navy's investments in the new and daunting technology of guided and ballistic missiles, one of great "magnitude." The plan was to turn "heretofore scientific oddities" into weapons systems. Squid meant reaction propulsion, just like the sea animal that moved by its principles. This included basic research into the "thermodynamic properties of new fuels" and their manufacture. It included study of "the basic phenomena of the combustion process and the conduct of materials at high temperatures," especially for their resistance to corrosion and heat. Above all, it meant applied research, a "project for the planning and execution of research and development for propulsive devices."[17] Here were von Kármán's new horizons in action, except for the Navy, not the Air Force.

Project Squid's aims were ambitious. Its initial research plan was for simultaneously balancing three categories of research: the known, the unknown, and the "totally unforeseen." Its foundation charter likened this to inventing the horse and buggy, the automobile, and the airplane all at once. Or the telegraph, wireless, and radio—simultaneously. Such an ambitious plan demanded core participation by the universities and its professoriate so as to allow full "freedom for gifted scientists to pursue their 'flashes of brilliance.'" This was especially true for the totally unforeseen, in which it was "perfectly appropriate to have as many as ten different groups working on the same project." Competition and duplication were essential to give the Navy a decisive edge with surprise technologies. Squid was really about making brand-new and finished weapon systems.[18]

The initial plan was for an actual laboratory, a "propulsive devices laboratory," near New York City, within driving distance to Washington, DC, and the major East Coast industrial labs. It was to have "a central test area with buildings, shops, and some laboratory facilities for actual experimental rocket tests" and "office space, conference rooms, lecture rooms, library." These were to be in a location "sufficiently remote as to offer a minimum nuisance to neighboring areas." The planners soon scrapped this idea in favor of an ingenious laboratory of laboratories. It was to unite different universities, like a squid's many tentacles and arms, under one Navy head, which would take on the burden of management. This was a practical compromise, to use existing programs and facilities in order to gain a technological edge against the USSR, as well as the Air Force and Army. The universities would provide the technical talent. The Navy would provide the universities with "basic scientific problems," while assuring a "minimum of interference" with their "normal" life. The project contracts initially covered up to three years and up to $100,000.[19]

The Navy chose Princeton University as the headquarters early on, this to center the project between ONR offices in Washington, DC, and New York City. It was also one of the premier Eastern universities. Princeton's president, Harold W. Dodds, jump-started the project with a barrage of invitation letters to the Ivy League.[20] The Navy also placed Frank A. Parker at Princeton as project organizer, its point man to organize the new university network. Parker came straight out of the Navy's Bureau of Aeronautics in Washington, DC. He was the perfect choice, a propulsion engineer and natural diplomat. Parker's listening tour of interested universities in February of 1946 helped shape the joint effort.[21] At Project Squid's organizational meetings, held at upscale venues like New York University's "Green Room" faculty club, and at the conference room of Princeton's Frick Chemical Laboratory, Parker also cautioned his own Navy colleagues that the war was over and the military could no longer impose its "supreme authority" over the colleges. He promised that Bureau of Aeronautics plenipotentiaries would not take posts at each of the universities but instead rely on the faculty themselves to have "some say in the actual operation of the project."[22]

In the end, Yale and Columbia, with Rutgers and the University of Pennsylvania, chose not to participate. By late spring, the first five Project Squid members were in place: Princeton, Cornell University Aeronautical Laboratory, New York University, the Polytechnic Institute of Brooklyn, and one outlier, Purdue, whose participation came rather late in the organization process, once the others were already in place. By the very definition of the program, Purdue should not have even been involved. It was a Midwest school, simply too far away from the East Coast. Hovde's leverage, and the entrepreneurial drive of the PRF, along with the innate talent at Purdue, changed all that. Parker visited campus on 8 March 1946, a month after his planning trips along the East Coast, to negotiate the deal. G. Stanley Meikle, PRF research director, soon led several delegations of Purdue's engineers back East to finalize the agreements. He also handled security clearances, including negotiating the overhead contract rate and operating costs. The PRF made it easy. Purdue was to receive $150,000 a year out of the total Squid budget of $1,150,000.[23]

Thanks to Parker's negotiations, Project Squid was organized by a task force method of R&D. The main Policy Committee set the tasks. The Technical Committee judged them by their merits. Squid was designed to provide for order and accountability between so many participating universities, yet also maintain their freedom of action. It was a project of phases, which were organized by research topics, and which were sometimes duplicated between universities. The phases, in turn, contained a variety of problem assignments, which were really discrete laboratory tasks or experiments. Purdue's range of phases turned out to be among Project Squid's most numerous and diverse, spanning both fundamental and applied research. The closest was New York University with its five phases. Purdue's seven phases, including fuels research in the Chemistry Department, combustion studies in Mechanical Engineering, and corrosion studies in Physics, were "somewhat overlapping in interest but not in method of approach."[24] They revealed the university's interdisciplinary strengths. Squid was following the OSRD pathway, awarding contracts to individual faculty members through their institutions. As one engineer argued, "Research cannot be planned, only cultivated where it is," so "find a good man and support him strongly."[25]

## PROFESSOR ZUCROW AND PURDUE ENGINEERING

With these teams in place by March of 1946, President Hovde surprised the Navy with a wild card. He did not really know Purdue all that well when he started his job. But he knew rocketry, and its future possibilities. Purdue likewise promoted Hovde as the kingpin of rocket research. As he announced rather dramatically in a letter to Princeton's President Dodd: "I have one individual to suggest for consideration for the full-time position as chairman or director of the project. I am giving my suggestion to Mr. Meikle, who will doubtless throw it into the discussion at the appropriate time."[26] He had in mind Maurice Zucrow.

As Purdue was negotiating the terms of Project Squid, it was also negotiating terms for Zucrow to come back home, to the team of Andrey Potter, Harry Solberg, and George Hawkins, and to the new president, Hovde, whom he knew from Aerojet during the war.[27] It was a circling back. Zucrow joked, "I did nothing in twenty-five years but reach Lafayette, Indiana."[28] There was some progress. At age forty-six, Zucrow left a twenty-four-year-old boss at Aerojet, William Zisch, for a thirty-seven-year-old new president at Purdue. As head of Mechanical Engineering, Solberg tendered the official offer on April 9. Zucrow accepted on April 17, at a salary of $6,000 for a twelve-month academic calendar.

The administration set up the rather awkward arrangement for Zucrow to divide 70 percent of his time to Mechanical Engineering, with Solberg as his main boss, and 30 percent of his time to the new School of Aeronautics, with Elmer Bruhn as his second boss. The arrangement served Solberg's primary interests, which were mainly in gas turbines for utility and industrial plants. He also set the terms for Zucrow's academic title: "Professor of Gas Turbines and Jet Propulsion." The division was also meant to "prevent duplication of effort" and to "ensure the proper coordination of research and instruction in this very important field." The job description seemed written exactly for Zucrow: "For Purdue University to take the lead in this field, we need someone who is a good teacher, a first-class research man interested in fundamental research, and a man of vision."[29] Purdue's offer must have been extraordinarily appealing at this moment. Zucrow was ready to return to his "first love" of university teaching and research. His cardinal rule for the rest of his career was that "the main function of a university professor is the teaching function." He must have meant it. Zucrow took a pay cut to come to Purdue. His salary there was "less than 60%" of his pay at Aerojet.[30]

Zucrow did not have to leave California. He could have stayed on in the aerospace industry, or continued teaching at UCLA. But there were some foreboding signals. With the postwar contraction, as military orders for aircraft and related parts plummeted, many thousands were now unemployed in Los Angeles. One week after V-J Day (15 August 1945), Aerojet had to let 843 go from a workforce of 1,475. It began a new venture in indoor plumbing to employ those who remained. Military sales plummeted from $10 million in 1945 to $2.5 million in 1946.[31] By the spring of 1946, GALCIT-JPL (the Guggenheim Aeronautical Laboratory at Caltech [GALCIT] now joined to the new Jet Propulsion Laboratory [JPL]) was also in serious disarray. Homer Stewart called it all "some real honest-to-God poison around Caltech." Two suspected communists, Sidney Weinbaum and Paul Duwez, took on jobs at GALCIT-JPL in Zucrow's own specialty: designing rocket power plants with new heat- and corrosion-resistant materials. Three of the founding GALCIT members were under investigation by Army Intelligence and the FBI as communists. Before the war, they had participated in a reading and discussion group that served as a "front" for Professional Unit 122 of the Communist Party of the United States of America. Malina and Tsien were Communist Party members. Summerfield was part of the reading group, a "comrade." Each had a party identification and underground code names. Each soon fled the scene. Malina to France. Tsien to Communist China. Summerfield to Princeton.[32]

The move could not have been easy. The Zucrows were leaving a home in Altadena, at the edge of the San Gabriel Mountains, just a few blocks from the Angeles National Forest and Caltech, and midway between the Aerojet headquarters and the JPL. He was coming to a Purdue campus and city overflowing with new students, the economy in difficult straits through 1946, racked by inflation and shortages. The economic downturn stretched Purdue resources thin. R. B. Stewart reported that the average Purdue professor made $5,619 a year, $823 less than an Indiana University professor. His people were "underpaid and overworked." In terms of teaching staff, graduate students taught 25 percent of all classes.[33] The housing situation at Purdue was precarious, with few comfortable places to live. Enrollments surged, 3,356 students in March 1945, to 11,472 in October 1946, then peaking with 14,674 students in 1948. Students shared attic rooms and cramped dormitories, slipshod temporary barracks, and a series of trailer parks around campus. Quonset huts were used as apartments and classrooms, though not at the same time. As one student "Ode to the Quonsets" exclaimed: "In each petite hut the classrooms two / Resemble the monkey house at Columbian Park Zoo." And "Classes in the Quonset huts are open to all / Providing you are not over five feet tall." Students baptized the campus as "Boom Town." War veterans dominated the scene, seven thousand at their peak, over half of them in engineering on the GI Bill. They organized an advocacy group, the Purdue Veterans Association, for better housing. They built playgrounds for their children. In one famous episode, which received national coverage, they boycotted the local barbers for price gouging and charging $1 for haircuts.[34]

Few veteran students knew that one of Purdue's own, R. B. Stewart, was among the architects of their good fortune. The Truman administration had asked him, based on his work for the Navy–Army Board, to serve as chair of the advisory committee to the administrator of Veterans' Affairs (1946–1958). This was one of the mechanisms for deploying the GI Bill, and it dispersed $13 billion by Stewart's distinctive principles of fairness and equity. "Every regulation that was written governing the processing of G.I. through any phase of education," said R. B., "passed under our supervision." It was hard work. R. B. and the committee, once again, had to rein in and discipline both the profiteering large schools and "small operators" trying to milk the program.[35]

As the GIs came and graduated, campus life returned to undergraduate normalcy. For new students from out of state, young people like Neil Armstrong from nearby

Ohio, Purdue was "Hoosierland," a kindly term that set its culture apart. Yet everyone became Boilermakers, or more commonly Purdueites or Purduvians, as their college careers progressed. There were many more men than women on campus, the majority of them in engineering. Naturally, the engineers referred to this imbalance as "the Ratio." It grew over the years, to their consternation, as for example 4.66 to 1 in 1947, and up to 5.98 to 1 in 1948. Purdue may not have had many women, but it did have cars. Lots of them. Both were a status symbol in the culture of the day. Students who were featured in the *Purdue Exponent* commonly shared mentions of their girlfriends and their cars, and even their grade point average (called the index), if it was worth it.[36]

Zucrow came to teach these young men and women, mostly as graduate students. He later remembered the "obligation he felt to the youth of America who had sacrificed their time and their lives during World War II." There was really no one like him in the country. Anywhere. He was one of a handful of Americans who understood the new jet propulsion as both engineering science and R&D experience. Solberg's primary concern was that Zucrow come to campus ready to create and teach four new graduate courses: Steam and Gas Turbines (ME 139), Gas Turbines and Jet Engines (AE 115), and Principles of Gas Turbines and Jet Propulsion I and II (ME 281 and 282, cross listed with AE), supplemented by specialty seminars and thesis and dissertation research. Solberg also assigned him to renovate a series of undergraduate and graduate courses: Fluid Mechanics (ME 55), Heat Transmission (ME 100), Advanced Fluid Mechanics (ME 106), Advanced Heat Transmission (ME 203), Hydrodynamics (ME 207), and Advanced Thermodynamics (ME 211). His task was to fortify them as engineering science and thereby "build a sound unified program." Zucrow agreed. Before experimentation in the laboratory, he pursued experimental work in devising more scientific courses and curriculums in the new fields.[37]

Purdue was bringing in Zucrow to share the best of Caltech and Aerojet, to re-create "special fundamental and applied courses in the field of high velocity turbine-rocket-jet aircraft." The need for these courses was acute. As of the summer of 1946, twenty young men had already applied for graduate work in jet propulsion at Purdue. They were, as a rule, already working at Wright Field or NACA Lewis, or in one of the Navy development sections. They took leaves of absence from their engineering jobs to prepare for the future in "the high velocity field." Zucrow was the answer.[38] A photograph captured him lecturing on jet propulsion to mostly older graduate students out at the airport, a variety of propeller models hanging on the wall. Wright Field Air Force officers, meanwhile, came to campus to negotiate for another cadre of graduate students. They met with Potter and Bruhn, then went to the Purdue–Indiana football game. The next academic year, the public relations officer at Wright Field returned the favor, bringing Captain Chuck Yeager for a visit to campus, soon after his historic flight breaking the sound barrier in his X-1 rocket plane.[39]

Zucrow's fall 1946 courses were innovations. The closest any other school came was Caltech, which offered closed courses for the military and a few elite engineers (beginning in fall of 1944), and Stanford University, which offered a special advanced course for thirty-four graduate students on gas turbines and jet propulsion in spring of 1947. Johns Hopkins taught a series of lectures for the military and industry called the Guided Missiles Familiarization Course (February–March 1947). Ohio State University, thanks to its Cryogenic Laboratory, offered courses on jet propulsion in 1948, supported by the Wright Field Air Materiel Command (AMC), which in turn by 1949 was offering courses on rocket propulsion at the Wright Field Graduate Center.[40] All of these courses were team taught. Zucrow taught his courses alone.

With Zucrow on board, Dean Potter remarked that Purdue was well on its way with the "development of our most important field." It is tempting to read backward and define this as rocket propulsion and aerospace engineering. But Purdue's top priority at the time was the graduate program, one that Solberg called Purdue's "big field of expansion," to raise its reputation as an engineering education leader.[41] He set out to dismantle the old "stagnant groups of courses" and raise the engineering sciences.[42] Purdue needed Zucrow in 1946 the way they needed him twenty years before: then to help build a brand-new graduate program as the first PhD student, and now to revive that graduate program for engineering as a full professor. Hovde spotlighted the imperative when he spoke with the faculty in the Hall of Music on Monday, 7 January 1946, late in the afternoon, his first official business day as Purdue's president. Graduate research was Purdue's "most important field." "Let us not be afraid to experiment," he said. Hovde raised "basic research" as Purdue's new "educational function." We must invest in "the development of the latent talents" of students, he said, by way of "the laboratory of the productive creative staff scholar of the modern university." Meikle similarly promised to abandon "rigid maxims, the rule of thumb routines, and the fixed dogmas"; to make "scientific sorties into the relatively unknown." For this Purdue needed "the foresight and philosophical power of understanding of the alert, intelligent, productive, creative thinker."[43] They were both describing Maurice Zucrow.

By several national surveys, Purdue was primed for academic success. In one, covering 362 colleges and universities and based on alumni listed in "Who's Who in Engineering" between 1931 and 1948, Purdue ranked fifth behind MIT, Cornell, Michigan, and Illinois. In another, on the collegiate origins of the "younger American scholar," that person primed for recognition as a scientist or PhD, in part thanks to their education in the liberal arts, Purdue ranked in the top twenty, with schools like Caltech, Chicago, Swarthmore, Johns Hopkins, Oberlin, and MIT. With them, Purdue was a place of remarkable "concentration of scholarly activity." At his desk after a year on the job, Hovde confirmed Purdue's promise in a short note to Vannevar Bush: "There is lots to do and lots to do with," he wrote, here among "my six thousand engineers."[44]

Zucrow's calling card in all these endeavors was his textbook. The original title, reflecting his expertise and interests, was *Principles of Jet Propulsion*. Solberg asked for a change to match his new job title: *Principles of Gas Turbines and Jet Propulsion*. Zucrow compromised by putting jet propulsion in the title first. The manuscript was completed by June of 1946 but only published in January of 1948. Military censors required Zucrow to delete classified materials from his Aerojet experience, hence the focus on basic principles. They also delayed the peer review. Such was the price of being first; "the first textbook on the subject in English," in three printings.[45] Zucrow wrote it for engineering students and professional engineers, in an approach he called the "'pedestrial quality' of a college textbook," to walk them through the material so as to teach them to walk on their own. This required "the slow, more fundamental, and theoretical approach." The textbook sold in 11,697 copies between 1948 and 1965. Along with his several later contributions to the top engineering handbooks, it transformed Maurice Zucrow into America's premier "jet and rocket engineer."[46]

Zucrow devoted chapters 1 to 4, fully one-third of the text, to the fundamentals of engineering science. He covered dimensional analysis and the nature of matter, the momentum and energy relationships of fluids (conservation and continuum), and the thermodynamics of gases and air. He translated the classic formulas of the great pioneers (Ludwig Prandtl and Adolph Busemann, William Durand and Theodore von Kármán, Aurel Stodola and Gustav Flügel) into terms accessible to the average undergraduate and graduate student. Chapters 5 and 6 covered the aerodynamics of the piston-engine propeller airplane, with reviews of the forces acting on the airplane (drag, lift, thrust) and the performance equation, as well as a detailed study of the airplane propeller in terms of momentum and propulsion, angle of attack, and torque. NACA achievements figured prominently here.

In chapter 7 Zucrow addressed the gas turbine power plant, and most of its fundamentals, tied to his brief period at the Elliott Company (1941–1942). There he mastered the details of the Lysholm compressor and began his first forays into high-temperature and high-stress metals, both of which took up a significant part of this chapter, as well as the textbook's appendix, a study of high-temperature metallurgy. Chapters 8 to 11 surveyed both propeller-driven and turbojet engines. They were all reaction engines, following Newton's third law of motion. He gave each its due but focused on the turbojet engine for aviation, with in-depth surveys of air compressors and turbines and the combustion process.

The book peaked with chapter 12, "The Rocket Motor." Zucrow offered detailed equations and surveys on the thermodynamics of rocket propulsion, especially the crucial thrust-to-weight ratio, ensuing that the power plant and propellants produced enough thrust to carry the whole. He valued the rocket for employing a thrust "practically independent of its environment." It was a machine of the "simplest means for converting the thermo-chemical propellant combination (fuel plus oxidizer) into the kinetic energy associated with a jet of flowing gases." Zucrow paid special attention to the "gas turbine–driven pumping units" in the V-2 and the Hellmuth Walter engines of the Messerschmitt 163. He was fascinated with the Walter HWK 109-509, breaking down its weight components: framework, piping, pumps and turbine, motor, motor gear box, electrical apparatus, control levers, propellant control valves, and steam generator. It weighed 365.8 pounds but provided 3,300 pounds of thrust. "It is of interest to note in passing that ratio of the landing weight (4,620 lb.) to take-off weight (9,020 lb.) for the Messerschmitt-163 airplane is 0.51," Zucrow said, adding that "this was the first successful rocket power plant used as the sole propulsion means of a piloted plane." He was not allowed to mention his own contribution to America's near equivalent, the test Northrop Flying Wing of 1944.[47]

Zucrow's *Principles* was a foundational event in the history of aerospace engineering. There had never been anything like it before. There was nothing like it then on the market, except for a few imperfect texts, filled with misinformation, or what Harvard's Lionel Marks called the "play of fantasy."[48] Zucrow's book had one serious competitor: Hsue-shen Tsien's *Jet Propulsion* (1946), based on his "closed" engineering course at Caltech in fall of 1944.[49] Zucrow and Tsien wrote parallel manuscripts. Both were based on von Kármán's "original plan of instruction" for Aerojet and Caltech. Both addressed what Tsien called the external and internal "aerodynamics of high-speed flows." Both culminated with chapters on rocket motor design as the power plant of the future. Both texts were at first classified (restricted), Tsien's written on contract for the Air Technical Service Command, Zucrow's written for the Engineering, Science, and Management War Training (ESMWT) program. Both surveyed the history, Tsien reaching back to the ancient Chinese and Greeks, more recently to the French and Germans. Tsien spotlighted Caltech's pioneering role, but completely ignored anything that Robert H. Goddard ever did. Zucrow paid his homage to Goddard as the ultimate pioneer. Tsien, the brilliant theoretician and mathematician, edited his volume, a detailed mass of eight hundred pages, with a team of authors. Zucrow wrote his handy textbook all on his own.[50]

Tsien's text was in high demand through 1945 and 1947, as the military and government, industries, and universities prepared their officers and engineers to learn the new technologies. The market was huge. The Cornell Aeronautical Laboratory needed it for its contract to build pilotless aircraft for the Army and Navy. Aeronautical Engineering Departments at Stanford University, Virginia Polytechnic, and Pennsylvania State College needed it to create new courses. General Electric and Shell needed it for their training programs. Bell Aircraft asked for it in April of 1946 for several of its rocket projects, just a year or so away from the X-1 rocket plane. In one of the more ironic requests, even the chief engineer of the Propeller Division of the Curtiss-Wright Corporation asked for it. For all the demand, Tsien's text proved unworkable in the classroom. John Wiley, America's premier publisher of engineering and science texts, canceled its publication plans, turning instead to Zucrow's *Principles*.[51]

Zucrow's other rival, as far as textbook authors went, was George P. Sutton, his former colleague at Aerojet, author of the seminal *Rocket Propulsion Elements* (1949), with generous references to Zucrow. Sutton's first draft of the book was actually typewritten lecture notes (in three-ring binders) for the Engineering Extension course at UCLA, XL-198, Rocket Propulsion Elements, which replaced Zucrow's ESMWT course in the spring semester of 1947. Here Sutton revealed his unique talent for summarizing complex material, the stuff of rocket science, with powerful descriptions and explanatory terms, joining mathematics, physics, chemistry, and engineering. The notes also read like a superb history of the Aerojet Corporation, including accurate descriptions of its rocket engine laboratories, test cells, apparatuses, instruments, and procedure. No one had ever done this before. Sutton was quite proud of Aerojet's achievements, which had "perfected" many of GALCIT's initial inventions, including Zucrow-inspired advances in "liquid propellant JATO units and one of the first successful stainless-steel rocket motors." Sutton's was a bold move, to take the culminating chapter of Zucrow's wide-ranging *Principles* and turn it into a book of its own. Zucrow's attention to the full range of jet propulsion fundamentals may have actually held him back, framing him as the mainstream engineer. Sutton seized the glamour to become the most authoritative voice of purer rocket science, with nine editions of the book to 2016, selling over sixty-five thousand copies. Sutton's book made its mark. We know that two of Purdue's first astronauts, Neil Armstrong and Eugene Cernan, read it for their graduate studies. Elon Musk also read the book as he eagerly prepared for one of his great adventures, founding a commercial rocket enterprise called SpaceX.[52]

For all their differences and rivalries, Tsien, Zucrow, and Sutton, each in their own talented ways, educated the first generation of engineers in jet propulsion and expanded the reach of the new aerospace field. Zucrow's contribution, more than the other two, was for higher education. Back at Purdue, these were exciting times for its brand-new School of Aeronautics, founded in July of 1945 with programs in aeronautical engineering and air transportation. The school experienced dramatic growth and dynamism in its first years, with growing enrollments and a new graduate program.

## PHASE 7: HIGH-PRESSURE COMBUSTION

When Zucrow joined Project Squid in fall of 1946, he started out with nothing, "no staff and no research facilities," so Harry Solberg remembered, just a desk and a chair, and his reputation from Aerojet and UCLA. As part of his hiring negotiations, Solberg made clear that Zucrow's first responsibility was to gather financial and material support from military, government, and industry sources. Zucrow approached Wright Field in Dayton to procure the latest German and American intelligence reports on gas turbines and jet propulsion, and to get some surplus equipment. He especially wanted the General Electric I-16 and I-40 turbojets, America's most advanced models; the Westinghouse Yankee 19B, which had only recently powered both the McDonnell experimental Phantom jet and the Northrop experimental Flying Wing; and the vanguard Messerschmitt 163 rocket power plant.[53]

Zucrow did have one support staff, someone who stayed with him for the rest of his career: Cecil Warner, who had transferred to Purdue from Tri-State College (Angola, Indiana) in 1937, receiving his BSME in 1939 and PhD in 1945. An expert in thermodynamics and heat transfer, he worked on these issues for "air-cooled radial engines" in association with George Hawkins's Army Ordnance project. He and Zucrow set out for Wright Field in November of 1946 to seek financial support for their investigations in "various thermodynamic and associated problems encountered in the field of aircraft power plants." Their pitch: to establish "the first such laboratory devoted exclusively to instructional research" in the varieties of jet propulsion. The Army Air Forces' T-2 Technical Intelligence Branch offered $21,000 for this jet propulsion research for the 1946–1947 academic year.[54] It was one of Zucrow's first big breaks, part of a national effort of nearly two hundred Air Force R&D "long-term projects" with industries and universities to train its personnel in the new technologies. The T-2 Analysis Division was in critical need of civilian expertise like Zucrow's, to acquire the "technical knowledge and experience in all phases of aeronautical science and engineering, plus imagination and ingenuity of a very high order."[55]

The Air Forces also donated an important war trophy to Purdue in 1947: a Walter HWK 109-509 rocket engine of the Messerschmitt 163, the world's first operational rocket plane, able to climb at 16,000 feet a minute to a maximum of 600 mph, with 3.75 minutes of full thrust engine time. Zucrow had first studied one at Aerojet in June of 1946, part of its contract for testing the B model and providing technical data. Richard F. Gompertz of the Wright Power Plant Lab brought one of the Walter engines to Purdue for Zucrow's study and use. It was, in a sense, the first Purdue rocket lab. They examined and fired it in a field adjacent to the airport in "a truly low-cost test operation," according to Gompertz. Their work together, he remembered, was a truly "humble beginning." Gompertz remained ever thankful for Zucrow's "accomplishments to teach others to be successful." He meant himself, too. Gompertz went on to direct the Rocket Propulsion Laboratory, part of the huge new Air Force Experimental Rocket Engine Test Station at Muroc Field (the future Edwards AFB). He also later worked on the Apollo program.[56]

Zucrow's early research and teaching also covered the turbojet engine. As one of his students highlighted, "It's easy to forget how revolutionary new jet technologies were in 1946, how radical and inspiring Zucrow's lectures were on Jet Engine Design and Performance," how all the novel information about "turbine engine parameters" ("pressure ratios, rpm, net thrust, fuel flow") became "mainstream."[57] Zucrow inherited Phase 2 of Project Squid, which studied combustion and heat transfer in turbojets, set in the Gas Turbine Laboratory at the East Hangar and adjoining Quonset hut of the Purdue Airport. There, six Allison V1710-109 engines drove twelve superchargers to provide the high-pressure air for the gas turbine experiments, whose results defined critical data sets and efficiencies for the new jet aircraft industry. He knew parts of the setup well: items from his old Internal Combustion Engine Lab (ICEL) were now reconfigured for jet propulsion.

Zucrow's graduate students, with Navy, Air Force, and NACA support, completed their theses on turbojets and moved on to significant aerospace careers, for example in the military services and at Pratt and Whitney and NACA Lewis. Richard Levi Duncan wrote the first PhD dissertation in aeronautical engineering, a study of the superior performance of air-cooled hollow turbine blades. As Robert Glick remembered, eventually "the program's objective was to improve the performance of radial inflow turbines by cooling the turbine wheel with a forced air flow over the structural part of the turbine." In the tradition of the Zucrow Rocket Lab, classmates maintained common research tracks, with mutually complementing and reinforcing studies, as in Ernie Petrick and Donald William Craft, who worked on measuring "heat transfer in air-cooled radial inflow turbines."[58] Zucrow, in turn, published several pieces on how to teach jet propulsion to undergraduates and several provocative studies on the technical limitations of the turbojet engine.[59]

For all these new approaches, Zucrow's primary interest remained in liquid-propellant rockets. In his *Principles*, he laid out the key challenges for improving the rocket motor: how to adapt the design and function to higher chamber pressures, and thereby higher thrusts and specific impulses. In the early days, the average chamber pressures for liquid-propellant rocket motors hovered at about 300 psi. He promoted the higher chamber pressures for more "effective" exhaust velocities. The trouble, of course, was the resulting higher heat loads on the chamber walls. The "adequacy of a rocket motor depends almost entirely upon its ability to perform without damage at high temperatures for the operating duration," he wrote. New materials to absorb or resist the high heats were one solution, as was regenerative cooling, using one of the cold propellants to circulate in the system to reduce heat. Zucrow focused on internal film cooling: "based on forming, over the inside walls of the chamber and nozzle, a complete film of liquid which by its evaporation keeps these surfaces cool."[60]

Sensing something of his own advantage, and risk, Zucrow understood that he had to compete with his university rivals, had to join what he called a rocketry "race" among the engineering schools. This was a race for surplus items, technical documents, and all-important military funding. Several schools were even trying to build whole missile systems. Some were small: the Cornell Aeronautical Laboratory and the Lacrosse missile, the University of Maryland and the Terrapin and Oriole sounding rockets, and the State University of Iowa Rockoon project. Others were big. Very big. Like Project Meteor, a Navy Bureau of Ordnance and MIT partnership, including five departments at MIT, fifty faculty, and six labs, to design and build a whole guided missile from start to finish. Johns Hopkins University and its Applied Physics Lab were doing the Bumblebee. JPL was building the Corporal. The University of Michigan was developing Thumper and Wizard. These weapon systems were so new and complex, the military and industry needed what Air Force historian Mary Self called the "high powered professors" to get them done.[61]

In a rare lapse of judgment, Zucrow and Purdue joined this race in December of 1946 with a proposal to make a guided missile all their own. The Schools of Aeronautics and Electrical Engineering proposed it as a new phase for Project Squid: to design and develop a guided missile in its "aerodynamics, propulsion, housing, guidance and control." John Weske, a visiting professor in aeronautical engineering, was to direct the "structural study and design of the missile" and "high velocity aerodynamics." Roscoe George, inventor of the George cathode-ray oscillograph, was to do the "electrical control equipment." Zucrow would handle "high velocity thermodynamics." The PRF was to manage "the supervision of its construction by contractors," though it was unsure about process and costs, since no one had ever done this before. It was an expensive program: $100,000 in several dozen increments for a four-year program, and a grand total of $2,328,965.[62] This was a lot of money, even for the time. Purdue lost the bid. Properly chastised, Purdue's engineers from then on mostly stayed on the small side.

Zucrow's real rocketry breakthrough came in March of 1947, during his second semester at Purdue, when the Navy's Project Squid approved a new phase for Purdue: Phase 7, "Investigation of Rocket Motors and Liquid Propellants at High Chamber Pressures." Zucrow was project leader. He was assigned an initial $18,700 by June. The approval came in an unusual place, at the famed White Sands Proving Ground near El Paso, Texas. Frank Parker had organized a trip of the Technical Committee there to witness a V-2 firing, and for consultations with the German Peenemünde team and Wernher von Braun.[63] It was a fitting place for the start of Phase 7. White Sands was where several of Zucrow's power plants, like the one for the Corporal missile, were first successfully tested. It was where he served as a consultant on the technical staff (1950–1960), where he mingled and debated with the scrappy tribe of American rocket scientists at the very far frontier of new knowledge. These were colleagues like the engineers Robert Truax and Milton Rosen, mathematicians and scientists like Homer Newell and Ernst Krause, dedicated to the US civilian program in the Viking rocket. White Sands was also soon to be named an Integrated Range (1952) for all the military services: the US Air Force, the US Army Guided Missile Program, and US Naval Ordnance. Zucrow's Rocket Lab, in time, conducted research contracts for them all.

This Squid phase assignment on high chamber pressures became the hallmark rocketry project at Purdue for twenty years. Andrew Haley, Zucrow's old boss at Aerojet, later paid homage to Phase 7: recognizing "high mass flow, high energy, and high pressure" as the keys to "large thrust." For "high performance, or specific impulse," he wrote, "a rocket engine requires high chamber temperature and a large expansion ratio (that is, the ratio of chamber pressure to exhaust pressure)." G. Harry Stine, a physicist and project engineer for the Viking and Aerobee rockets, who helped build White Sands, also remembered the historical role of Phase 7, if in a more creative way. He wrote it into one of his science fiction short stories (under the pseudonym Lee Correy), an homage to the joys of rocket pioneering out in the New Mexico desert. Efficient rocket combustion and higher thrusts required higher chamber pressures. From his imagined distant future, he wrote, "Zucrow pointed it out years and years ago."[64]

Phase 7 funded the Purdue Rocket Lab, dedicated to "basic and applied research." It was a modest facility at least in terms of location and funding. Zucrow built it on the cheap. Purdue donated the land, in a set of farm fields and prairies next to the airport. Zucrow applied some of his first Air Force grant and all of the Navy Project Squid grants, about $40,000 in total, to construct the first lab. For instruments and equipment, he raided the surplus war items from the Joint Army and Navy Machine Tools Committee of the War Department. These included a thirty-kilowatt diesel-electric generator for power, a Chevrolet half-ton stake truck, and a "hand cranked sound powered field telephone." Two Purdue aeronautics students, Signal Corps veterans from the war, rigged the telephone up to connect the lab with the PRF offices, the wire partially hung along the airport fences. "We needed communications," remembered Zucrow, "because experiments in rocketry can be dangerous."[65]

Purdue had little choice but to keep the Rocket Lab within small budgets. Project Squid's own budgets were slashed in half by the Truman administration defense cutbacks in 1946–1947. Of all the Squid universities, only Purdue survived the cuts with all of its phases intact, if reduced in monies.[66] These kinds of financial stresses wore at the project phases, at least until the Korean War, when new monies appeared. As a rule, the Squid administration at Princeton held to a strict accounting of funds. The universities, for example, needed official approval to use Navy-financed equipment for purposes other than Squid's, as when Purdue requested to use one of its labs for another contract, with the Allison Division of General Motors. Project Squid went to von Kármán himself at the Policy Committee for permission. At other times, President Hovde actually represented Purdue's Project Squid when it came to the most important requests. The standards were high.[67]

Zucrow's standards were high as well to meet Purdue's and the PRF's primary rule of "educational function." The Rocket Lab (aka Rocket Test Facility) was meant to serve students. Walter Hesse's master's thesis, for example, helped inaugurate it. Hesse came to the lab from Purdue (BSME 1944, MSME 1948, PhD 1951), where he was in the V-12 program and a first-string baseball star. He served in the Navy as the assistant engineering and electrical officer aboard the submarine USS *Blenny*. As a graduate student, he taught mechanical engineering and studied aircraft propulsion and rocketry. His thesis was the first such tract, anywhere in the country or world, on how to build a rocket lab. He and Zucrow took a road trip in the summer of 1947 to get started, visiting some of the country's first rocket testing facilities: at Reaction Motors Inc. (Lake Denmark, New Jersey); the M. W. Kellogg Company (Jersey City, New Jersey); several US Navy Bureau of Aeronautics facilities on the East Coast; and the US Air Force facility at Ohio State, the Herrick L. Johnston Cryogenic and Rocket Research Lab. Their conclusion: design one building, with one control room for two adjoining test stands, and with spare rooms for a machine shop and office.

Hesse's study, sponsored by the Navy's Project Squid, covered the detailed design parameters of the Rocket Lab's test cells, test stands, control room, and instrumentation, all to meet the crucial standards of "accuracy, safety, and simplicity." Purdue's site was already "isolated," one key requirement given the noise and dangers. Cement slabs and concrete block walls provided the "high strength" to meet the "high energy" of rocket propulsion. He recommended two test cells attached to one control room. Each cell was to be fifteen by fifteen feet and twelve feet high, with three reinforced concrete walls (one and a half feet thick), a ceiling with one-foot reinforced concrete, and one wall with an "overhead roller metal door." Each cell was to have two "air-circulating fans" and a ceiling exhaust blower; "explosion-proof lighting fixtures;" and two twelve-by-twelve-inch windows, one for observation and one for the television camera, each with four-inch-thick bulletproof glass on the test cell side, and quarter-inch safety glass on the control room side. He added a workshop with "lathe, drilling machine, tool grinder, power saw, welding equipment, hand tools," a "small chemical lab" for studying "bi-propellant ignition and other problems," and office space for "test data analysis."[68]

Rocket science was not easy. "Setting up the rocket motor could take days," wrote Hesse, but test-firing the "rocket motor" took only thirty to sixty seconds. He described a normal firing: "The fuel and oxidizer pass through control valves and enter the combustion chamber through injector nozzles. In the combustion chamber, the propellants react at elevated pressures and temperatures. An ignition system is not shown because various propellant combinations can be utilized that will spontaneously ignite upon mutual impingement. The high pressure, high temperature gases expand through a properly designed de Laval nozzle and escape to the atmosphere at supersonic velocity. The reaction of the ejected gases produces a thrust on the rocket motor." He also gave a warning: "It is anticipated that some fires and explosions will occur in the test cells because there is no experience available on testing at the high pressures which are to be used." In the case of an accidental explosion and fire, a "quick-opening valve" from the control room would surge water by way of perforated pipes above, and a sloped floor to drain below. Propellant tank cubicles for the hypergolic and cryogenic oxidizers and fuels were also separated from the test cells and control room by concrete block bulkheads. For extra safety, the motors were fired in the direction of the "prevailing seasonal winds," aimed toward the southeast, in the direction of the Purdue Airport and School of Aeronautics. Slanted earthen bunkers, laced with stone, were placed 50 feet away (136 feet long by 30 feet wide and 12 feet tall) to buffer the noise and exhaust the gases into the air. A storage tank held five hundred gallons of water to battle fires. A ventilation system, with a capacity of 2,730 cubic feet per minute, drew the toxic gases from the test cells. "Flush-type showers" were also available for anyone coming into contact with the toxic chemicals, especially the nitric acid, which would eat away a person's clothes and skin and, if breathed, begin to eat away at the lungs.[69]

If Hesse wrote a master's thesis on how to build a rocket lab in the present, Edward C. Govignon wrote one on how to outfit and operate it for the future: "A Study of Some Possible Methods for Improving the Performance of a Rocket Propelled Missile" (MSME June 1948). Sponsored by Air Force Technical Intelligence at Wright Field, and supported by the Kellogg Company, it was a "design analysis" of a guided missile for a five-hundred-mile range and a fifteen-thousand-pound payload. He was looking ahead to improving range and payload, not with the current norm of chamber pressures at 300 psi, but a Zucrow-style liquid-propellant rocket power plant using 1,000 psi. Higher combustion pressures "give improved performance, but introduce new problems. The reaction temperature is increased, the buckling tendencies of the combustion chamber wall are aggravated, and the rate of heat transfer to the combustion chamber wall is augmented." That was the rub. He basically outlined the coming research agenda for the Zucrow Rocket Lab. Yet the higher pressures (1,000 psi) were worth the troubles for the resulting higher specific impulses (up to 20 percent) as a measure of thrust power, and they were one way to save on propellant weight (although increased pressure also meant "stronger and heavier" rocket motors). Govignon's thesis, not surprisingly, was classified as secret, given that it proposed a "military rocket powered vehicle" and used secret sources of its own, from Caltech's JPL, the Glenn Martin Company, Aerojet, the RAND Corporation, and the Air Force.[70]

Both Hesse's and Govignon's master's theses spotlighted the heart and soul of the lab: instrumentation. This was, of course, a Zucrow specialty from his old days working on combustion experiments and power plants in the 1930s and research in jet propulsion at Aerojet in the 1940s. The Rocket Lab needed much more: Thermometers for temperatures reaching beyond 5,000°F. Pressure taps connected to electric strain or reluctance gauges to measure chamber and injector pressures. Thermocouples to measure temperature rises at inlets and outlets and at the chamber–nozzle junction.

Slide-wire devices to capture the "distance and rate of travel of the propellant control valves." Oscillographs to record much of this physical and electrical data. Motion pictures to capture catalysis experiments to determine the "time intervals" of ignition delays. In terms of hydraulics, hydrometers measured the specific gravity of propellants; potentiometers measured valve-opening rates too. Thrust was determined by way of a camera shot of a pressure gauge. Regulator, tank, and chamber pressures were recorded by hydraulic lines connected to pressure gauges. A Bolex sixteen-millimeter movie camera captured the flame patterns in the first seconds of the test. Other cameras captured gauge readings as well.[71]

Zucrow and his team built the Rocket Lab between 1947 and 1949 to these specifications, as Project Squid "Problem Assignment #1" for the "Design, Construction, and Instrumentation of a Rocket Test Facility." His lead faculty partners were Cecil Warner from mechanical and George Palmer from aeronautical engineering. Warner's role at the lab was essential. An author in his own right, he taught undergraduate and graduate courses, wrote research proposals and reports, maintained budgets, and supervised the everyday work of the lab, including the graduate students. Two of Zucrow's old colleagues from Aerojet also came for short periods to help design and build the Rocket Lab: Marvin Stary and Irwin Weisenberg. To this extent, with Zucrow's experiences as well, they helped shape an Aerojet extension laboratory for the Midwest.[72]

Half of that first year (May of 1947 to May of 1948) was spent awaiting construction approvals and enduring delays. Surveyors prepared the land in March of 1948 and construction of the test pits and buildings began over the summer. They were complete by November, with a six-inch layer of gravel poured at thirty feet on all sides. It took another year before the two test cells were fully ready and operational, though Zucrow wisely outfitted and tested the first one before he started the second. Faculty and student teams prepared the delicate web of propellant lines and instrumentation devices and panels. They created a unique "tank weighing apparatus" on which to hang the high-pressure propellant tanks. Project Squid praised all of this. Zucrow's team revealed "good foresight in planning their work." As Govignon's thesis confirmed, they had already prepared the instrumentation, and begun experimenting with it on various research problems at other Purdue locations, for immediate installation once the lab was ready. This allowed them to complete initial thermochemistry, heat transfer, and ignition lag studies using four Aerojet 25-ALD-1000s sent by Wright Field and their own regenerative-cooled rocket motor made at a machine shop on campus. Phase 7, concluded Squid reviewers, "will provide much valuable engineering and scientific data for the rocket field."[73]

Purdue's Phase 7 was the only one of its kind in all of Project Squid. The only one devoted to the actual building and outfitting of a complete rocket laboratory. The closest thing were several expensive wind tunnels at Princeton. By 1949–1950, now renamed the High Pressure Liquid Rocket Program, Phase 7 was third among the top-funded nationwide phases in terms of operating costs at about $77,000 (out of Purdue's overall allocation of $631,000) and was slated for more, $12,358 to be exact, for "an accelerated program for the motor-performance phase of the high chamber pressure project." It was just the kind of "applied research" that Project Squid was looking for, and it had a surprise benefit, something altogether unique: the education of graduate students through sponsored research.[74] This educational dimension was a troublesome one at other universities. New York University, for example, conducted well-financed liquid-propellant rocket work and heat-transfer studies, much like Purdue, but suffered from disorganization. It simply could not keep qualified and motivated professors or students. Mentorship was everything. A project leader was crucial to anchor and educate young people as part of a research team. A faculty leader and teams fully "integrated into an academic setting" were essential to "guarantee enthusiastic interest of a large group."[75] Zucrow was proving his worth.

By fall of 1947, Zucrow was not just in charge of Phase 7 but had "general technical cognizance of all SQUID work at Purdue." He was the Squid technical director on campus and its representative on the national Technical Committee. He also hosted the Applied Research Panel at Purdue (9–10 November 1948) with government and Navy guests, including his old friend Robert Truax. The panel meetings were organized the Zucrow way by two of his graduate students, Stanley Gunn and Clair Beighley.[76] In another classic Zucrow move, he did not announce the completion of the Rocket Lab at this time. One of his undergraduate students did, a senior in aeronautical engineering, Eldon "Al" Knuth. Knuth was a top scholarship student; US Army hero at the Battle for Metz in the Second World War; president of the Engineering Honorary Society, Tau Beta Pi; and junior editor at the campus newspaper, the *Purdue Exponent*. He featured the Purdue Rocket Laboratory in an issue of the *Purdue Engineer*. Knuth described the sophisticated affair of the Rocket Lab. The two test cells to observe and record the static firing of rocket power plants; the control room and instrument room, machine shop and chemistry lab. Project Squid's Phase 7 was geared to help "propel rockets and similar missiles" farther, faster, and safer, to breach the far boundaries of "high-speed long-range flight." No one had ever experimented with such high pressures before. The dangers were legion. Students tested propellant mixtures in "miniature glass rockets," just a few drops at a time, to discover their "ignition characteristics" and their proper combinations, before using them in real motors. The Rocket Lab was doing extraordinary things on the heat transfer and liquid cooling of rocket engines. All at the edge of a farmer's field in central Indiana.[77]

Zucrow's Rocket Lab turned out to be a rather small niche in a vast American landscape of industrial, military, and university laboratories. Premier among them were the large-scale multimillion-dollar organizations like the Atomic Energy Commission's labs at UCLA, or the Argonne Lab at the University of Chicago, the Lincoln Laboratory at MIT, the Applied Physics Laboratory at Johns Hopkins, the Ames Laboratory at Iowa State. These were the vaunted Federal Contract Research Centers, doing top-secret work and thereby "attached to rather than integrated with the universities."[78] Cornell's Aeronautical Laboratory, a Project Squid member, was a wealthy cousin to these centers, the whole lab a gift from the Curtiss-Wright Corporation to the university in 1946. It was a wartime research lab worth $4 million, with a new wind tunnel and with $675,000 donated by aviation manufacturers as a "working capital fund." As represented in its glossy and finely edited journal, *Research Trends* (1954–1958), the lab specialized in airplane structures and aerodynamics, but soon amassed a wide range of further expertise in radar and electronics, computer modeling, economic forecasting, aviation safety, weapons systems, and jet propulsion and combustion. Purdue alumnus Clifford Furnas was its director.[79]

As for rocket labs in particular, none could compare with Caltech and Princeton. By the end of the war, Caltech's JPL was a campus all its own on the west bank of the Arroyo Seco, at Devils Gate Dam. Caltech also had a second jet propulsion laboratory, known as the Jet Propulsion Center (JPC), a gift of the Daniel and Florence Guggenheim Foundation in December of 1948. It was the second such gift, to complement the original GALCIT grant from the 1920s. The Guggenheim also funded a second such JPC at Princeton, which was already the headquarters for Project Squid. Harry Guggenheim called the two centers his "Rocket Institutes." Thus, Caltech and Princeton enjoyed the boon of two jet propulsion units each. Guggenheim endowed each "Center" with $250,000 to fund the Robert Goddard professors and $2,000 annually for each of the three Daniel and Florence Guggenheim Fellows at each university. The money provided for two years of study for the PhD, with national searches for "unusually promising graduate students in jet propulsion." Slick, well-designed posters announced the fellowships, framed by rockets in flight and parabolic arcs into space. They would ensure America's "leadership in the development of rockets and jet propulsion, with particular emphasis on peacetime uses." G. Edward Pendray, Guggenheim's main adviser, claimed that it engaged in a "nationwide study of various universities in which rocket and jet propulsion work is now going on." That was a publicity stunt. The foundation focused exclusively on "seeking a suitable institution in the East, and another in the West, in order to provide a truly national program of jet propulsion training and research."[80] Some national search, leaving out everything in between.

Places like Purdue were handicapped from the start by this elitism. Zucrow made it a point of pride, renaming his Rocket Lab as the Jet Propulsion Center in the summer of 1951, after he added several new buildings and programs, without a penny of Guggenheim monies. Doing so was bold, to say the least, raising a poorer and isolated Purdue to national prominence. Rolf Sabersky at Caltech even thought that, because of the new name, and surely because of Zucrow's fame, Purdue was actually one of the Guggenheim centers. But Zucrow had built his JPC on his own. The leading professional journal *Missiles and Rockets* eventually named it one of the "big three" such centers, alongside Princeton and Caltech. Martin Summerfield at Princeton did not take kindly to the slight, listing Princeton and Caltech as first and second among the "advanced educational programs in rocket propulsion," then mentioning the others, at "Purdue and elsewhere."[81]

Zucrow was defiant about his new Rocket Lab and Jet Propulsion Center. He and Warner laid claim to national "leadership" with their broad course offerings in jet propulsion, compared to the seven faculty at Caltech and five at Princeton, all of whom taught more specialized courses.[82] His relations with Princeton were cordial, but at times strained. Professional differences got personal. In one instance, Summerfield was informed by a friend, who overheard some gossip at an American Rocketry Society (ARS) national convention in the "heat of the argument," that Zucrow was unhappy with Summerfield as editor of *Jet Propulsion* for giving preferential treatment to his friends and Princeton associates. A rift ensued. In another instance, Summerfield and the ARS published a piece by Princeton's Luigi Crocco questioning some of the methods and conclusions of Purdue's work on combustion oscillations. The case became a cause célèbre at the Rocket Lab, a mark of its identity and pride. As J. Michael Murphy remembered, "They couldn't understand how Purdue could write a paper on combustion instability that proposed things differently than they had. And I think that was where things started to get kind of hot, because Summerfield should've stayed neutral as the editor, and he didn't."[83]

Biases and conceits appeared in the sources at times, as when James Conant spoke of his Harvard University as one of the few privileged centers of "wisdom" in the land; or when one former Purdue engineering dean admitted that the folks at MIT only trusted "their own" as the best graduates in the field. No other schools warranted.[84] The popular Hollywood film *Titanic* (1953) parodied this elitism, referring to the collegiate tennis star "Giff" Rogers (played by Robert Wagner) as a "fine, healthy bumpkin." Passengers on the doomed ship are confused by the big "P" on his collegiate sweater. "Oh, this is for Purdue. It's a college out in Indiana. Everybody thinks it's Princeton." In real life, the astrophysicist Louis Friedman once chose a fellowship from Cornell over Ohio State because of his admitted "elitist New York attitude." After all, he was "not a football player." A Caltech colleague also once commended Theodore von Kármán's European bearing. He was no "Joe Dokes from Indiana." In *The Right Stuff*, Tom Wolfe later portrayed Purdue's and Indiana's own Virgil "Gus" Grissom as just such a country yokel, the goofy Hoosier who did not deserve to be in the ranks of the New England Yankees who graced the test pilot and astronaut corps. It was a sad calumny, replicated in the popular movie.[85]

Zucrow resented these attitudes, and the prestige of places like Caltech and Princeton. He proudly admitted to being a Harvard man, but muted any pretension: "I have had students from every place. I find that a good man is a good man." Elite schools were no better. In another case, Zucrow protested MIT's submissions for the ARS Graduate Student Award. Places like MIT enjoyed the huge "advantage" of having "elaborate and expensive facilities at their disposal," with the guidance of teams of mentors. The advantage was patently unfair, so he argued. The Graduate Student Award ought to go to more deserving and innovative students, those with the creativity and "imagination" to write shorter seminar papers (not massive theses) all their own.[86] He meant people and places like Purdue.

## ORIGINS OF AN INFORMATION AGE

All of these episodes, threaded between the Second World War and the Cold War, also spotlight a nexus between jet propulsion and computer technology, between the space age and the digital age. Vannevar Bush, who advanced guided missile research by small steps in the war, was an early promoter of the new information technology. He even predicted that someday we might "reduce the size of the encyclopedia to the size of a matchbox." He foresaw that "the man of the future will sit at a desk on which are translucent screens, a keyboard, push buttons, and levers." With time, our bodies and minds would fuse with "cogwheels, thermionic tubes, and strands of wire." He also imagined photographic cameras pinned to our foreheads, like coal miners digging through rolls of microfilms, 1970s style. He did not get all of this quite right. Gen. Gladeon Barnes, promoter of an American V-2 missile, was more successful in result, named "the father of modern-day computers." He commissioned the Electronic Numerical Integrator and Computer (ENIAC) at the University of Pennsylvania, the country's first viable digital computer.[87]

Project Squid contributed its own advances to this rising information age, at least in the way the military and university shared new knowledge. Its administrative structure was flexible, vertical in its application of military authority, yet horizontal in the freedom of action for university scientists. The Navy set the research agendas, the projects and phases. The universities proposed modifications, but otherwise fulfilled their commitments with personnel and facilities, experiments, and reports. The results and records were the heart of it all: information. The Princeton administration offered a central repository of resources and rules and a clearinghouse for technical publications. The whole point was to create an archive of jet propulsion, for communication and sharing, for learning and application. A number of ad hoc panels offered flexibility, joining experts from different universities to resolve challenges. All these units wrote project and phase reports, approval reports and rebuttal reports, quarterly reports and technical reports. It was an intricately designed information network.[88]

From the start, Project Squid was relentlessly demanding in terms of cooperation and competition, but also surveillance. The ONR imposed a strict funding regime, at first for three-year contracts but later extended to five years. It applied contracts based on R. B. Stewart's OSRD and Army–Navy models, which formalized and disseminated the role of the peer review technical committee. Joining university and industry participation, peer review was an essential check and balance on the research project. Project Squid was also frugal. No scientist or engineer in it could hide behind pretense or doctored results. Project administrators denied the Polytechnic Institute of Brooklyn funding for "unwarranted" equipment purchases in the study of high-temperature metals and alloys. Reviewers, including experts from the leading universities, were stringent. They compelled Zucrow to rewrite his initial proposal for a study of rocket combustion chambers in far greater detail.[89] They reprimanded the combustion work at Purdue (on the "turbulence of flames" as related to ramjet studies) for failing to work smartly and cooperatively with other universities.[90]

The ONR made regular on-site inspections or contracted intermittent outside reviews of the projects and phases. The Atlantic Research Corporation offered critical evaluations of each university in terms of scientific and technical quality, fiscal restraint, personnel issues, and even their "cooperative" spirit. It made recommendations for funding reductions or cancellations. In a "serious criticism," its investigators scolded the Cornell Aeronautical Laboratory for padding its Squid projects with irrelevant researchers, "a buffer" simply to keep them on the payroll. The evaluators criticized New York University for sloppy accounting and personnel policies, for assigning researchers "indiscriminately" to any number of phases, each without its own budget, resulting in over-expenditures of $27,000. They targeted the Polytechnic Institute of Brooklyn for lack of "cooperation" and weak "technical reporting." They even found Princeton itself guilty of sloppy procedures and over-expenditures. In terms of administration, Purdue largely avoided these kinds of fault lines. It was the only member of Project Squid to use a research foundation (the PRF) to administer the contracts and accounts, each with a distinct number and budget to function efficiently within cost limits. The PRF worked.[91]

There were problems among the sponsors. The military services competed fiercely with each other for strategic resources and influence, each creating R&D networks within their own branches. Nor did the military liaisons always meld well with the academics. There were anti-intellectual biases at play, as for example against those who "figuratively wear the class ring," the men of the "university group." Officers sometimes exhibited the public distrust of "longhairs" and "eggheads."[92] Intellectuals. Or as one general put it, the odious "cult of scientists-know-best."[93]

For all these various challenges, and although a Navy initiative, Project Squid engaged in wide-ranging cooperative work, both within the military services and between the military and civilian contractors. At times, the Air Force supplemented Squid budgets and lobbied for new phases or projects to serve its special interests as one part of Squid's "inter-Service activity." Project Squid worked closely with other independent partners engaged in jet propulsion work, like Army Ordnance and the Battelle Institute. Agents at the lowest levels sought each other out for technical cooperation on issues like exhauster capacities, water-cooling systems, or the prices of dynamometers.[94]

The universities held both "informal meetings" and "formal symposia" on such basic topics as "fluid dynamics, fuels, combustion, materials, instrumentation." Zucrow and his people actively organized and participated in them. At times, these symposia became contentious. At the Conference on Problems of Heat Transfer in Rocket Motors (Princeton University, 2–3 December 1949), Martin Summerfield even questioned a basic premise of the whole Project Squid approach. Zucrow had opened the conference with a talk on his premier early findings based on "the reaction between WFNA and different hydrocarbon fuels at high chamber pressures." Though recognizing the still imprecise and rough data and the need to solve the key "operating problem" of cooling the rocket motor, Zucrow offered this striking claim: "It is apparent from the curves that a substantial improvement in performance can be realized at the high combustion chamber pressures." Summerfield was skeptical. He ended the conference wondering "whether a wholly analytical approach to the problem of cooling rocket motors is at all possible." Zucrow's Rocket Lab was about to prove him wrong.[95]

Project Squid administrators and researchers were keen on the "free exchange" and "flow of information" in all directions: between universities, with the military and industry, and back again.[96] Princeton printed and shared the technical reports of the universities and their phases. The Navy provided indexes of foreign publications on rocketry and jet propulsion. The Air Force shared the authoritative reports of its Wright Field Document Center. Squid also published a confidential bulletin, *Technical Library Abstracts*, for its members, listing descriptions of new articles and publications from around the country on all technical aspects of jet propulsion and rocketry.[97] Over time, Project Squid morphed into such offices as the Rocket Propellant Intelligence Agency and the Solid Propellant Information Agency. Both had headquarters at the Johns Hopkins University Applied Physics Laboratory, initially funded for a mere $20,000, like Zucrow's Rocket Lab. These later became the Joint Army–Navy–NASA–Air Force and Chemical Propulsion Advisory Committee, designed for the exchange of classified materials between the academy, military, and industry.

R&D experts depended on these compilations of data and new findings, but they also needed to learn more about the new and still mysterious science of jet propulsion. By the summer of 1949, Princeton began planning for its aeronautics publication program. It eventually became Progress in Astronautics, also known as the Princeton Series. It had nine volumes by 1960, with one hundred experts participating, funded by the US Navy and Air Force. Princeton claimed a grave need for this "series of texts of high scholastic caliber" to "cover the basic theory underlying all jet propulsive devices." This was to be a series of "theoretical and practical" monographs for the scientist and engineer, with "far reaching implications in the stimulation of research and in the education of men for advanced work."[98] Princeton made these boasts at the very moment Zucrow published his *Principles of Jet Propulsion and Gas Turbines* (1948), without even mentioning it.

Yet Zucrow's textbook and courses left their legacy. Neil Armstrong, who attended Purdue between 1947 and 1955, with time in the US Navy and Korea between 1949 and 1952, remembered Zucrow with a special nostalgia: "I chose Purdue because at that time there were only two rocket scientists in the country, Wernher von Braun and Maurice Zucrow. I came to Purdue because of Dr. Zucrow." This influence, with that of his other engineering professors, helped to crystallize Armstrong's vocation to become "first and foremost an aeronautical engineer."[99] There is still more to this story. Armstrong studied at the Naval Air Test Center, Patuxent River, Maryland, in 1953 and 1954, just when Zucrow's PhD graduate and Project Squid alumnus Walter Hesse served there as chief engineer and academic instructor for power plants (1949–1955). Hesse completed his Purdue dissertation at Patuxent, where he outfitted a "project airplane" to study turbojet engines with instrumentation designed and made at the Zucrow Rocket Lab. He also taught Neil Armstrong and three of the future Mercury astronauts there in the Test Pilot School. The course covered the mechanics and operations of advanced aircraft.[100] Another Project Squid and Zucrow veteran played his part in the Armstrong saga. Robert W. Graham, back on the Purdue campus to "recruit" for NACA, helped convince Armstrong to answer the ad "for a test pilot for the X-15."[101] Armstrong applied and got the job. Graham also worked on heat-transfer studies of the ammonia–oxygen engine for the X-15 rocket plane, which Armstrong eventually flew. Here were examples of collaboration on rare personal and professional levels, perfect unions of two Purdue strengths: power plants and people. They are also emblematic of Zucrow's own character and success, his complete devotion to teaching, in that two of his best students taught Armstrong.

No matter how cooperative, little of Project Squid's information was "free." Security classifications encompassed it. Squid started out fully unclassified. The goal was to produce as much basic research and fundamental knowledge as quickly and openly as possible. Only the "new process" and "finished propulsive device" would be classified.[102] That was the aim, but not the result. The constant tension in the program was toward development. Researchers found the most thrill and prestige in those "engines of a type unforeseen at present," in the "specific propulsion devices." These included Joseph Foa's pulse jet at Cornell, Abe Kahane's intermittent ramjet at Princeton, and Maurice Zucrow's liquid-propellant rocket at Purdue. These actual devices had to be classified. It was inevitable that Project Squid would become retroactively classified too. Most of its sources were deemed so already. The stakes were simply too high and the system already primed for secrecy.[103]

Project Squid very quickly fell into a hierarchy of classifications, the secretive world of the espionage laws, written into the US Code as Title 18, USC, Sections 793 and 794 (1948–1950). They prohibited the "transmission or revelation" of "information affecting the national defense of the United States" to "an unauthorized person." Actual documents often contained this stamped statement. For offices and labs, the "Restricted" category meant that foreign aliens needed clearance, that visitors needed approval, and that finished work was not allowed in the "published journals." The "Confidential" and "Secret" designations were much more "serious," needing clearance for all personnel, an "official letter" for visitors, identification badges of some kind, "strict accountability of all materials published," along with "fences and guards." These cost about $25,000 every year for any facility so designated. Clearance and access to Squid libraries could take up to a year to receive, with paralyzing delays. The headquarters at Princeton was fully "SECRET," as one description had it.[104]

Research scientists and engineers in the most advanced fields of R&D lived and worked in a mostly paperless world, at least in terms of what they could read at home or write about and publish in public forums. Everything in the top secret world, and in the Q and Omega clearances for atomic physics, was subject to sensitive compartmented information (SCI). One of Zucrow's students, Paul Petty, remembered that "the work was absolutely compartmented. Closed off to everybody else. Isolated. Information went in, nothing came out. If you went to a meeting and took notes, at the end of the meeting, someone gathered up your notes and stamped them 'Classified.'" In another case, Zucrow student Elliott Katz, just starting his career at Convair on the Atlas missile, wrote an analysis, "and then," he said, "they took away what I wrote and classified it as Secret. Because I didn't yet have my Secret clearance, I could no longer read what I wrote. That's not unusual, by the way."[105]

There were some strange phenomena when it came to secrecy. NACA Lewis had strict protocols for secret materials at its Northern Ohio centers. There were "secret places" to hold materials. There were special handlers for them, called the "classified document control clerks," who carried them between offices. There were special envelopes and color-coded forms to sign. In one dramatic scene, Walter Olson, chief of the Fuels and Combustion Research Division at NACA Lewis, dutifully reported back to the Research and Development Board that the NACA security officer had properly "burned" three secret files, numbered and logged according to procedure. One NACA Lewis employee suffered under the burden of compliance. Bruce Lundin, as chief of the Engine Research Division and then of the Propulsion Systems Division (1951–1958), received at least seventeen notices about rather minor security infractions. He left classified documents in an unlocked drawer, on top of his desk, in his outbox, and in unlocked file cabinets. He once left a confidential document on a conference room table. He blamed the "stress" of his job and the extraordinary "thoroughness" of one of the guards. The only reason we have them is that Lundin himself, probably more entertained than ashamed, saved them for posterity.[106]

Zucrow's world was filled with security classifications. He had top secret and "atomic energy clearance" between 1949 and 1952 when he worked for the government. He and his leadership team (Cecil Warner and Bruce Reese) had secret clearances for their Purdue work associated with their administrative functions and their own research projects. Most graduate theses and dissertations were either open or restricted, and only a very few were confidential or secret. Zucrow had the authority to declassify them by three-year intervals. The Purdue Library employed a security regime for the transportation, storage, and lending of these kinds of classified materials. The Rocket Lab secretaries, among them Wanda Eleene Boyd and Sandra Lee Law, also received secret clearances, as they were the typists and archivists for classified documents. With the opening of Chaffee Hall in 1965, they even had a much talked about "secret vault," though it was really little more than a small locked room at the back of the main office.[107] These women were some of the guardians of the Rocket Lab's secrets, otherwise protected by the chain-link fence that separated it from the airport, by the gate and long gravel road that led into the lab, and by the miles of grasslands and farm fields that surrounded it.

CIRCLING BACK TO THE INTRODUCTION, WE TEND TO THINK OF BUSH'S PROposed National Research Foundation as a failure. Congress finally passed legislation to fund it in 1950, but only as a much-diminished National Science Foundation (NSF). As a member of its National Science Board, Andrey Potter confirmed its value and purpose in terms Bush would have approved: it had "the responsibility of building up a stock pile of basic scientific research" as "the foundation for the technology of the future, and for the preservation of the national health and security." Potter saw nothing to fear in its claims to "government control." Purdue's experience with the Morrill and Hatch Acts and the Agricultural and Engineering Experiment Stations, and its relations with NACA and Wright Field, proved otherwise. The US government and military had a track record of getting along with both industry and the universities.[108]

The Truman administration offered Hovde the directorship of the NSF. He declined, likely because he knew where the real power was. In the military services. He had helped to put it there. By this time, the services had parceled out significant preserves in the universities and industries for their own weapon systems. Project Squid was one such preserve, a windfall for Purdue, positioning Zucrow and his Rocket Lab within a strong and sustained national network of laboratories, engaged in sometimes competitive, more often collaborative sponsored research. With Hovde as its local patron, Project Squid meant a vindication of the Bush ideals for "the free play of free intellects" and the relatively free flow of information too.[109] At least while Hovde and other Bush acolytes were in positions of authority, their experiences from the OSRD ensured mutual respect. Hovde praised the Navy for joining "the scientific forces of the universities" and its own "technical establishment." It was a marriage of equals.[110]

Not all observers of these dramatic changes in academia were so optimistic. The radical poet W. H. Auden parodied the militarization of college campuses, now filled with professors and students cultivated in the high purposes and profits of R&D. He wrote of the new university man, "Pompous Apollo," for whom "Truth is replaced by Useful Knowledge; / He pays particular / Attention to Commercial Thought." Auden counseled instead: "Thou shalt not worship projects nor / Shalt thou or thine bow down before / Administration." Conservative thinker Russell Kirk also decried the federal government's new projects and contracts flooding the universities, "the American worship of the great idol Panacea."[111]

Purdue alumnus Clifford Furnas, then director of the Cornell Aeronautical Laboratory and soon the assistant secretary of defense for research and development under President Eisenhower, offered a more practical and cynical perspective. Research, so he defined, was "the observation and study of the laws and phenomena of nature and/or the application of these findings to new devices, materials, or processes, or to the improvement of those which already exist." R&D meant, in no uncertain terms, "applied research: pursuit of a planned program toward a definite practical objective—a preconceived end result. It takes the results of fundamental or exploratory research and tries to apply them to a specific process, material, or device."[112] The weight of the Cold War moment dictated a complex array of academic research, military planning, and industrial production for these purposes.

# 5

# PROJECT STATE

## Guided Missiles for the Cold War

MAURICE ZUCROW'S FIRST TEN YEARS AT PURDUE WERE BUSY ONES, starting with his commitments to Project Squid, his renovation of the engineering curriculum, and also his work for the Research and Development Board, as we explore below. World events dictated the pace, events like the National Security Act of 1947; the Soviet tests of their atomic bomb in 1949 and hydrogen bomb in 1953; the rise of National Security Council directive no. 68 in spring of 1950, defining the ideological nature of the Soviet threat; and especially the outbreak of the Korean War in the summer of 1950. These events quickened military spending for a possible war with the USSR, along with all the newest and best technologies to support it.[1] "Modern warfare," so the Second World War had proven, depended on "sudden scientific invention" and technological innovation.[2]

These were dramatic years, to say the least. Purdue University played its role, through its president, Frederick Hovde; its business officer, R. B. Stewart; and its new professor of gas turbines and jet propulsion, Maurice Zucrow. They made up an extraordinary team that contributed to the administration, funding, and technical expertise of federal R&D in the earliest years of the Cold War. They contributed to what I term a rising American project state. Among historians, Aaron Friedberg and Carroll Pursell have called it a "contract state," spotlighting either its positive or negative business dimensions, the money side.[3] This distinction is worth making. Contracts tend to spotlight the dominant patron–client relationships, the military and industrial angles, often working within the free market and democratic principles, sometimes not. In the worst-case scenarios, they spotlight the rise of what Harold Lasswell called the "Garrison State" or what Daniel Yergin labeled the "National Security State." They critiqued a political and economic system to serve war, not peace. To serve the state, not the people.[4] Projects, on the other hand, focus more on the nuances of science and engineering as the arbiters of military and industrial advantage. The project state defines the optimism and pragmatism of the age, the power of the project to compress time and manage engineering potential. Projects were, by definition, future oriented. They were about calculating the new and achieving results. They inhabited a realm of the mysterious and unknown. Yet they were also supremely practical, offering the military and government a format to coordinate teams of scientists and engineers in the resolution of a specific problem, or in the achievement of a discrete aim. These teams were small, local, and self-governing.

A whole new category of engineer, the project engineer, arose to manage projects in industry, to guide them along their pathways of design, fabrication and testing, procurement and production. In the university, sponsored research projects created the new positions of project manager and principal investigator, with wide vertical and horizontal power, linking them to profit margins and professional power outside of the traditional university structure. Projects gave professors leverage. They required small mountains of paperwork, in proposals and peer review, contracts and subcontracts, reports and revisions. But it was paperwork for prestige and a higher purpose.[5]

R&D projects are a forgotten phenomenon of the Cold War era, probably because they were ubiquitous, hidden everywhere in plain view. The ultimate projects included the Manhattan Project that built the atomic bomb by 1945 and Project Apollo that landed on the Moon in 1969. But there were myriad varieties in between. Like Project GALCIT or Project Squid for jet propulsion; like Project Charles or Project Hartwell for aerial and underwater surveillance. The US Code and National Security Act of 1947 formalized and dispersed projects throughout the defense and national security establishment. The National Security Council and the Central Intelligence Agency (CIA) had them. The National Advisory Committee on Aeronautics (NACA) had them. Projects could be small- or large-scale technical experiments, veiled in secrecy or displayed in public. They could be isolated to specific labs, like the Purdue Rocket Lab; or extend to whole countries, like one for "psychological operations in Thailand"; or cover the whole globe, like the network of communication antennas orbiting Earth, the infamous Project Needles.[6] The Army had one of the most notorious projects of all: Project Paperclip. It was the secret endeavor that brought German scientists and engineers to the US, "under contract," initially in "protective custody," to help America with its new aerospace projects.[7] The 1950s became the classic age of the "missile project" in the US, in the words of one vice admiral. The very thrust factor in jet propulsion became a metaphor for the project plan: just as propulsion threw a weapon forward in physical space to reach a target, the military needed to organize the project on an "accelerated" schedule for a "target" deployment.[8]

## THE RESEARCH AND DEVELOPMENT BOARD (RDB)

As the war wound down, the Office of Naval Research (ONR) began to rebuild American R&D from the bottom up. Vannevar Bush's teams disbanded the Office of Scientific Research and Development (OSRD). He was left to manage one last redoubt, the wartime Joint Committee on New Weapons (JNW), part of the office of the Joint Chiefs of Staff (JCS). It survived the war and had staying power in various iterations. In June of 1946, the secretaries of the war and the Navy first transformed it into the Joint Research and Development Board (JRDB) of the JCS, chaired by Bush. The JRDB was a compromise between the military services and the holdover civilian science leaders from the OSRD. Bush put it together hastily, meant to continue the cooperation and advise the JCS. He believed that the JRDB should have the power to "allocate and reallocate" military responsibilities and funds for "research and development in specific fields." The "swift technological advances" of the Second World War and early Cold War demanded this. He wanted the JRDB to be "revolutionary" in character, "the first time in history that a nation has attempted to set up a defense program in which research and development are fully integrated with strategy." Bush wanted in on strategic decision-making. These were goals he never achieved. But the JRDB did have the significant power to advise and recommend, and served as a forum for the "resolution of differences," for the "elimination of duplication" and gaps, and for the "coordination of facilities, project, and fiscal data." It was a place where civilian experts and military officers continued to learn to communicate.[9]

The US government transformed the JDRB once again, largely in name only, when it became the Research and Development Board (RDB) of the Office of the Secretary of Defense, established in Public Law 253 of the 80th Congress. It was part of the transformational National Security Act of 1947, the package that established the Department of Defense, the US Air Force, and the CIA. The RDB is the one we forget because it only lasted five years. Its mission, much like its earlier versions, was to review R&D projects and report on them to the Department of Defense. This was an authority to evaluate and advise, not direct and manage. The RDB also inherited the organizational structure of the JRDB, with Bush as chair, along with a 250-person staff and committees of some 2,500 technical advisers from both civilian and military offices. The committees were devoted to technical cooperation in fields like the geophysical sciences, or electronics, or guided missiles, and each had their own subcommittees and panels. The Programs Division gathered information and data about "research and development projects" in the military services and the CIA, knowledge that was "essential to an intelligent consideration of present and projected programs." The RDB's committees and panels had as their key task to create "integrated national programs in their fields."[10]

In late 1947, Bush introduced a creative flourish to the RDB, a bridge institution between it and the JCS called the Weapons Systems Evaluation Group (WSEG). It was his attempt to compensate for the structural weaknesses of the JRDB/RDB with a group of technical experts to influence policy and strategy. They were to offer "professional" and sophisticated technical "analysis and evaluation" of weapon systems, including their possible combat use. They were to offer the secretary of defense "an impartial, supra-Service perspective." Secretary of Defense Forrestal approved it in December of 1948, after a favorable recommendation from an ad hoc committee of scientists (including Frederick Hovde and Lloyd Berkner). Still, the WSEG was threaded by controversy. As with the RDB, the JCS and military services saw it as a threat to their prerogatives. Every step Bush took was met with obstruction and defeat.[11]

Historians have not been kind to the RDB. According to one judgment, "The RDB, paralyzed by conflicts among the services, came to naught, neither effective in its coordination of military research nor viable as a mechanism for bringing civilian influence to bear on military planning."[12] Several other critics highlight how the RDB represented a defeat for civilian control of the military and its budgets, opening the way for corruptions and abuse. It fell into a trap of military domination, or was "fragmented" by its committees and panels and even a weak WSEG. It was ignored by the JCS and military services, unable to control "waste and duplication of staggering proportions."[13] Nor did it have a monopoly on advice. It was not the only R&D institution in town. The Air Force, for example, created Project RAND (for Research and Development) in November 1947 as a joint initiative with Douglas Aircraft, and it soon became an influential nonprofit corporation, in part as a maneuver against the RDB.[14]

For all these deficits, the RDB achieved some success. Its committees and panels brought civilian and military technical experts together in "active association," as Don Price phrased it. Political and strategic coordination may have been lacking above. But technical coordination happened below. Most scientists also brought the democratic spirit of the university to their work, along with a "self-confident, free-wheeling, and frank collective approach."[15] In these ways, the RDB was foundational. It helped establish the merger of "technical and policy analysis," defining precise scientific and engineering methods and findings for better informed political deliberations. Civilian scientific "analysis and review" of weapons and weapons systems became a central tenet of Cold War defense procedures.[16] The RDB's achievements were informational.

The RDB also formalized the R&D project. It instituted an administrative and statistical system for use by the military and government, by industry and academia, in the form of the "project card." There was no new weapon or weapons system, no novel aerospace technologies, without a project, and there was no project with a card. This was one of the RDB's self-defined mandates: the "accumulation of complete data on all existing Army and Navy research and development projects" and the "standardization of project data." It all came in the form of the compact information on the one-page cards. The RDB even created a project to develop the card. As Bush confirmed, "The accumulation of such data has become a project in itself because of the volume involved."[17] Like many of Bush's designs, the project cards were ideal in form but not in practice. At the start, the RDB received eighteen thousand of them, although roughly five thousand arrived incomplete and unreviewable.[18] Still, the RDB project card carried value. The Armed Services Procurement Act of 1947 allowed project contracts to be exempt from competitive bidding and to be privileged with liberal overhead rates, as long as they were "for experimental, developmental, or research work" related to national security.[19]

R. B. Stewart, chief business officer of Purdue University, played a signal role in all of this. He was one of the architects of the R&D project as sponsored university research contract, an outgrowth of his work for the OSRD and the military services. In this capacity, along with his directorship of the GI Bill, R. B. oversaw two of the largest transfers of national wealth in American history. From the government to the private sector, that is. His magnum opus was published, under the auspices of the War and Navy Departments, as the *Principles for Determination of Costs under Government Research and Development Contracts with Educational Institutions* (1947), sometimes known as the Blue Book. In it, R. B. established a common set of guidelines for the administration and accounting of sponsored research. He provided the formula for computing the direct and indirect costs for R&D projects and contracts between the military and educational institutions. His cost principles were based on actual costs (salary and wage expenses) and a "campus-wide average," which were then calculated to meet the specific project. This provided for both flexibility and parity between universities.[20]

None of this was easy. R. B. spent months negotiating to find a common ground between the military services and universities. His high standard was a "maximum assurance of fairness." This meant reducing government costs and ensuring government equity in the disbursement of funds. It meant keeping universities responsible for the monies they received, and fully able "to recover all costs, direct and indirect." R. B. knew his *Principles* were not perfect. They merely "reconciled" differences and set the terms for "fair distribution and agreement." Within a year of their publication, a committee was already at work to improve procedures and paperwork. The military services still needed convincing that overhead was not a "subsidy" or gift to universities, but an accounting of the "total costs of the projects." This was especially difficult in that the universities were in a rush to profit, with sloppy accounting and hidden fees. According to the OSRD veteran Lawrence Hafstad, everyone wanted in on the "bandwagon" of "the billion-dollar annual program of research and development."[21]

R. B.'s overarching concern was to protect the freedom of the university, whose researchers needed "wide latitude" from the government in order to support their work and pursue their visions. The quality and content of the "research project" depended on this. Otherwise, the government was wasting its money. Quality research also demanded fair reimbursement for overhead, he argued, because the government was not just buying into a lab but into the whole university, "a whole organic structure including libraries, students, recreational facilities, administrations." R. B. fashioned a novel philosophy of overhead. For "the university is not a place where a scientist works with his head buried in a laboratory; it is a community, an environment, a growing organization." He emphasized "the homogeneity of the entire university organization, and the inseparability from the total, of a single research laboratory and a project."[22] R. B.'s interventions were no small thing. From 1940 to 1954, federal funding for university R&D rose from $20 million to $285 million. Universities like Purdue were beneficiaries, since 52 percent of it went to engineering departments and 22 percent to physics, among other physical sciences. In all of this, the Department of Defense, with the National Science Foundation and most federal agencies, along with universities and colleges, followed "the most equitable means" formula of R. B.'s Blue Book. By 1954, the "median rate" for higher education overhead reached as high as 42 percent for private, 38 percent for public schools.[23] Quite a windfall.

## THE GUIDED MISSILES COMMITTEE (GMC)

Purdue president Frederick Hovde was also deeply involved in the National Military Establishment (the earlier name for the Department of Defense), receiving his travel orders to report for meetings at the Pentagon, enjoying secret and top secret clearance to review intelligence and classified records. Hovde resigned his civil service appointment as chief of Division 3 of the OSRD on 31 December 1945. Bush appointed him its new chief (without compensation) on 6 January 1946. He was already at work as Purdue's president but went to DC every second week to finish closing down the division.[24] He was also Bush's handpicked chair of the RDB's Guided Missiles Committee (GMC) from February 1947 to April 1949.[25]

Purdue's campus knew the rough outlines of this work. University news releases and biographies spotlighted his wartime work as "chief of the rocket ordnance division," now the "guided missile expert." On 6 September 1947, as chair of the GMC, he was the only civilian to witness Operation Sandy, the launch of a V-2 missile from the carrier USS *Midway*. It rose to twelve thousand feet before it exploded. Hovde took four days to get from Purdue to a point in the Atlantic just south of Bermuda. In November of 1947, he also traveled with Hugh Dryden and the Special Committee on Self-Propelled Guided Missiles to visit NACA facilities at Langley, Virginia, and then on to the Wallops Island Pilotless Aircraft Research Station to view missile launches.

It was not all hardship. When in DC, he still visited his old haunts of the elite Cosmos Club, or the officer's mess at the US Naval Gun Factory, for cocktails and meals. *Time* magazine named him, not Robert H. Goddard, or Theodore von Kármán, or

Maurice Zucrow, as "the rocket man," one of the first such titles in the public record. It also called him a "good-natured, beer-drinking, chain-smoking, bridge-playing educator."[26] Amid all of this, Hovde ran Purdue and made his way around Indiana too: for a Poultryman Association event, a veterinarian convention, and a Wheat Improvement Association dinner.[27] Purdue students parodied his rocketry fame in the *Rivet*, a student-run satirical magazine for fall of 1949. The faux *Time* cover featured Hovde, framed by a guided missile, and roasting him in a myriad of articles for spreading his Purdue duties just a bit too thin.[28]

During Hovde's tenure, the GMC was still very much in a preparatory stage, finding its way amid the challenges of the new technology, with the goal of providing more coordination of the nation's guided missile program. Bush lectured Hovde about his duties as the guardian of an "enormous military program of research and development amounting to half a billion dollars a year." Hovde needed to maintain civilian authority over the military, and unity over division. Bush was hoping Hovde would continue to be his man.[29] Hovde was an attentive and loyal student, mostly. He regarded his GMC work, based on his wartime experience, as providing crucial "civilian leverage" over the military's self-interests, an essential balance in American democracy.[30] The intentions were noble, the realities rather less so. Hovde's chairmanship of the GMC was handicapped from the start by a number of factors.

Bush, for one, did not make Hovde's job easy. He was Hovde's immediate boss at the RDB, at least at the start. They were both known quantities, having decided against an American V-2 late in the war. It was a decision that now seemed ill-advised amid a rising missile race with the USSR. Bush remained rather equivocal about the guided missile. He understood its "all-important" role in warfare. Yet he was also wary of its efficacy as a weapon of offense, partly based on the failed German experiences with the V-1 and V-2. He trusted his own record with more successful weapons of defense like radar and anti-submarine technologies.[31] Bush was also fearful of what Mary Self has termed a "missiles 'mass hysteria'" between 1945 and 1947. It was creating rivalries and duplication between the military services, and a race for tens of millions of dollars. Both were unbecoming of a free market and democracy. Bush thus offered a preview of the very warning that President Eisenhower gave, some ten years later, about a rising "military-industrial complex."[32]

Hovde was also hampered by what he had inherited from the Guided Missiles Committee of the JNW, also known as the Dewey Committee for chair Bradley Dewey, whose "national program for guided missiles" had set weak civilian oversight and allowed widespread military interservice competition. By the fall of 1945, the Dewey Committee defined guided missiles as those "unmanned vehicles moving above the earth's surface whose flight path could be altered by a mechanism within the vehicle." It established the following R&D priorities in scheduling and financing: first were air-to-air missiles, second were air-to-surface, third were surface-to-air, and only fourth were surface-to-surface missiles. Within this last category, long-range cruise and ballistic missiles (with ranges of five thousand miles and beyond) were relegated to the long term (ten or more years). Their technical challenges remained formidable. As historians agree, between 1945 and 1952 the US simply lacked the proper rocket science and intellectual "manpower" for "intercontinental rocket missiles or satellites."[33]

These various constraints were made even worse by the Truman administration defense budget reductions, beginning with the "black Christmas" of 1946 and continuing into the next few years. Hovde's GMC was forced to coordinate a series of "amputations," often most ruinous to projects in their earliest and most promising phases and to the very costly facilities being planned and built. The JCS did not make any of this easier, with a "complete lack of adequate guidance" as to "requirements, time scales, and relative priorities" for guided missiles.[34] Furthermore, by demobilizing the OSRD, Hovde had participated in the diminution of his own authority at the GMC, where the military services enjoyed full "cognizance" over their own missiles, both those "in being" and those "to come." In principle, the GMC could only offer "broad guidance." It could not assign "development projects" of any kind.[35]

Historians have raised serious criticisms of the GMC's work under Hovde's direction. They have found a poorly functioning set of panels and groups, vulnerable to pressures from competing Navy and Army and Air Force priorities. They have identified a "lack of vigorous leadership," otherwise inconsistent and aimless.[36] Hovde once conceded to Bush that the new "national program of guided missile research and development" was "extremely difficult to define." The work was frustrating and exhausting. His fellow member, Rear Adm. D. V. Gallery, felt that the GMC was a "rubber stamp." It was just "going through the motions." At one meeting, "in one hour we disposed of a stack of 188 sheets of paper, covering 53 projects which totaled $46,620,925."[37] Hovde complained most about the military services, each bound up in their own "self-interest" and "inter-Service" rivalries. He struggled to keep the services in proper balance as they lined up guided missile contracts with US companies, all in competition for scarce funds and R&D talent. The services also experimented, often with a frustrating pace and high costs, with various propulsion platforms, from turbojet to ramjet, from solid-propellant to liquid-propellant rockets, at both subsonic and supersonic speeds. In one premier case, the Air Force and its Matador guided missile, in contract with Glenn Martin, clashed with the Navy's Regulus, in contract with Chance Vought Aircraft. The Snark and Navaho programs were also torn by both technical and strategic indecision about the optimum propulsion modes and stages. This all amounted to what engineers called the very imperfect nature of the "missile art."[38]

For all these rather negative appraisals, Hovde enjoyed some organizational and technical successes at the GMC, as in his creation of its Technical Evaluation Group

(TEG). This was actually an early test case for Bush's higher and broader WSEG. Hovde first applied it in the GMC before Bush later proposed it for the JCS. But whereas Bush's WSEG mostly failed as a technocratic gambit from above, Hovde's TEG succeeded as an initiative of scientific/technical experts from below. The TEG had its origins in a remarkable group of planning consultants. These technical experts included Lawrence Hafstad, Robert Gilruth, L. J. Henderson, Clark Millikan, W. A. MacNair, and H. Guyford Stever. They had been called upon by Hovde to advise the GMC on the potential for a long-range intercontinental ballistic missile (ICBM) with a five-thousand-mile range. Along the way, the consultants realized the immensity of the task. How the country's guided missile program was "almost entirely in the study and experimental stage," with fifteen of the sixteen actual missile projects all less than a year old. There was a dire need for an "experimental development program" of "building, equipping, and firing the test vehicles." And there was a real need for a TEG to offer a "wide latitude" of technical expertise in basic and applied research. The planning consultants set the contours of guided missile technical procedure, with their priority for "comprehensive, logical stepwise engineering development." This was the only way to properly test and prove new kinds of propulsion, aerodynamics, and guidance. Their inspiration was a project report out of the Convair company, the main contractor for the ICBM, which admitted to its "extraordinary difficulty." Success meant "a series of intermediate steps of successively increasing difficulty. Each step will be an extrapolation of the previous one. When confronted with the problems of each step, the engineer will have the benefit of experience acquired in the previous one." Engineers and policymakers had to proceed much like the rockets themselves, in first and second stages of increasing sophistication and capability, like their very test vehicles of "stepped" ranges, weights, and thrusts.[39]

By 1948, the TEG became a small group of experts, circulating up from the various panels, who advised Hovde and evaluated the work of the GMC based on what they termed project management or systems engineering. The reasoning was that "a guided missile is essentially a combination of many technical skills into one overall operating system, or vehicle." This required the "integration of missile projects at the technical level" given the "profound interaction among the various components in the design of a guided missile system." One contractor was essential to "make engineering compromises between the components in order to optimize the performance of the whole system." The complexity of the guided missile also inspired the creation of the ten GMC panels, in place by December of 1947 and which answered to the TEG. They represented the many facets of the guided missile system: propulsion and fuels, launching and handling, aerodynamics and structures, warheads and fuses, guidance and control, test range instrumentation, test and training equipment, countermeasures, target drones, and facilities.[40]

The GMC did not invent this project or systems approach. One early TEG report actually drew its provenance back to the turn of the century: "It may be pointed out that the Wright Brothers operated on a project basis in order to develop the airplane." They were the first of the pioneers to define the technical problems, acquire engineering data, procure the parts, and design and develop the whole aircraft. The Wrights invented aerospace R&D. The TEG also borrowed the process from the aviation industry in the Second World War, which used a single contractor to select and integrate aircraft components and make needed compromises for the whole. It allowed for both economy and speed of effort.[41] Hovde's TEG applied these stepwise methods to develop reliable missiles on budget and on time via the "parallel advancement" and "close integration of all components." It transformed Bush's cramped and vertical project card into a more streamlined, stepwise, horizontal project schedule. These early systems approaches were, in fact, an elaboration of the project system. Engineers needed to fulfill the project contract promptly and successfully by coordinating all the complex technical details and balancing the technical expertise of the university and industry for military needs. Hovde called this the "integration into one coherent working unit of a larger number of basic scientific techniques than any weapon so far conceived."[42] He and the TEG had created the early institutional foundations of systems engineering, what the Air Force later called "concurrency."[43]

Hovde believed in the project system, especially its "associational" struts of civilian and military advisers. The civilians were a formidable group, the very people who were helping create the country's missile and rocket institutions. Among them were William Bollay of North American Aviation, head of its Rocketdyne division; R. W. Porter of General Electric, in charge of building the Hermes, an American version of the V-2; Homer Stewart of JPL, in charge of the Corporal missile; and Ralph Gibson at the Applied Physics Lab, head of the Bumblebee surface-to-air missile. They were not pushovers. The GMC delegated to them and depended on them. The military liaisons usually deferred to their expertise. The TEG was busy. Its members visited military and contractor sites "to make technical analyses and assessments," to define the competing missions of the services, and to share "constructive criticisms and recommendations." Their criticisms were comprehensive. They found too much waste with booster production and too much duplication in ground-to-air missiles; too many plans for "large government facilities" and not enough dispersion of contracts to industries and the universities.[44]

As of the summer of 1948, after three months of work, the TEG had submitted technical estimates to the GMC for twenty-two out of the thirty-three reviewable missiles then in military R&D.[45] Hovde's GMC sent these on to the RDB, along with reports on priority issues like countermeasures, wind tunnels, and service budgets. It offered technical advice to the military services on foreign developments. It created a glossary of guided missile terms, a six-hundred-page brochure on test range instrumentation, and a survey of scientific personnel in guided-missile R&D. It submitted an impressive study of the "integrated test range program," which created shared sites for the services and contractors at White Sands Proving Ground, New Mexico; the Experimental Rocket Engine Test Station at Edwards Air Force Base; and the Cape Canaveral Atlantic Missile Range. The GMC also made some weightier boasts. It had "obtained confirmation that the Committee has not only the right but the responsibility to comment on strategic and tactical considerations" as related to guided missiles. It had established a system of semiannual reports covering guided-missile progress in the services (supervised by its Ad Hoc Subcommittee on Periodic Reports). And with most pride, the GMC "to date" was "able to settle its problems and act upon recommendations before it without reference to higher authority." It had solved technical disputes without taking them to the Secretary of Defense.[46]

Hovde's ultimate achievement as chair of the GMC was the report "National Guided Missiles Program" (1948). It fulfilled his essential RDB mandate to "prepare an integrated national program of research and development" by way of balancing "the related scientific fields of aerodynamics, propulsion, guidance, control, warheads, fuses, launching, range instrumentation, etc."[47] The program touted the GMC's organizational achievements, which were a model for information sharing and communication between government agencies, the military services, the industries, and the universities. The GMC also established the principle "that civilian scientific groups should play the major part in missile development." Their technical expertise was crucial, so as to "continue, under its directive, to determine and approve major project assignments." This meant that the GMC should weigh such "factors" as the military "application" of any missile, the "technical competence" of the contractor and all "component research and development," along with "manpower, facilities, and funds." The emphasis here was on projects, as essential scientific/technical rationales for the new weapons systems. The engineering took precedence over the interests of any one contractor or military service. This was the ideal, at any rate.[48]

Another standard in the national program had to do with achieving a "balance of effort" over the "scientific and technical problems" in its charge. This meant "avoiding unwarranted duplication" and "filling gaps." As for duplication, most of the GMC's work involved gathering information and disseminating advice. For example, it called for cost reductions and "consolidations" in several duplicate Navy projects: the Kingfisher, Lark, Rigel, Regula, and Triton guided missiles. It warned about duplications between the Army's Jupiter and the Air Force's Thor missiles, and between the Army's Nike and Air Force's Bomarc. As a rule, though, Hovde and the GMC, perhaps taking a cue from Project Squid, tolerated duplication as a necessary evil, especially given the "technological and strategic uncertainties" in early missile R&D projects.[49] Forging into the unknown often required parallel paths. The services and industries also accepted this standard, so long as duplication did not become wasteful.[50]

As to gaps, there were plenty of them, and they were even more worrisome than duplications. Some were internal. The GMC was frustrated by the conflicting "fiscal and budgetary practices" between the services. Committee members and their contractors did not have enough technical data about supersonic wind tunnels, about "warhead and fuse design," about solid-propellant booster development, or about "high-altitude meteorological observations" (the Earth Satellite Vehicle Program). The most troublesome gaps were external. The GMC and its TEG most feared the looming Soviet advances in long-range surface-to-surface missiles, raising regular warning to both the RDB and the services. As the GMC argued, intercontinental and submarine-launched ballistic missiles were bound to replace America's fleets of bombers as delivery systems for nuclear weapons. These were still experimental technologies at best, but that was all the more reason to invest in them in 1948, in order to achieve results by 1958.[51] The work of Hovde's GMC was foundational.

All this talk of duplications and gaps was really about preparing for the future, attending to the horizons of new knowledge, and maintaining readiness in the arms race. Hovde's GMC sometimes called this the "balance of efforts," balancing the old and the new, utility against obsolescence, expectation against surprise. He and his colleagues also referred to "shifts of emphasis," breaking the balance in order to face threats and leap ahead with new projects and systems. Both of these phrases came straight out of Vannevar Bush's lexicon.[52] Hovde participated in one such shift of emphasis shortly after resigning as chair of the GMC. In June of 1949, he joined an ad hoc committee at the Department of Defense, with Lt. Gen. J. E. Hull, director of the White Sands Proving Ground, and N. E. Bradbury, director of the Los Alamos Scientific Laboratory "to study the technical feasibility of combining atomic warheads with guided missiles." They recommended in favor of the "marriage," as they put it, favoring implosion-type bombs that necessitated still bigger and smarter guided missiles.[53]

Aside from travel and per diem expenses, Hovde only received a pro forma "one dollar a year" salary. He served out of a sense of national duty. That was a patriotic reward. He also used his position at the GMC to represent the universities, to serve their interests. Throughout his work, he argued for the promotion of basic research as that most fertile ground for the exploration and discovery of the new and unknown. He lobbied for it in the industrial laboratories, which were motivated by the "stimulus of competition," but especially in the university labs, driven by the "constant flow of young people into and out of them."[54] Under Hovde, the GMC's national program recommended that the government "emphasize fundamental research on liquid rocket engines so as to secure design information for large rocket motors." This meant that it needed to better fund "educational institutions with propulsion and fuel laboratories" and provide them with the best "graduate research facilities" and student support.[55] Part of the reward for chairing the GMC was to do the Purdue Rocket Lab's bidding.

As Hovde completed his work on the GMC, returning to campus full time, some Cold War controversies came with him. Government-sponsored research in the universities was conflicted. As a rule, most scientists and engineers avoided political controversy. Their clearances demanded it. In their professional lives, many were able to keep their political views compartmentalized from their technical work and funding proposals.[56] But politics intervened as higher education addressed the challenges of a divided Cold War world. Pressure mounted from within the academy for conformist anti-communism. The National Education Association recognized that this created "awkward choices and calculated risks." Still, the schools needed to educate for democracy, for "patriotic citizenship" and the "American way of life," against "communist advocacy" and "totalitarianism."[57]

These tensions peaked with the anti-communist Red Scare in the immediate postwar years. It all began in 1947 with the Hollywood and espionage hearings of the House Un-American Activities Committee (HUAC). It continued in 1949 when Whittaker Chambers named the Soviet spy Alger Hiss, who was eventually convicted of perjury. It peaked with the anti-communist attacks by Senator Joseph McCarthy in 1950. Amid all this, in the spring of 1949, HUAC sent letters to eight-one colleges requesting their textbook lists for select liberal arts and social sciences courses, this to review them for pro-communist ideas and to spotlight the suspect professors.[58] This was a line that Purdue's president Hovde, a conservative Democrat, refused to cross. He was no mere pawn of government power. True, he framed the Cold War as a moral battle between Communism's false equality and democracy's real freedoms. Yet he also rejected extremes. The HUAC's textbook requests were the "most serious threat to intellectual and academic freedom that has yet occurred in our nation." He gave an emphatic no to such politics of "fear."[59]

Purdue's R. B. Stewart, a conservative Republican, was also a fierce defender of academic freedom in these years. Like Hovde, he cautioned against the "totalitarian" urge in the US for "government control, government subsistence, government employment, government regimentation." He resisted the "moblike actions of pressure groups" and "the vacillating policies of public officers."[60] R. B. faced his own separate textbook controversy. It was actually far worse than Hovde's. The Truman Veterans Administration, backed by the attorney general, was attempting to review textbook choices, as well as curriculums and polices, for any educational institution operating within the GI Bill. President Truman was looking to close any institution guilty of "communist doctrine" and "subversive" content. R. B.'s governing committee for the GI Bill, backed by the administration's solicitor general, refused Truman's order and won, seemingly against all odds. "We were that close to having a dictatorship in the administration set up to tell the GI's where to go to school," said R. B.[61]

Zucrow turned his attention to these controversies in a talk to Purdue students that he titled "The Engineer and Democracy" (8 November 1950), mostly relating to McCarthy's campaign against spies and seditionists within the US. It was not a very optimistic assessment. "The modern world," he began, was "a sick one, a very sick one!" It was "torn by economic strife, political strife, inflation, high prices, fear, and possibly a third world war." Then came the surprise: "There is no simple remedy for the disease from which the world is suffering." Science and engineering could offer little. "The argument is frequently advanced that because the engineer is skilled in the art of applying scientific thinking to solving technical problems he is particularly well-qualified for solving the problems of society." That was "specious," he said, because the scientist worked with "certain invariant natural laws and the simplifications he introduces" in experiment. "Social and world problems are not susceptible to such simple treatment." Most of what the engineer needed to do was study "the book of history and in an unbiased, critical manner" be sure "to learn the lessons contained therein." He was advocating the historical method: balancing facts, evidence, and interpretation. In answer to the political pressures of the day, Zucrow advised that "the engineer has the social responsibility as an educated man to make a real effort . . . to distinguish fact from fancy and refuse to accept an unfounded opinion or propaganda. He must be independent in thought," even if "the conclusions reached do not agree with the popular clamor of his times." The engineer had to be a "non-conformist" and "independent thinker." A person who could "examine in an unprejudiced manner the wrongs, inequities and injustices in our own society and then proceed to improve that which needs improvement." There was no "perfect democracy," only the pathways toward it. Zucrow's lodestar was the Bill of Rights and its guiding principles of "freedom and liberty based upon the dignity of the individual."[62] For these kinds of talks, and his mentorship of the honorary societies, Zucrow won the Leather Medal Award in 1952, given by Sigma Delta Chi to faculty and staff for promoting and boosting Purdue University.

Not everyone took the Red Scare so seriously. Purdue's satirical magazine for students, the *Rivet*, once featured a parody quiz, "Is You or Is You Ain't Subversive?" Students were to answer twenty multiple-choice questions to check on their "Communist inclinations." The questions were inside jokes, requiring some background knowledge, as for example that rival Indiana University's school color was red, and that the standard reference source for engineering students was the *Mechanical Engineer's Handbook* (1951) by Lionel Marks. The quiz asked: The Red Menace is: "the latest in lipstick" . . . or "Indiana University?" Karl Marx is: "Harpo's brother" . . . or "the author of Marx's Handbook for Mechanical Engineers?" Both were subversive, it turns out. The highest score meant that "the FBI will see you soon"; the lowest, that you were a Republican.[63] Most Purdue students probably scored low.

## THE PANEL ON PROPULSION AND FUELS (MPF)

While navigating the communist threat and defending free inquiry at Purdue, Zucrow also went "national" through his participation on the Guided Missiles Committee (GMC) as chair of its Panel on Propulsion and Fuels. The panel's acronym was MPF, for "Missiles: Propulsion and Fuels." It was a relatively small role, but it was influential. Zucrow was one of some twenty-five hundred experts working for the RDB. Within the GMC, Zucrow's MPF was one of its ten panels. With the TEG and various ad hoc subcommittees, these panels reported directly to Hovde and the GMC. All this meant that Zucrow, like Hovde, was present at the foundation of America's guided-missile establishment.

Zucrow's service on the MPF encompassed five years (1947–1952), all of them during the Truman administration. Like Hovde, he was motivated by a combination of national and self-interest. Based on his wartime work and participation on Project Squid, the MPF gave him the opportunity to canvass the US defense establishment for jet propulsion, an invaluable experience for both him and his students. They would all need jobs someday. He became, once again, the technical rover and fixer, the jet propulsion plenipotentiary and ambassador of the long-range missile. There were potential conflicts of interest. As chair of the MPF, Zucrow initially worked for Hovde, who was also his boss at Purdue. While on the MPF, he also evaluated the workings of the M.W. Kellogg Company, the Technical Analysis Division of the Air Forces, and Aerojet. These were the very places where he was already an adviser or contractor. Like his colleagues in the business of R&D, balancing these interests with integrity was simply part of the job, and a measure of his professional character and reputation.

Early on in his tenure on the MPF, Zucrow requested and was granted, by approval of the RDB, sole and leading authority on questions of jet propulsion research and development. His earlier work for Aerojet, as for example on the Corporal and Nike power plants, was his calling card. Zucrow's MPF "assumed primary cognizance, regardless of the end purpose, of projects" in pulsejet, ramjet, and rocketry engines and fuels. It was a bold move, and Zucrow made some bolder claims during his tenure, recommending ever more "expanded" R&D for both ramjets and especially for liquid-propellant rocket engine research.[64] There were limits to what exactly the MPF could do. For example, during the economy measures and budget reductions after 1946, the MPF did not have the authority, but could only "strongly recommend" that the GMC more equitably redistribute project funds. Many of Zucrow's civilian colleagues in the GMC, moreover, were part-time, often serving less than a week a year, meaning that the full-time military staffs held the advantage on the panels. In the case of impasses, the only recourse was to petition the TEG, which everyone seemed reluctant to do.[65]

The charge of the MPF was to "implement" the directives of the GMC. The panel conducted evaluations and recommendations, usually in negotiation with the military services, by way of conferences and reports. The three civilian members were balanced by one military member each from the Army, Navy, and Air Force. Their precise mandate was to focus on coordinating long-range defense R&D so as "to assure continued and increasing progress" and "an optimum balance of effort." The MPF rarely met in any one place, but traveled around the country to see and hear, share and review, with the actual installations and offices. These were places like White Sands Proving Ground; North American's Santa Susana Rocket Test Station; Aerojet at Azusa and the Jet Propulsion Lab at Pasadena; the Applied Physics Lab in Silver Springs, Maryland, and its subcontractors, Marquardt, Wright Aeronautical, General Electric, and the Continental Aviation and Engineering Corporation; General Electric in Schenectady and Bell Aircraft in Niagara Falls, New York; the M.W. Kellogg Company in Jersey City, Reaction Motors Inc. in Rockaway, and the Naval Air Rocket Test Station at Dover, New Jersey; Redstone Arsenal in Huntsville, Alabama; and the new Air Force Tullahoma center in Tennessee. Zucrow navigated these sites with ease. He chaired the panel in the Pentagon or the National Defense Building in Washington, DC. It met at the Power Plant Laboratory of the Engineering Division at Wright Field, Dayton (which became Wright-Patterson Air Force Base in 1948), and at the NACA Lewis Flight Propulsion Laboratory in Cleveland. It also met at Purdue University, as for example the official MPF meetings on 24–25 February 1949, 8–9 May 1950, 1 November 1950, 2--21 September 1951, and 1–2 May 1952. Sometimes the Pentagon literally came to Purdue.

Some of these visits were to receive briefings, others to give them. Its members met about once a month, though not always together. They formed working groups to confront specific questions. The MPF requested and received top secret briefings on Soviet developments, as for example from the Air Materiel Command at Wright Field, or from the Office of the Secretary of the Air Force. Among its most active military representatives were Robert Truax from the Navy's Bureau of Aeronautics, and C. M. Hudson of the Army Ordnance R&D Division. Zucrow made a special effort to draw in his closest and best civilian colleagues, like Weldon Worth, chief scientist at the Aero Propulsion Lab at Wright Field; Milt Rosen and Homer Newell from the Naval Research Lab; Frank Parker of Project Squid; and especially Addison Rothrock and Walter Olson from NACA Lewis. Olson remembered how Zucrow plucked him out of NACA in 1950 and "managed to get his hands on" him to serve in the MPF. It was an easy choice for Olson. Zucrow had a "national name" for his textbook and for "having fathered some of the men who are now administering guided missiles." The panel gave Olson access to "all sorts of secret projects."[66]

The MPF acted as an essential conduit and communicator. It took an active role in the accumulation, distribution, and sharing of information. Based on the Project Squid model, it recommended new standard practices for the distribution of reports, as well as the standardized nomenclature in them, all "in the interests of conformity." Besides traveling to investigate various sites, the MPF created extensive mailing lists with both the "prime contractors" and "component contractors" involved in guided missile projects. These were to acquire information, but also to share its findings and any important details. The MPF offered an invaluable service. It was a unifying force. In time, once the early disputes were settled, Zucrow reported favorably on cooperation with other GMC panels and working groups. He also applauded the military service representatives for agreeing on technical issues "in all major cases."[67]

In these ways, the key achievement of Zucrow's MPF was arbitration, applying its scientific engineering authority to settle technical disputes between the military services and industrial contractors. These were among the panel's most important contributions. No grand political or strategic deliberations here. Just the small technical advantages that could lead to big rewards, as for example in missile performance; or that could lead to big savings in time and money, two key "measures" of the "rate of development" of a new weapon. Zucrow and panel members may have only had evaluation and recommendation power, but their expertise doubled and redoubled it.[68] Their technical evaluations were substantial. For example, the MPF defined cobalt, columbium, chromium, and nickel as critical materials for the making of propulsion systems, this a service for the RDB Committee on Geophysical Sciences. It evaluated designs of auxiliary power plants for guided missiles and new parts for various kinds of turbines. It advised new methods of pressurizing rockets and ways to derive rocket fuels from petroleum products. The MPF approved technical details, worth hundreds of thousands of dollars, for the country's major guided-missile systems.[69]

Zucrow also revised the RDB project card with new formats. He created a detailed questionnaire that requested all of the essential technical information, seemingly bland and perfunctory, but otherwise very useful for engineers. He also created a project summary form. It transformed the vertical look of RDB project card into a horizontal box, reduced to the basic data parameters necessary for engineers and managers, spotlighting the time schedule and "estimated prototype missile completion date." Zucrow further offered a "rating system" for MPF jet propulsion technical estimates, to predict "hardware within two years." He based it on the kinds of graduate school grades he assigned at Purdue: at one end, "an 'A' rating indicates a probability of 70–100%, whereas a 'C' rating indicates a probability of 0–30%." Even these grades had their limits, Zucrow warned, since every one of these technical estimates represented an "oversimplification of the difficulties encountered in developing the types of engines considered." Rocket science was in its infancy. Any prediction was a "gross indication" of only "probable progress."[70]

Zucrow's panel enjoyed special success in organizing two symposia, also adapted from the Project Squid model, as a means to share information and solve problems. They were both classified as secret. One was on the aerodynamics of supersonic inlets (at Wright Field on 8–9 May 1950), with the key participation of Worth and Olson, along with the Fort Bliss and Huntsville "Germans."[71] The other was on the thermodynamic properties of gases, in conjunction with the RDB Committee on Basic Physical Sciences and its panel on chemistry, and drawing from Zucrow's wide network of military, industrial, and university contacts. Zucrow was fluent in these symposia fields, and in the related dimensions of guided missiles, as he sometimes displayed in masterpieces of improvisation. Once, in answer to an offhand query about methylal (dimethoxymethane) as an alternate rocket fuel, he picked up a tab of paper and took a minute to scribble his own calculations about its "expected performance." He then judged its specific impulse inferior to aniline or octane, noting the further disadvantages of its low boiling temperature and high vapor pressure, along with being very "difficult to handle in the field" thanks to its "upper and lower explosive limits." His summary judgment was to not waste any time with it.[72]

The MPF also allowed Zucrow to follow his special interest in ramjets, something Project Squid was not funding at Purdue. His *Principles of Jet Propulsion and Gas Turbines* even featured a ramjet aircraft on its front cover, hardly visible against the black background, but perfectly streamlined in an art deco style. For Zucrow, the ramjet was a visionary technology for high speeds in the atmosphere, a pure kind of power plant. It had a simple combustion chamber to compress incoming air and combine it with fuel for thrust, without all the more complicated parts of a normal turbojet engine. Its key limitation was that it needed another power plant to first accelerate it to its most efficient speeds. Zucrow sponsored several early Purdue master's theses on the ramjet, a means to teach his students and learn more himself about its performance specifications, potentials, and limitations.[73] Zucrow's MPF was a consistent patron of expanded ramjet R&D, as for example in the Air Force's Navaho MX-770 intercontinental missile and its X-7 "test vehicle," as well as in the Marquardt Aircraft Company's designs.[74]

In the winter and spring of 1949–1950, Zucrow and the MPF made the first survey of the national ramjet program, which it "initiated on its own" at a 1 December 1949 meeting. This was another bold move, a mark of the independence and integrity of experts within the RDB system. It was also comprehensive, based on detailed inspections and studies of the national sites. The MPF advised a series of symposia for the military and contractors to share information and work better together. The initiative culminated a few years later with the panel's contributions to the GMC's National Unitary Plan for ramjets (April 1951). They were based on a wider circle of site visits through 1952. Along the way, the MPF surveyed and investigated, consulted and reviewed, but also acted as ombudsman, settling disputes between the military and its contractors, lobbying the RDB for smarter investments. Zucrow also used the ramjet survey as a teaching moment, reminding the engineers he met along the way about the correct approach to R&D. All power plant work, he cautioned, depended on a "proper balance in translating the results of basic research into final products." A complete research spectrum included "(1) basic or fundamental research, (2) applied research, (3) engineering development, and (4) experimental design and test." These were lessons he had learned in person at Aerojet. These were lessons on the record from the failure of the German weapons program in the Second World War. Engineers needed to be "ever vigilant" to the right balance and coordination in the testing of engines and all their components, in both static and flight conditions, and with a mix of both optimism and skepticism. As Zucrow cautioned, "Optimism regarding the solutions of the different problems to be overcome must be regarded with a certain amount of skepticism until the solutions of the problems have been demonstrated under conditions simulating operational use."[75] He taught this to his students at Purdue in simpler terms: "Beware of optimism."

Under Zucrow's tenure, the MPF also advanced the possibilities of the long-range, liquid-propellant rocket, as both surface-to-surface missile for Earth and a sounding rocket for outer space. In February of 1949, responding to inquiries from the RDB's Special Ad Hoc Subcommittee on National Guided Missile Program Planning, Zucrow confirmed his panel's support for missiles of the V-2 and Hermes type. As a cautious visionary, he saw their "merits" in the "long term," but also their "attendant difficulties" in propulsion, needing more time and money for R&D. The MPF "strongly" recommended "design and engineering work toward a large single-stage or two-to-three stage rocket system," with up to two hundred thousand pounds of thrust. This was possible given all the "propulsion and structure developments to date," especially at North American with the Navaho MX-770. The current technical work proved that the US could achieve "substantially greater impulse per unit weight." The panel interjected that "at least one end use should be fairly clear." The US needed more of a strategic purpose, as that "the cost of a large first stage is principally in the propulsive system," and only by first defining its "design" could the government then proceed to the "building and testing" of "fuel pumps, tanks, and motors."[76]

Seizing the prerogative, over the course of a full year from 1950 to 1951, Zucrow's MPF made its way again across the landscape of US rocket propulsion test facilities, ranges, and industrial sites. Along the way, the MPF offered its robust endorsement of the Viking sounding rocket, a joint ONR–Reaction Motors project managed by Robert Truax, Milton Rosen, and Homer Newell for its improved high-performance fifty-thousand-pound thrust engine.[77] These visits and evaluations concluded with an important milestone thanks to Zucrow's MPF: its "Program Guidance Report" (December 1950), which was one of the first documents in the RDB system to go on the record "that it considers rocket development in this country, with currently available propellants, to be ready for the initiation of a long-range rocket missile," ideally with multiple stages, for a range beyond three thousand miles. Zucrow later called this a major turning point, when a critical mass of rocket engineers saw an American ICBM really as "feasible" in propulsion terms.[78] Backing their claims was the MPF's report titled "National Effort on Liquid Propellant Rocket Engines" (June of 1951), also known as the "Liquid Rocket Survey," a detailed technical review of what was working around the country and what needed improvement, and which the MPF supplemented with a study on the unitary plan for liquid rocket test stands.[79]

Zucrow's arguments and justifications for an American ICBM have remained, to now, a hidden transcript in the history of the early Cold War arms race. They were not overtly political. They traded in the arcane technical details of rocket science. But they buoyed higher, more authoritative voices in the policy debates, ultimately leading to the strategic decision to accelerate development of the Atlas ICBM. The new chair of the GMC, Clark Millikan, now offered a rising confidence in the "advances" of rocket motors of bigger sizes, beyond the V-2 threshold, with "higher performance efficiencies and considerably greater thrust." Reliability had "increased remarkably." Booster rockets were now "common practice." J. E. Lipp, representing the RAND Corporation and the Air Force Long Range Rocket Program at Wright-Patterson, was even more colorful. He offered a perfectly geometric study of the performance "curves" of the upcoming ICBM: weight and cost, flight and time, reliability and vulnerability, and including the "lethal radii" against the enemy target. All were precisely measured and calculated, plotted and graphed, with a ten-year timeline for development. Some Air Force officers balked at Lipp's expensive predictions as "unrealistic." He inspired "some deep, wrinkled brows" and even "a little bit of profanity" about the challenges of the new propulsion and guidance. Yet they decided to proceed with a "hardware analysis" rather than risk complacency about the "future years to come."[80]

Zucrow's contributions were not without self-interest. Throughout his MPF career, he advocated for more military and government investment in basic university research, especially with regard to liquid-propellant rocket engines. This was only natural, part of his scientific/technical mandate to promote basic research. As when he and the MPF created a handbook listing all the properties of liquid propellants. Or when the panel formed a working group (with Colorado State University and NACA Lewis) to confront the seemingly intractable problem of rocket instability. The MPF shared crucial information on the latest findings among the military services and their contractors. Zucrow also added instability as a research project at his own Rocket Lab. One crucial lesson of the MPF was that the US needed more "basic knowledge" and "engineering design data" to keep the "store of available technical knowledge" high. This justified "the technical benefits resulting from a judicious amount of duplication of effort for stimulating new ideas and approaches."[81]

There was nothing subtle about any of this. Circling back to Hovde's own justifications, Zucrow was constantly plugging for more money to support university and graduate student research. He and the MPF called for laboratory investigations and new performance data in combustion studies, high-energy propellants, cooling techniques, heat- and corrosion-resistant materials, ignition lag and "high frequency pulsation," and improved instrumentation. First and foremost, at least for Zucrow, was the need to design and develop the tanks, chambers and nozzles, and propellant systems for higher chamber pressures.[82] All of this was in the country's interest. These appeals also mirrored the strategic plan of Zucrow's own Rocket Lab.

As a complement to his work for Project Squid, the MPF also reveals a sideline career for Zucrow: as a committee member. He served on dozens of them, spanning government and industry, academic and institute committees. He attended many hundreds of such meetings over the years without any formal salary, just travel expenses and a per diem (in one case, $50 a day). Few of their contributions were fully transcribed. If they were, they remain buried in government or corporate archives, still classified with secret technical details.

The committee form remains an elusive object of study. It was an essential foundation of the R&D networks during the Cold War, what the US government once called a democratic forum for the "advice of the governed."[83] Committees, subcommittees, ad hoc committees, panels, and summer study groups—all served as forums for lively debate. By its nature, the committee preserved anonymity. Members could keep their opinions private. Yet the committee also aimed for unanimity, subsuming its discussions, however disputative, into blanket approvals that kept projects moving or improving. It normally did not have a formal procedure for an expert adviser to dissent.[84] There is also evidence that some committees were handpicked for success, to include those persons already primed to approve just what their sponsors wanted. Top officers in the Air Force spoke regularly of making sure "the right people," aka those who were "reliable," were on the most important committees.[85]

Zucrow raised his own committee participation to an art form. For him, the committee was a means for equitable peer review. The few transcribed minutes that survive reflect that he did not say a lot, did not dominate the conversations. Rather, he paid close attention, considered the issues, and finally intervened, perhaps with an insight that stopped the distractions cold, or one that moved along the discussions with fresh ideas. The committee was, for Zucrow, just another classroom. His fellow committee members testified to his mastery of the fundamentals and to his foresight. They were battle-tested military officers and hardened managers of million-dollar programs, not easy to sway. People like Donald Putt, who directed R&D for the Air Force; or Adm. W. F. "Red" Raborn, R&D patron in the Navy; or William Avery, who directed ram-jet propulsion at the Applied Physics Lab; or James C. Fletcher, director of NASA. All remembered Zucrow for his feisty and smart, sometimes combative, but more often "forward-looking" committee work. Fletcher remembered Zucrow's work on the Visiting Committee of JPL and on the Technical Advisory Board of Aerojet, as seminars of a kind. Zucrow taught them. W. H. Pickering, director of JPL, recalled that "you never stopped prodding and asking the right questions (to the annoyance of many lesser spirits), you kept things moving in the right direction."[86] In one of his classic witticisms, Zucrow judged committee work with this gem: "Nothing is so good it can't be better, and nothing is so bad it can't be worse." Committees were, on balance, a necessary good of the defense establishment.[87]

## AIR DEFENSE AND THE ATLAS ICBM

In these years of the country's first guided missiles, the US Air Force's priority commitment remained to airplanes in a culture of the long-range jet bomber, and to a lesser extent the short-range jet interceptor. The Strategic Air Command and Air Defense Command were the country's bulwarks of deterrence. Guided missiles mattered, but primarily as weapons for air-to-air offense and surface-to-air defense. They answered the US fear of another surprise attack like Pearl Harbor, this one nuclear, and likely fatal.

Early on in their work, Hovde's GMC and Zucrow's MPF investigated the promise of air defense missiles for use against Soviet nuclear bombers, within "an early warning system capable of acquisition of targets as small as one square meter at one thousand miles," for possible development by 1952.[88] One of the most promising was Project Nike, the US Army Ordnance Department's guided missile system for defense against Soviet bombers. As an extension of Hovde's Division 3 work in the Second World War, the Allegheny Ballistics Laboratory helped to develop Nike's "single solid-fuel booster rockets" and tested both the rockets and warhead in its wind tunnels. Aerojet built the "solid-fuel rockets for the multiple-booster combinations," as well as the primary "liquid-fuel rockets for the missile propulsion system," to which Zucrow had contributed improvements. Douglas Aircraft Company, the primary contractor, improvised with "unusually thin surfaces" to reduce weight and drag and a "canard configuration" for better steering. Bell Telephone Laboratories created "extremely accurate radar tracking and computer equipment," as well as a computerized "system tester" in order "to obtain statistical data concerning the effect of various parameters on overall kill probability." Hovde's TEG was impressed with the team effort: "In spite of the distance separating them, the prime contractor and the various sub-contractors and their auxiliary agencies appear to have worked expeditiously in a harmonious relationship during the various phases of the work to date."[89]

As Bell Labs was discovering, such defensive guided missiles really only made sense as part of a wider and deeper distant early warning (DEW) network. The idea was in wide circulation. Project RAND (MX-791) at Douglas was researching the possibility of a satellite for air navigation and traffic control. Vannevar Bush advocated a "national system of air navigation" for electronic communication and "air warning." The Air Force took on the responsibility to investigate the promises of DEW in its Project Charles and Project Lincoln over the course of 1951 and 1952, joining civilian and military cooperation and leading to the first investment in a computerized and coordinating Semi-Automatic Ground Environment (SAGE), a sophisticated system of electronic surveillance and response to identify and defend against a Soviet nuclear bomber attack. It was all very contentious. Air Force Strategic Air Command (SAC) officers resented the intervention of scientists in the DEW debate, fearing they would undermine priorities in funding for strategic bombers and their support networks. To SAC, air offense mattered more than air defense. The Air Force was also wary of the utility of the new and untested DEW and SAGE systems. Media coverage of the issues heightened public anxieties over the escalating arms race, questioning whether the US was safe enough from a likely Soviet nuclear attack from the air.[90]

In the winter of 1952–1953, Hovde joined these debates as a member of the Ad Hoc Study Group on Continental Defense, organized by the Department of Defense to join civilians from industry and the universities with representatives from the services and Pentagon. The mission was to study current measures and recommend improvements in the organization of the "defense of the North American continent against atomic attack."[91] The group's results are a case study in conservative committee work. The task was formidable: to find a middle ground, offend neither the SAC or DEW advocates, nor contradict the several other committees studying the issues. In the end, the group carefully navigated the labyrinth of countervailing forces and filters in the defense establishments. It advised steady investment in DEW, convincing the new Eisenhower administration "to proceed more deliberately with continental air defense planning." By the summer of 1953, proposed funding for DEW and SAGE rose to $25 billion in five years.[92]

These initiatives for surface-to-air missiles and the DEW network quickened the momentum for the military services to invest in longer-range guided missiles to match the new defensive systems with new offensive ones. This brings us to the Atlas ICBM (MX-1593). It had a long career, going back to 1946–1948, centered at the Convair company. The Air Force had shelved it for a few years, awaiting advances in liquid-propellant power plants of the very kinds confirmed by Zucrow's work in the MPF and Lipp's work at RAND. North American Aviation boosted the Air Force's confidence as well with its superb new engines for the Redstone and Navaho missiles, which adapted and vastly improved upon the V-2 model.[93] By the summer of 1951, the Department of Defense, in league with the Air Force and the RDB, approved "Atlas" as the official public name for MX-1593. Engineers were quietly moving the project along.[94]

Into the spring of 1952, Zucrow's MPF confirmed that the US was "ready for the initiation of a long-range rocket missile," beyond three thousand miles in range, by way of focused "engineering research and development." Based on this advice, the GMC subsequently approved Atlas as a bona fide "Strategic Rocket Project."[95] With such authoritative technical advice, the Wright Air Development Center advanced the project within the Air Force hierarchy. Rocket engineering had readied more powerful engines to launch a nuclear-armed ICBM into the edge of space. Innovations in transpiration cooling, originally designed to cool power plants, and one of the specialties of Zucrow's Rocket Lab, now also ensured that an ICBM warhead would survive the extreme heats of reentry. Col. R. L. Johnston, chief of weapons systems development at Wright, saw all of the advances in propulsion "packages" and "clusters" and stages as a new threshold upon which to facilitate the ICBM, if "in a fairly modest way," especially given the need for further work in structure and guidance and control.[96]

Zucrow's MPF contributed another boost for the Atlas into the fall of 1952, one even more compelling, in that it addressed the rising threat of a lethal "technological surprise" from the USSR. On receiving secret briefings about accelerated Soviet developments, Zucrow validated the threat and the fear "that the United States may be lagging in the art of rocket jet propulsion." The Soviets and their German engineers were already developing rocket engines with higher thrusts and longer ranges. These included designs for lighter and more efficient spherical thrust chambers, with higher-energy propellants, that could bear higher chamber pressures. American research needed to answer in turn.[97]

Much like the debates and disputes over the DEW network, the various defense lobbies did not receive these kinds of technical recommendations, however pristine, without challenge. Zucrow's own RDB thwarted his appeals. An RDB staff report blunted the conclusions of the MPF, recommending instead that the US military proceed with restraint on the ICBM, this to avoid "outstripping" other weapons systems in place. The RDB advised the bare minimum: "support should be slightly increased."[98] The Air Force was also mired in disagreement, a story that historians John Lonnquest and Chris Gainor have told in detail, centered on the struggles between the Air Research and Development Command in favor of the ICBM, and the Air Staff and Strategic Air Command against, rallying instead for its investments in bombers and interceptors.[99]

It took the Scientific Advisory Board of the Air Force to break these organizational and strategic stalemates, when on 17 September 1952 it appointed the Millikan Ad Hoc

Committee from within the RDB-GMC to offer a final authoritative scientific/technical opinion. Chaired by Clark Millikan, Hovde's replacement as head of the GMC, it included the country's top experts on the ICBM as a weapons system: H. W. Bode, M. U. Clauser, C. S. Draper, G. B. Kistiakowsky, G. F. Metcalf, H. J. Stewart, and M. J. Zucrow. The committee was a turning point. The top civilian authorities in the Air Force marshaled the leading civilian experts within the RDB to advance Atlas. The Scientific Advisory Board and RDB charged the Millikan Committee to work through December to make an "additional detailed study" of the "technical estimates." These included costs, timetables, and the crucial R&D "solutions to the problems encountered in the guidance correction, composite propulsion system, and the re-entry and terminal phase of the trajectory." Zucrow had to make yet another travel circuit along his well-worn MPF trails. Besides receiving Air Force briefings, the group visited the major contractors: Convair on the missile, North American Aviation on the propulsion system, and Bell Aircraft on guidance and control.

The "unanimous" decision of the committee was not without some drama. Millikan prefaced it all by saying their work began "during the week of December 7, 1952." Atlas, in other words, might prevent another Pearl Harbor. With the guarded optimism of experienced engineers and scientists, the committee voted to "retain" project Atlas as "an intercontinental ballistic missile to carry an atomic warhead." It was gratified by the recent advances in a lighter warhead, with more killing power, and hence a wider target area, and by Convair's innovative approaches to the missile's structural design. Yet it also advised a "step-wise flight test program" to confirm Atlas's structural design, propulsion system, guidance challenges, and warhead reentry issues. It recommended ten years. There should be no rush into the "immediate construction of a very large test vehicle." Convair, the main contractor, needed to conduct the maximum ground testing of "all component parts and subassemblies." Most of the document covered engines, echoing Zucrow's meticulous work on the MPF, advising reliance on the promising first-stage liquid-propellant engines of the Navaho, along with careful consideration and coordination with the Viking and Nike projects for possible multistage options.[100]

Official Air Force histories tend to diminish the Millikan Committee's contributions to Atlas, not to speak of all the work that came before it. Several historians argue that "ballistic missile technology languished in the United States," at least until 1953, when Lt. Gen. Bernard Schriever and a cabal of like-minded defense strategists transformed Atlas into a true "weapon system development and acquisitions program." Schriever certainly remembered himself as the tipping point. How "most of the scientists" recommended a step-by-step program of testing with smaller missiles, all from a "technical and scientific standpoint." How he answered instead that, "it's important that we shoot for the moon."[101] Similarly, Dr. Hans Friedrich, director of research and technical development at Convair Astronautics, and a former member of the Nazi von Braun team at Peenemünde, once celebrated Schriever's Atlas ICBM not as the product of the small steps of rocket engineering, but as a "giant step toward Outer Space."[102] Both were deservingly proud of their breakthrough strategic technology.

Yet the Millikan Committee contributed a significant vote of confidence to advance the Atlas ICBM. As of January 1952, Atlas was still at the Power Plant Laboratory at Wright-Patterson, hampered by limited funds as a "state of the art project." By the summer of the next year it was moving along a "development directive" for testing.[103] The engineers and scientists who served recalled their own contributions in 1952 as technical experts from the RDB-GMC system. Homer Stewart called it the moment when "we really caught up fast on the Russians and the big ICBMs." George Kistiakowsky said the Millikan Committee was "the first of its kind" to secure Atlas's future and freely volunteered how Zucrow "played a key role." Zucrow, for his part, remembered how the Air Force rejected his call, through the Millikan Committee, for even larger thrust engines (750,000 pounds) to match the Soviets. He had wanted even more.[104]

Telling the story from Zucrow's perspective allows us to turn from the heroic back to the prosaic, to value technical preparations as essential groundwork for Schriever's strategic decision-making.[105] The Millikan Committee represented, for Zucrow, the culmination of five years of intense work in the MPF, during the rough and ready pioneering days of early rocketry. He was one of that "dwindling band of happy engineers," in the phrase of fellow pioneer E. N. (Ed) Hall. Among them were Robert Goddard and Robert Truax working for the US Navy and William Bollay and Sam Hoffman of North American Aviation (Rocketdyne). They had labored through the "chaotic, halcyon days," the "long years of risky and largely surreptitious effort." Their proving grounds were places like Hall's own Wright Field in "Mudville, Ohio, USA., on the banks of the Mad River." They were mostly mechanical engineers in the propulsion tradition, engrossed in the "development engineering" of laboratory research, static test stands, and trial and error testing. Their "subtle reasoning and intuition" and "analytical and experimental work," argued Hall, had created a vast archive of knowledge, the foundations of the "philosophy" of the ICBM. To his despair, these free-wheeling pioneers were eventually downgraded by Schriever's "institutionalization" plans. They were replaced by new generation of "supermen," the Electrical Engineers at Ramo-Wooldridge and their heady new science of systems management, experts who just did not understand that "electrons obey linear differential equations" but "reacting gases do not."[106]

Hall's point here was that the slow and steady work of propulsion engineers over many years had enabled Schriever and company to speed things up in the matter of a few. Rocket engineering required a cautious approach: "to establish compatibility of the major elements of the device"; to resolve problems with propellant feeds and control systems, with ignition delay and combustion instability; to monitor and calibrate with "extensive and elaborate instrumentation."[107] Here were the elements of the new school of "design reliability," as confirmed by one of Hall's colleagues in Army Ordnance. "*All* components, down to the last relay, valve, or even soldered joint, are vital." The aim was to reduce malfunctions to "not more than one unit out of ten thousand." The guided missile, as "the brainchild" of its designers, was only as strong as its smallest and various parts.[108] The eloquent Ralph Gibson, Hovde's old pal from Division 3, and now director of the Applied Physics Laboratory, distilled these principles in his creative "philosophy of the guided missile." He fancied the guided missile as an avatar of the human being, as a complex organism joining propulsion (circulation and muscles) with aerodynamics (structure and form) with guidance ("the brains"). Together they gave it speed, range, accuracy, and "damage potential." Gibson inspired his engineering audiences, especially at Purdue, with "man's attempt to endow his guided-missile creatures with more-sensitive, more-versatile and more-responsive intelligence and with more reliable and stronger sinews to translate this intelligence into action." His primary source for the smart section on propulsion? Maurice Zucrow's *Principles of Jet Propulsion*.[109]

These thrills and trials of developmental engineering circulated into the realms of popular culture. A group of young people in the Detroit Rocket Society turned them into romantic notions about the guided missile as a kind of living being, with "radar vision." Donald Ritchie, one of its founders, wrote that "rockets today are becoming quite human, even so far as having binocular vision and psychosomatic difficulties in their operation." G. Harry Stine, the engineer-writer laboring away at White Sands Proving Ground, one of Hall's happy band of pioneers, portrayed the guided missile as a "beautiful monster," a mermaid of interplanetary flight, whose engines were "iron maidens." He was enamored of all "the little relays, the pieces of wire, the lengths of tubing, the bolts and nuts," all the "unaccountable tiny parts, each indispensable." He wrote in fascination of the "parabola of revolution of her nose cone down her sleek, unbroken sides," of "the graceful curve" of the rising missile in form and function, the gift of America's rocket engineers to the arsenal of futurist technologies."[110]

THESE EARLY YEARS OF THE GUIDED MISSILE, SPOTLIGHTING PURDUE'S VARIED contributions, open a new perspective onto the history of the Cold War. The RDB did not operate for long, only the five years between 1947 and 1952. During that time, it actually survived several imposing critiques. First, from President Truman's President's Scientific Research Board came the report chaired by John Steelman, *Science and Public Policy* (1947). It mostly applauded the government's project system of R&D, embodied in the efficiencies of the "research triangle" of government, industry, and academia. They were far superior to the rather bloated military labs. Second came the much more critical review from Herbert Hoover's Commission on the Organization of the Executive Branch (1949). Its investigations of the National Security Act found an RDB embattled by the more powerful military services. It found "disharmony and lack of unified planning," as well as "extravagance in military budgets and waste in military expenditure." The "military cliques" threatened "democratic traditions," economic "efficiency," and "true national security."[111]

In response, the Eisenhower administration decided on a major overhaul. The Defense Reorganization Act of 1953 converted the RDB into a new Office of the Secretary of Defense, with a pair of assistant secretaries: one for defense R&D and one for engineering applications. This was not a repudiation, more an adaptation of the RDB's best practices: to centralize and unite policy and budgetary planning, fortify technical advice about projects, and dampen the interservice rivalries.[112]

One critic was not so charitable. Vannevar Bush came to abhor the RDB, his own creation, for its fixation on "the project idea." He shamed it and the military for valuing applied over basic research, engineering over science. "This may be a reflection of our cultural immaturity," he warned. "As we proceed, there may be more among us, highly successful in affairs, anxious to serve humanity, who will wish to look at the stars, or delve into the earth, or probe for the secret of life, not because it will add to the comforts or reduce the hazards of existence, but because it may render us a more dignified and understanding race with greater satisfaction in living."[113] Bush remained the patron of knowledge, not things.

This controversy entered a wider public debate, as the National Science Foundation and American Council on Education (chaired by Alfred Sloan of General Motors) promoted the Bush line for the freer pursuit of scientific knowledge. *Fortune* magazine, no less, decried our country's neglect of "gifted" individual scientists. We had become instead a "gadget-happy nation," obsessed with "technological end products" like laundry machines, televisions, and guided missiles. The culprit was "the governmental policy of support via *projects*," or what Bush called the "excessive subsidy of the mediocre." Because "the temptation in this country is almost overwhelming to justify every human activity in terms of its immediate usefulness or efficiency." In other words, "our wealth of scientific gadgets and our vast organization of scientific projects are in heavy disproportion to our depth of scientific thought. We research 'the hell' out of everything; we *contemplate* very little."[114]

The academics who served in the RDB were not so negative. Zucrow expressed his support for the "project system" between 1945 and 1965, as did Lee DuBridge, president of Caltech. They and their colleagues were able to sell their research on contract and educate new scientific and technical elites.[115] Edward Baldes, a biophysicist in aerospace medicine at the Mayo Clinic, remembered how the RDB committees and panels got "a great deal accomplished in those days." Cooperation was the rule. Clifford Furnas and Fred Whipple agreed, especially in the early years. Homer Stewart of the Jet Propulsion Laboratory, who served on the GMC, applauded its "technical auditing function," one "outside but paralleling the technical command function" of the military. "This was a very instructive and fairly influential activity. I think it helped considerably to speed our rate of development in the guided missile era." He also recommended the project system. "In those days military projects were set up with universities in mind." The R&D project helped unite the university with the military and industry, all in the "national interest."[116] Thus spoke the technical experts.

The interweaving stories of Hovde, R. B. Stewart, and Zucrow reveal something often neglected in the Cold War: the importance of the technical over the political. Military strategy and government policy were not everything. Project engineering and technical expertise counted too. So too did the personal honor and professional values they brought to their government work, multiplied many times over by their friends and colleagues.

The many contributions of both Hovde and Zucrow to the national defense establishment were significant. During his tenure at the GMC, partly in response to the challenges of the Korean War, Hovde also served as a member of the RDB Committee on Program Review, which recommended a scientific adviser to the president and an OSRD-style body to better coordinate research and development. Hovde later served on the Air Force Scientific Advisory Board and was a member and chair of the Army Scientific Advisory Panel. As a consultant on the President's Science Advisory Committee, he helped plan federal science policy in 1957, before and after Sputnik. At the request of the Atomic Energy Commission, he went on to advise the Argonne National Laboratory, receiving the coveted Q clearance, one of the highest top secret classifications. There he helped George Hawkins to oversee a research contract on nuclear reactor design to study "heat transfer to boiling water at pressures between 2000 and 4000 psi."[117] The Boilermakers struck again.

One of Hovde's more interesting contributions came in 1959 as a member of the ad hoc panel for the evaluation of the Bomarc IM-99B missile. It was a recapitulation of Hovde's rocketry work, a reflection of the defense world he had helped create. An old Purdue alumnus and colleague, Clifford Furnas, now chancellor at the University of Buffalo, was chair on this panel called by the Office of the Director of Defense Research and Engineering and its Guided Missiles section, part of the Office of the Secretary of Defense. The panel featured two university presidents, along with representatives from General Motors, the Philco Corporation, the Institute for Defense Analyses, and Hughes Aircraft. It met in the weighty New York City offices of the Institute of the Aeronautical Sciences. It visited the Lincoln Laboratory and MITRE Corporation at MIT, the Boeing Corporation in Seattle, and the Pentagon. It enjoyed a few suppers at the Cosmos Club in DC. It heard reports from the US Air Force Air Defense System Integration Division, with briefings on all of the complexities of rocket science: power plant, airframe, airborne guidance and fuses, radar, and data links and processing. The guided missile had come of age.[118] Hovde had too.

For his part, Zucrow did not participate at such high levels again, preferring to advise the military services on more technical issues. He was a member of the Technical Consulting Group (1950–1953), and then a consultant on the Aeronautics Panel (1953–1958), both in the Office of the Assistant Secretary of Defense for R&D. He served as a member of the Ordnance Scientific Advisory Committee of the Department of the Army (1959–1962). His more meaningful contributions came in a smaller, more important setting: his Rocket Lab back at Purdue and his education of a generation of rocket engineers.

# 6

# SATELLITE VISIONS

## The Dilemmas of Space Exploration

As Maurice Zucrow completed his busy five years as chair of the Panel on Propulsion and Fuels (MPF) and as a roving consultant to the burgeoning guided missile industry, he also made a vital contribution to the US civilian aerospace program. During his busiest days working for the Guided Missiles Committee, Purdue remained his home, where he continued building his Rocket Lab. Note the emphasis on the rocket. It was a unique place in the US, smart and small and set apart, brimming with confidence in basic and applied rocket science, applied to both guided missiles for war and civilian rockets for outer space.

Purdue University's contributions were part of T. A. Heppenheimer's influential American "main line," a core of jet propulsion and rocketry engineers who helped to shape rocketry R&D.[1] The main line remains something of a lost chapter in the history of rocketry, often overshadowed by the German rocket team under Wernher von Braun, those men secreted into Fort Bliss, Texas, under Project Paperclip, at work at White Sands, and after 1952 at the Redstone Arsenal in Huntsville, Alabama. American popular culture focused on their German V-2 weapon as the emblem of the space age. This was unavoidable. There was no mistaking that it was the world's first ballistic missile, that the Nazi regime and its rocket team had mastered the mechanics of liquid rocketry. The V-2 was a paradigm shifter. It changed the way people thought and planned, inspiring a cascade of research and design effects. It was also quite attractive, that pointed and streamlined shape, standing nobly by its tail fins. For American advertisers, the V-2 was the machine about to reach beyond Earth, fathom "the Great Undiscovered Ocean of Truth."[2]

The popular film *Destination Moon* (1950) featured just such an Americanized V-2. Its promise reached into popular idioms, with such catchphrases as "countdown," "blast off," and "go ballistic." G. Harry Stine featured such a V-2 in his story about the Southwestern Rocket Society at White Sands, set in the year 2018, building a civilian rocket to the Moon: "With the echoing cry of *Zero!* flame spattered across the concrete and the giant cylinder began to move, its voice snarling and thundering in a cascading maelstrom of sound of all frequencies." As the rocket took flight, "a cottony white trail spewed from it as it disappeared against the sky."[3]

American main line aerospace engineers were already reaching beyond the V-2 model, learning from it, improving it. Some of the more magnificent and historic adaptations came in the form of rocket planes. American engineers made the V-2 fly, in a manner of speaking, giving it wings and pilots in the shape of the great American X planes. Edward Heinemann, famed Douglas Aircraft designer of military combat planes and National Advisory Committee for Aeronautics (NACA) research aircraft (like the D-558), featured them in a classic study of jet propulsion. The X-1 rocket plane, powered by a Reaction Motors liquid-propellant engine, was about to break the sound barrier. Supremely confident, Heinemann paraded machines of ever higher and faster flight trajectories, from airborne jets to rockets for outer space. It was a vision shared by Bell Aircraft in one of its company advertisements: the graceful lines of aircraft and rockets rising through expanding S curve spirals above the earth.[4]

Through all of this, American rocket engineers began to enjoy a new confidence, built upon their rising professional community in the American Rocket Society (ARS) and their technical expertise for the military and industry. Prestige and profit, along with personal achievement and fulfillment, were now at stake. Engineering science as rocket science was coming of age, as were the spaceflight dreams of the ARS and NACA. This wave peaked in 1952, when both professional and political trends joined in a fascinating moment of convergence for long-range civilian rockets. Once again, Zucrow was at the heart of it. A convergence of rockets not just for intercontinental war, but also for peaceful exploration.

## AMERICA'S SATELLITE DREAMS

For all his interventions in the defense field, Zucrow did not forsake his spaceflight utopianism. He maintained it in his work with Project Squid and the MPF. One inaugural moment came in June of 1947 at a talk before the ARS and the American Society of Mechanical Engineers (ASME) in Chicago, widely reproduced in the newspapers, when he expressly predicted the advent of spaceships, in his words, "man-carrying rocket ships capable of shooting beyond the pull of gravity." He predicted that "multi-step" rockets, with chemical propellants, not even yet atomic power, would soon achieve speeds of 25,000 mph, or 36,700 feet a second, but the "high temperatures of more than five thousand degrees" in the engine combustion chamber required ongoing research into engine cooling. By his own research, film cooling was one new solution, what he described as using a "thin film of fuel in liquid form" through pores in the chamber wall, with the effect that "the relatively cool film of evaporating fuel protects the wall from the high temperatures of the burning gases inside."[5]

He was not alone, of course. He never claimed to be. Zucrow gave credit to those colleagues who, just before him, had pioneered the equations and techniques to achieve escape velocity, premier among them Frank Malina and Martin Summerfield.[6] They all shared roughly the same moment of inspiration, with one subtle difference. Zucrow's predictions were more realistic, grounded in the engineering challenges, the research and testing of the fuels and engines. He was the more cautious visionary.

One of Zucrow's more ingenious of students, J. P. Sellers, wrote his master's thesis on the topic. Titled "The Problem of Escape from the Earth and the Stepped-Rocket," it was soon republished in the journal of the ARS. Sellers was an ARS founder, by the way, membership card no. 246. He was funded by Phase 7 of Project Squid. Sellers was also a pilot and had worked for Pan American Airlines in Miami before he went on to Ohio State and Purdue to continue his studies. He displayed the supreme confidence of a newly minted engineer and spaceflight enthusiast. The rocket engine was, for him, perfectly achievable, just another example of "the controlled movement of a vehicle from a given point to another." But the rocket also had a special "place in the evolution of man's conveyances." It was the power plant ready to escape gravity, and his thesis offered a measured study of gravitational forces, rocket propulsion, "external ballistics," the detailed design parameters, and ultimately, escape velocity. Sellers gave due credit to Goddard's "outstanding" early contributions, his vision of "extreme altitude" by vertical flight. He also praised Zucrow, Goddard's heir, for the same "foresight" in advancing the science and engineering of "projection from the earth."[7]

These Rocket Lab interventions came at a time of high promise for jet propulsion. The Allison J33 turbojet engine, produced by the Allison Division of General Motors in nearby Indianapolis, helped make the P-80 the fastest aircraft in the world. The X-1 was approaching Mach 1 speeds in its test to break the "transonic speed curtain." Anything seemed possible now. United Press was reporting Kenneth Arnold's "flying saucer" sightings, of craft speeding beyond 1,200 mph near Mt. Rainier, Washington, followed by a wave of similar sightings from around the country at places like Joliet and Kansas City. For Purdue locals, the Army Air Forces R&D Section from Wright Field brought its war trophies to downtown Lafayette, at Fifth Street between Columbia and Main. On display were a Tojo Fighter (Nakajima Ki-44) and Northrop Flying Wing, a V-1 buzz bomb and Messerschmitt 163 rocket plane.[8] Rocket fever came to campus, thanks to the national media and Zucrow's news releases. At the aero modelers' shop in the basement of the Purdue Memorial Union, not far from the model train club, students built "rocket ships" using carbon dioxide cartridges from seltzer bottles. Cary Hall residents built "match rockets," powered by a match head wrapped in tinfoil and attached to a straw, or to lightweight balsa wood cars fixed to a fishing line, and complemented by sparklers.[9]

Accompanying these popular ventures in spaceflight were some high-level proposals, motivated by the ongoing V-2 flights at White Sands Proving Ground, to do the actual thing. Francis Clauser at Douglas Aircraft wrote the classic, and partially secret, "Research and Development" (RAND) treatise on the rocket as "man-made satellite." He defined it as a "conservative and realistic engineering appraisal," an achievable "project." Humanity was poised for space. "If a vehicle can be accelerated to a speed of about 17,000 mph and sized properly, it will revolve on a great circle path above the earth's atmosphere as a new satellite. The centrifugal force will just balance the pull of gravity," he wrote. All America needed was $150 million in five years to "design, construct, and launch" a rocket, with either a four-stage launcher with liquid oxygen and alcohol or a two-stage with liquid oxygen and liquid hydrogen. He and his team offered a total package of details: the calculus of reaching and maintaining orbit, mass ratios, power plants and fuels, airframe, launch facilities, guidance and control, telemetry, and even descent and landing to Earth. The rocket also had multiple uses, the perfect American device for surveillance of the Earth or the stars, for a communication relay, for an "intercontinental" missile, or even for a "man-carrying" platform.[10]

These kinds of serious proposals needed rockets, and rockets still needed significant investments in laboratory and test-stand research. To accommodate the science and manage the R&D at the same time, the RDB assumed the lead for the Earth Satellite Vehicle Program (ESVP) in 1946, handing over its management to Frederick Hovde and the Guided Missiles Committee, with Zucrow advising from the Panel on Propulsion and Fuels. The program included a wide and unwieldy umbrella of projects. The military services and Department of Defense, in cooperation with industry and universities, were investing in design studies and experimental payloads to study the upper atmosphere and even launch a satellite into Earth orbit. The venues included Douglas Aircraft (RAND), the V-2 Panel, the Caltech Guided Missiles and Upper Atmosphere Symposium, the Army's WAC Corporal rocket efforts, and the Navy's Rocket Sonde Research Section. The Navy's Project Viking, in cooperation with Glenn Martin and Reaction Motors Inc., became the most sustained and successful of all the satellite programs, and whose architects, Robert Truax and Milton Rosen, often received Zucrow's endorsements and favor.[11] Their fortunes rose when, in 1952, the International Council of Scientific Unions, inspired in part by some of the American scientists in the ESVP (James van Allen, Lloyd Berkner, and Fred Singer included) announced the International Geophysical Year, forthcoming for 1957–1958 to study the upper atmosphere and Earth, perhaps even by rocket and satellite.[12]

Zucrow had his own utopian moments, especially as an elected fellow and member of the Board of Directors of the ARS (1951–1955). He was also a member of its Publications Committee and chair of the Program Committee in 1952, responsible for organizing the ARS conferences, and he took care to balance panels on the technical dimensions of jet propulsion with the scientific and public interest in "the space travel problem." Under his lead, the ARS promoted stronger links with the International Astronautical Foundation.[13] At the fall meeting of the ARS and ASME in Chicago (9–10 September 1952) to mark the centennial of engineering, Zucrow's close colleagues Bernhardt Dorman and J. Preston "Pres" Layton organized the panels High Altitude Flight and Interplanetary Travel, featuring Milt Rosen and Homer Newell. Zucrow presented a technical paper with his graduate student C. M. Beighley, "Experimental Investigation of Performance of WFNA-JP-3 Rocket Motors at Different Combustion Pressures." He brought in Walter Olson and John Sloop from NACA Cleveland to speak on rockets and ramjet flight. He also fulfilled a personal request from Wernher von Braun to speak on his dramatic new obsession, the satellite vehicle.

Von Braun was already breaking into the mainstream of American popular culture at this time, what with his colorful articles in *Collier's* magazine beginning in the spring of 1952, as well as forthcoming books and media appearances, all celebrating him as America's wunderkind of space planes, space stations, and colonies on the Moon. He also raised a stealth project to build and launch a satellite to orbit Earth, secretly discussing its possibilities with Dr. Aristid V. Grosse, director of the Research Institute at Temple University and a close friend and confidante of President Harry Truman. Grosse had in mind what he called "the American Star," a "propaganda satellite" to impress the world, especially the peoples of Asia, with history's first artificial satellite to orbit Earth. Von Braun was especially interested in the "surprise" factor of the project, a definite advantage to whomever launched it first, and something very sellable to the military for a relatively low cost, also valuable as a "new idea in the field of military science which is virtually 'pregnant with dreadful potentialities.'"[14]

Von Braun called the satellite "our unusual project," a first step in his broader plan for human spaceflight. He mounted a small crusade through the summer of 1952 to marshal business and political support, to help make his satellite and space station "an engineering reality," and ultimately to conquer "the last frontier." He boasted of having convinced, among others, Juan Trippe of Pan Am, Lawrence Bell of Bell Aircraft, secretary of commerce W. Averell Harriman, and retired general Lucius Clay, a close adviser to Republican presidential candidate Dwight Eisenhower. On the technical side, seemingly aware of the new RDB and Air Force overtures for the Atlas missile, von Braun cleverly proposed a committee of experts to retool the missile into a space launcher. The experts included Clark Millikan of Caltech, James Lipp of RAND, the leading project engineers and managers of Atlas, and Maurice Zucrow of Purdue. Von Braun underlined that he had "solidly sold" Zucrow on the spaceflight plan and that his "reaction was *very* favorable." Zucrow's response to "Dr. von Braun," as he always called him, was sober, repeating twice that although a satellite was "technically feasible," it was still burdened by "several major problems" needing lengthy investigation and solution. At any rate, the plans faded by late fall, when Grosse wrote von Braun that "wise men" had cautioned him to "go slow," especially given the upcoming presidential election.[15]

These overtures also ended in a very public way at the annual ARS convention in New York City (December 1952), in a set of dueling speeches. Von Braun spoke in excited terms on the "satellite vehicle as an instrument of world peace," the rocket as spacecraft. Lt. Gen. L. C. Craigie of the Air Force countered with a speech on rockets as essential for air power and deterrence, the rocket as guided missile. He raised a powerful point of "contrast" with the von Braun plan. Said Craigie in no uncertain terms: "Our enemies as we know them are not located on another planet or on the moon." Properly chastised, von Braun took the opportunity later in the conference to apologize for his exuberance and promised not to doing anything "half-cocked" but rather defer to the military authorities.[16]

Although at times restrained about moving too fast with any spaceflight venture, as with the von Braun project, colleagues remembered Zucrow as a forthright protagonist for the peaceful uses of spaceflight. He guided the ARS, for example, to help "build a decent space program in this country." James Harford remembered how powerfully he "came to the defense of some of the risks that we took in the old ARS days."[17] Zucrow often reminded his fellow committee members about the promise of rocketry to reach into outer space. On the Aerojet Technical Advisory Board, Zucrow fought against the "sterile chants of the competing technical liturgies" and the "industrial parochialism" for defense contracts. He had a gift for returning discussion to balanced perspectives about missiles and rockets, with "moderation, compassion and concern."[18] In 1952 he won the ARS's G. Edward Pendray Astronautical Literature Award for his textbook, his leading scholarly articles, and his work on the editorial board of *Jet Propulsion*. The award was a small sculpture in the shape of an elegant half-moon, traversed by a V-2 rocket.

Zucrow became America's spaceflight emissary to Europe. He attended the First International Congress on Aviation and Rocketry Materials (28 March to 1 April 1955), held at the famed Hall of Chemistry in central Paris and at the Hotel Claridge on the Champs-Élysées. There he shared the stage with one of his aerospace heroes, Eugen Sänger, at a conference dedicated to the memory of another, Robert Esnault-Pelterie. The French also paid their proper respects to Konstantin Tsiolkovsky, Robert Goddard, and Hermann Oberth. Zucrow took on the role of their successor, with premier billing as the American professor "from a city far away that carries the illustrious name of Lafayette."[19] Back at Purdue, with four hundred attending in the Memorial Union, as winner of the Sigma Xi award in 1956, he "paid tribute to the pioneers in the field of rocket research who forged ahead," these "dreamers" of the space age.[20]

Zucrow brought his students into the spaceflight wave, largely through the ARS. The "Indiana Section" may not sound like much, but it was effectively the ARS student chapter at Purdue University's Rocket Lab. Its first meeting was on 22 February 1950, when 250 Purdue faculty and students met to elect officers. They then held a "launching banquet" at the end of the semester on 8 June 1950. Zucrow filled the Indiana Section of the ARS with his graduate students and made it a vanguard ARS section, one of the most active, well before there were any official ARS student chapters. He tapped into his network of national contacts to bring powerful speakers to campus. A group from Bell Aircraft and Reaction Motors Inc. came to lecture on the propulsion successes of the X-1 supersonic rocket plane. Lt. Col. Marvin Demler, chief of the Power Plant Laboratory at Wright-Patterson, spoke on campus; so too did the secretary of the Navy, Dan Kimball. These speakers often brought recently declassified movies of America's secret rocket launches: JATOs at work on the B-36 bomber, V-2 launches at White Sands, and the first tests of the NACA X planes. Several speakers for the Purdue ARS were native sons made good. In February 1956, Stanley Gunn, from the new Rocketdyne Division of North American, spoke on its Santa Susana Rocket Test Station for "large thrust engines." In October, Robert Graham came from his new position at NACA Lewis to discuss his paper "Aerodynamic Heating and the Reentry Problem." The Indiana Section had created its own history.[21]

Zucrow and Cecil Warner were the faculty advisers. Warner did much of the work as director of the Indiana Section (1951–1955 and 1959–1963). Through the 1950s, he was also a member of the ARS National Program Committee, an associate editor of the journals of the American Rocket Society, and a member of the Technical Committee on Propellants. At Purdue, Zucrow and Warner expected the graduate students to run the show. With a fancy letterhead all their own, they served as presidents (Delbert Robison, James Bottorff, and Phil Diamond) and vice presidents (Jacob Botje and R. K. Rose) and had their own board of directors (Cecil Moore, W. L. Gilliland, David Elliott, and Helmut Wolf). With Zucrow's advice, they planned panels and trips to the various ARS conventions and organized joint sessions with the ARS and Institute of the Aeronautical Sciences. They published their technical papers in the ARS journals, either coauthored with their mentors or with graduate student partners, or on their own.[22] In December of 1956, Zucrow took a band of his best graduate students to the elite First International Congress on Rockets and Guided Missiles in Paris. J. R. Osborn spoke on pressure transducers, and Al Graham spoke on rocket cooling. Coauthored with Zucrow, theirs were two out of the fourteen international papers on propulsion. Zucrow thereby launched the careers of his students from the platform of his own myriad research tracks.[23]

When Andrew Haley at the ARS main office wrote to ask the Indiana Section for a short history of its work, he was surprised to learn that students ran it. He was so used to working with established aerospace PhDs and veteran engineers. Phil Diamond, then president, apologized that he really had very little to share, given the constant turnover in its membership. After all, students came to Purdue to graduate within a few years. He sent along the short résumés of the student leaders, this in the interests of memorializing "the pioneer student section of the ARS" and "preserving our name in the annals of rocketry." Don Crabtree, still an undergraduate assistant at the

Rocket Lab, did little more than list his high school career and an honorable mention scholarship.[24]

The Indiana Section of the ARS premiered rocketry on Purdue's campus for faculty and undergraduates, the staff and general public, to hear and see. Beginning in fall of 1950, graduate students Bruce Reese and Clair Beighley set the trend with talks on jet and rocket propulsion. By 1953 the section had grown by 20 percent, with six open meetings on topics ranging from ramjet engines to nuclear power plants for aircraft. The section offered a public lecture series in 1955 and 1956 on such topics as "rocket propulsion, astronomy, celestial mechanics, escape from the earth, step rockets, interplanetary flight, and interstellar flight." At the annual Purdue Activities Carnival in October of 1955, the ARS tables and exhibits featured a movie spool replaying clips of test firings at the Rocket Lab. On display were an Aerojet 38 ALDW 1500 unit and the Messerschmitt 163 Walter engine. The Rocket Lab's students also built a portable rocket test stand for public display, called the Purdue Rocket Laboratory Demonstration Rocket Engine, featured in the *ARS Journal* and at campus events. It cost $10 to meet thirty-nine operational checkoffs in order to fire the rocket and its exhaust plume some six feet long.[25]

Purdue Rocket Lab students won the ARS Student Award six times in the first eight years it was awarded. It was for "best student paper on rocket and jet propulsion." Beighley and Eldon Knuth won in 1950. David Elliott in 1951. Richard W. Foster won in 1952 for his paper "The Theoretical Design for a Rocket Engine Utilizing Uranium 235." He won it again in 1955 for the rocket test stand and was featured in the *Journal of Jet Propulsion* under the rubric "evolution of a missile engineer." Don Crabtree may have had the most interesting story. He won the award in 1956 for a paper titled "Design Evaluation of One Type Nuclear Propulsion System." The award was presented to him at the ARS national convention in New York City, where he shared the stage with other award winners like Bernard Schriever, William Pickering, and Hermann Oberth. He even "helped take Oberth to his room when he became a little tipsy." Crabtree had first met Zucrow when they sat next to each other at the National Science Fair held at Purdue in 1954, where he won second place in the physical sciences. Zucrow offered him a "test cell monkey" job at the Rocket Lab. Crabtree spent twelve years there, with a year's internship at Rocketdyne. He went on to become an engineering professor at Purdue. He later remembered his comrades in the ARS group "who during that brief period of time seemed to put together a place where a young eighteen-year-old kid could fulfill a dream."[26]

Zucrow's promotion of his own students in the ARS paid off handsomely with legacy and recognition. His former students Bernie Dorman, Clair Beighley, Stanley Gunn, and J. P. Sellers later became active members of the Southern California Section. Sellers's son remembered them, from the eyes of his own childhood, as "young engineers enjoying the good life in L.A. while making plans for the moon."[27]

## THE DILEMMAS OF SPACE EXPLORATION

The US military services were actually well-tuned to spaceflight dreams in popular culture and among rocketry enthusiasts. Aviation was becoming aerospace, said Rear Adm. D. V. Gallery. He loved American slogans like "the sky is the limit" and "out of this world." We would soon create "transoceanic missiles with pinpoint accuracy. Buck Rogers and Superman do it every Sunday," he said. Lt. Col. Marvin Demler at Wright Field kept a massive file of newspaper and magazine clippings of rocketry and spaceflight stories from the popular press. For him, it was a valuable marketplace for ideas and aspirations. In their fascinating newsletter, the *Air Force Missile*, Purdue Air Force ROTC students recognized outer space as a pathway for guided missiles and as a new zone for space research, a place for humanity to experiment and adapt in zero gravity.[28]

The military services could afford these extravagances because they still enjoyed a dominant position in congressional funding and public support. The ARS was not so fortunate. It was largely reliant on the military and industry and could ill afford too many distractions with spaceflight visions. Its postwar leadership retooled the ARS as a professional association for jet propulsion engineering and the guided missile. The growing aerospace industry was a complex market of interests and lobbies, including the International Association of Machinists and the Aerospace Industries Association. But in the landscape of guided missiles, only one organization intersected all of its moving parts: the ARS. Its directors achieved remarkable success in building a whole new "technical society" through the early 1950s, with a full-time staff and with "attractive modern offices on Fifth Avenue." Its clear priority was in "technical progress in the guided missile field." A "practical, realistic national program leading toward the first step into Space" came second. This hierarchy worked. President Richard Porter, most famous for running the American V-2 program, reported a dramatic rise in membership: from 919 people in 1951 to 3,207 in 1955. Subscriptions to the society's journal, *Jet Propulsion*, increased at the same time from 394 to 1,017.[29]

Perhaps most telling was that there were only seven corporate members in 1951 but sixty-four in 1955, including the Ramo-Woolridge Corporation and the Hughes Aircraft Company, General Electric Corporation and Esso Research and Engineering, and the Atlantic Research Corporation and American Bosch Arma Corporation. They paid ARS corporate memberships to supplement the individual dues paid by their own engineers. The ARS was also bringing in record funds from corporate advertising pages in its journals. It took great pride in its "first annual Guided Missile Equipment Supplement" (in *Jet Propulsion*, spring 1955), a whole separate issue for "experts" to write promotional materials for their guided-missile products. The ARS charged a $150 fee. Advertising and revenue policies tended to drive board meetings through the 1950s, more so than any actual spaceflight issues. By 1957, the ARS hired a former Reaction

Motors Inc. account executive, William Chenoweth, to help run the advertising program and better exploit the "missile market." [30]

By 1955, the ARS was a lynchpin of the rising network of military, academic, and industrial investments in guided missiles, what one insider called "the Ordnance-Science-Industry team."[31] Prominent regional sections (thirty in all by 1955) were located at the country's premier laboratories and test stands: places like the Douglas Aircraft Company, Lockheed's Guided Missiles Division, Rocketdyne at Canoga Park and Santa Susana, Reaction Motor's new Rocket Engine Plant in northern New Jersey (worth $4 million), and the Naval Research Laboratory. Bell Aircraft had its very own ARS Niagara Frontier Section. The Air Force Experimental Rocket Engine Test Station, the Naval Ordnance Test Station at China Lake, and the aerospace industries at Palmdale and Lancaster had their own California Desert Section. The ARS now accorded new honors and membership status to its specialists, dubbed "missile engineers." They often met at fashionable sites like the Pasadena Athletic Club, the Chicago Engineer's Club, and the National Press Club.[32]

These developments marked an end to the high-spirited pioneer days just after the war. Between 1955 and 1958, the ARS transformed itself into even more of a professional society, codified in a new structure of technical panels, originally for solid rockets, liquid rockets, ramjets, propulsion and combustion, instrumentation and guidance, and spaceflight. The panels were a supreme mark of specialization. They were a mirror image of the technical panels within the Research and Development Board's Guided Missiles Committee (RDB-GMC). They expertly set the priorities for conferences and publications. Martin Summerfield wisely saw them as a means to build membership and subscription ranks to integrate ARS members from the military, industry, and academia by their expertise. By 1958, as a member of the Policy Committee, Zucrow voted against one initiative to expand the technical committees, now to include such topics as flight mechanics, materials and structures, and magnetohydrodynamics. He was opposed to such multiplications and specializations. He was also probably upset with the loss of his own charismatic authority as one of the ARS pioneers. Committees of new experts were now replacing him and his old informal network of friends and colleagues.[33]

In these contexts, too much talk about "interplanetary flight" threatened the new corporate identity of the ARS. Spaceflight enthusiasts like G. Edward Pendray, Andrew Haley, and Frederick Durant lobbied for more coverage of "space travel" in the ARS journals, beyond the current one-third of pages devoted to it. But they lost the battle. Haley and Durant were further troubled after their return from the Second Astronautical Congress in London (1952), stung that Europe was taking on the mantle for peaceful human spaceflight. There, the International Astronautical Federation took the lead in advocating for the technical possibilities of multistage rockets, orbiting satellites, and even spacewalking astronauts. They were also embarrassed that the most famous representative of the ARS at the congress was not even an American. It was none other than Wernher von Braun, promoting his famous plan for a multistage "satellite rocket" with reusable lower stages and a winged orbital vehicle.[34]

To co-opt some of these utopian drives and futuristic visions, the ARS created the Ad Hoc Space Flight Committee in 1952, a means to give voice to the younger and more visionary engineers. The conservative members of the board of directors tended to limit its work. The board's "rocket engineers" preferred a restrained approach, with a focus on the "engineering problems" and rocket and jet propulsion for aircraft and guided missiles. They stood for "defense mobilization," leaving possible spaceflight, satellites, and orbital stations for the longer term, only after successful "specific technical development." In the pull between the technical and the interplanetary, the technical usually won out. Zucrow also cautioned "restraint" in late 1951. Spaceflight was not yet "of a scientific nature." We should not "wander too far from technical problems."[35] So the board consistently rejected specific appeals for spaceflight ventures, as for example Rollin Gillespie's Project Arc for a two-passenger "circumlunar rocket." ARS president Charles W. Chillson (at Curtiss-Wright Corporation) criticized Robert Truax's proposal to "enter the space flight field" and lobby for a "pseudo-serious" space rocket, "satellite space stations and interplanetary voyages." He was backed by Robert Porter at General Electric and Chandler Ross at Aerojet. Industrial and military concerns remained the priority, especially the all-important clearance and security issues that defined them.[36]

In sum, a striking investment of federal monies and national resources were at stake for the ARS and its members. The missile age from 1951 to 1958 saw federal funding for guided missiles and associated technologies rise from $21 million to $2.5 billion. As Andrew Haley boasted, "The rocket industry, including missile systems and associated gear, is now the largest industry in America both in dollar volume and size of labor force." Industry responded with structural adaptations in their own organizations: as for example Republic Aviation and its Guided Missiles Division, Raytheon and its Missile Systems Division, and Lockheed Aircraft Corporation and its Missile Systems Division. There were already nearly seventeen thousand subscribers to *Missile Design and Development*, a leading professional journal.[37] Missiles were a perfect growth industry, binding defense and the national economy. As one analyst noted, "Breakthroughs in missile technology are continually threatening the whole defensive or offensive apparatus of one side or the other." The pace of obsolescence and innovation guaranteed new contracts and more profit for years to come.[38]

The military services were the primary sponsors for all this spending. There was no better proof of their vast landscape of R&D establishments than the booklets and newsletters of Federal Procurements Publications Inc., which circulated to professionals as *Federal Procurement and Subcontractors Daily* and the *R&D* newsletter, offering a clearinghouse on the expanded project system, what its editors called the "Industry-Science Complex" and its contracts for military services and products. There were nearly $3 billion a year in funds to leverage, both out of patriotism and for profit. Its guided-missiles directory revealed a growth industry, now spending $40 million a week: for "typical" projects or "special" ones, for large projects or small, for "laboratory" projects or "systems" projects. The outfit publicized the new contracts, along with directories and contacts for the military agencies and manufacturers. The Air Force's Arnold Engineering Development Center, Office of Scientific Research, and Research and Development Command were just waiting for clients to help them plan for the aerospace future.[39]

With so much activity and money circulating in industry and the military, academia had no choice but to participate. Harry Solberg, head of Purdue's Mechanical Engineering, had a ready form letter to send to President Hovde whenever one of his faculty needed a leave of absence or sabbatical for outside work. As he wrote in the case of Bruce Reese, off to the Allison Division of General Motors, and later in the case of Cecil Warner, requesting a sabbatical at Aerojet, they both needed to immerse themselves in the active business of classified military contracts because "the field of jet propulsion is developing very rapidly and security regulations often retard or restrict the dissemination of technical information." Faculty made the same justifications in their routine requests for "permission to engage outside activity." The "public service" of committees and boards, without compensation, allowed them to receive the latest scientific/technical advances and possible research contracts.[40] Zucrow made this one of his mantras. To teach well, his staff needed to circulate back into industry every few years, given the rapidity with which developments were taking place and the pace of new literature and techniques.[41]

Warner's sabbatical at Aerojet offers a revealing case study about the importance of these cooperative ventures between academia and industry. He spent the 1953–1954 school year there, working as a project and development engineer for R&D in the Liquid Engine Division and Thrust Chamber Department, alongside Zucrow's old colleagues Chandler Ross, Marvin Stary, and Ray Stiff. Warner mostly worked on the "booster rocket system" for the F-99 Bomarc antiaircraft missile. He made redesigns of the existing gas generator and thrust chamber based on a new propellant mixture. He made theoretical and experimental studies of propellants, ignition characteristics, and "new rocket starting and shutdown equipment and procedures." He was also responsible for establishing and conducting the flight approval tests for the booster rocket system at the proving grounds. As a member of the new project group system, he supervised the full range of contract specifications, including "extensive development testing, pre-flight approval testing, flight approval testing and finally flight acceptance testing" of the complete rocket system. Warner also took back two crucial insights for his Purdue teaching. First, "the importance of the ability of the engineer to get along well with all classes and types of people." At Aerojet, he took every opportunity to mingle with all of the mechanics and technicians, to "check personally daily with the Shops, Proving Grounds, and Hydraulic Laboratory." Second, "the necessity of an engineer to be able to express himself when preparing reports." The Rocket Lab needed to teach its engineers how to write and how to speak with precision and clarity, even some eloquence.[42]

## A MIDWEST RESEARCH TRIANGLE

Purdue's Rocket Lab was entangled in this labyrinth of visions and dilemmas of applied research and corporate profits. It was a university unit eager to promote peaceful spaceflight, but also dependent on and supportive of the industrial and defense establishments. By the force of his achievements and the good luck of location, Zucrow networked a unique research triangle to balance these interests: between Purdue University in Lafayette, Wright-Patterson Air Force Base in Dayton, and the NACA Lewis Flight Propulsion Laboratory in Cleveland. There was nothing quite like it anywhere.

Wright-Patterson and NACA Lewis were already closely aligned, with liaisons on all manner of aerospace topics. Wright-Patterson had its own university, the Air Force Institute of Technology, established in 1946, which sent student tours up to NACA Lewis and its best graduates on to Purdue and Ohio State, Caltech and Stanford, Cornell and Princeton.[43] NACA Lewis, in turn, worked closely with the Army Air Forces on several joint projects. One was the College Thesis Program, which joined them with the universities to share and sponsor student research. Another was a German–English dictionary project for new jet propulsion technical terms.[44] Beyond these explicit institutional bonds, propulsion engineers within this research triangle created a small network of cooperation and influence. They delivered lectures in each other's forums. They depended on each other for contract research and consultations. They met at academic conferences, like the Cleveland-Akron Section of the ARS, or the regional and national meetings of ASME.

Wright-Patterson Air Force Base was the country's leading center for military aeronautical science and engineering, the site of a massive physical plant housing laboratories, production and testing facilities, and aircraft and weaponry. Through the Second World War, no other government research facility, outside of the Manhattan Project, compared. Its Air Technical Service Command (ATSC) and Air Materiel Command (AMC) had institutionalized what Lt. Gen. B. W. Chidlaw called the bond of "cooperation and integration of science, industry, and the National Military Establishment."[45] Zucrow parlayed his many contacts with Wright Field engineers into lasting relationships. The Air Force sent its officers to his Rocket Lab for graduate work and funded their research projects. In fact, the Air Force often preferred working with the universities, a quick way to avoid all the red tape demanded from NACA.[46]

Zucrow's relations with the Air Force were close. In the summers of 1947 and 1949, Wright Field's Intelligence T-2 sent him to Europe for inspection tours through England, Sweden, France, Switzerland, Italy, and Germany. He visited R&D sites for gas turbine, turbojet, and rocket power plants, consulting with experts at the Brown Boveri Company, Rolls Royce, the De Havilland Corporation, and the Royal Aircraft Establishment (Farnborough). For Zucrow, as for others who made the circuit, this was a learning curve and cautionary tale. European jet propulsion remained well ahead of America in designs and developments, even by several years. It was time to catch up.[47] In terms of rocket propulsion, the Air Force, much like the Navy, appreciated Zucrow for his Aerojet experiences and his focused studies of liquid propellants, engine design, and the advantages of higher chamber pressures. The Air Force understood that there appeared to be but "two methods by which substantial improvement in specific impulse over present rocket performance may be obtained: by the choice of new high-impulse propellants and by operation at higher chamber pressures." Rockets flew better and farther with these methods, but a recent survey found a "surprising lack of orientation of thinking" on them in industrial labs around the country. There remained a "real need for experimental data." For that the Air Force now relied on the Purdue Rocket Lab and NACA Lewis.[48]

Zucrow's relationship with NACA Lewis in Cleveland was more complex, and significant. The Purdue Rocket Lab and Lewis enjoyed a remarkable research symmetry: in turbojet, ramjet, rockets, and nuclear propulsion. NACA was more established and prestigious, but still depended on Zucrow for advice and cooperation. The trouble was that Zucrow was geared to development; NACA Lewis was not. True, during the Second World War, NACA had engaged in a series of "developmental" projects in answer to urgent aeronautical goals. NACA opened its labs to industry engineers and expanded its technical reports for their particular needs, all to promote immediate solutions and improvements. NACA designed high-end radial engines, superchargers, and new fuels, all to meet wartime demands. Yet immediately after the war, chairman Jerome Hunsaker (1941–1956) and executive secretary John Victory rededicated NACA to its more traditional role as a research institution, one in "service" to the military and industry and careful to respect their developmental preserves of power and profit, as well as their patrons in Congress. Hunsaker called this a noble pact of "teamwork between science, military, and industry."[49] The politics behind all of this was a bit meaner, though. The Aircraft Industries Association, for one, made demands that in its "over-all distribution of effort," NACA dispense with the development of any actual machinery. Industry would build the engines. There was money to be made. NACA was willing to do the basic research in answer to industry's pressing questions and obediently established a new Industry Consulting Committee to resolve disputes.[50]

NACA Lewis conformed to these demands, approved by headquarters in Washington, DC, by renaming its organizational units according to "fundamental fields of research" rather than developmental "engine types." Groups that studied turbo jet engines now studied aircraft compressor research. Units that studied rockets now studied combustion research or high-pressure combustion. Both were now subsumed under even grander conceptual rubrics like the Fuels and Thermodynamics Division. Even the name of NACA Lewis changed: from the Aircraft Engine Research Laboratory to the Lewis Flight Propulsion Laboratory. NACA stripped itself of any hint of developmental prowess.[51] Abe Silverstein, soon the director of research at NACA Lewis, confirmed that he and his laboratory heads were "heartily in sympathy" with all of this. Time and again they reaffirmed that NACA was a "service organization," providing basic research for the military and industry. Or as Silverstein later wrote to his organization, the "truth" was that NACA was "not in the hardware business." It did not do "procurement of end items."[52]

By these arrangements, Hunsaker trapped NACA in a web of relationships that weakened rather than empowered it, and that limited its future prospects. He was constrained by his own traditional thinking, by his preference for basic research, and by his preserves of East Coast power. His forte was in the distant past. He had been MIT's first aeronautical engineering PhD. He had written the very first chapter on aeronautics for *Marks' Standard Handbook for Mechanical Engineers*. Both dated to 1916, when Zucrow was still in high school. Hunsaker had helped to design the Navy Curtiss NC-4 biplane, the flying boat that was first to cross the Atlantic in 1919; as well as the dirigible airships, the *Shenandoah*, *Akron*, and *Macon* (all of which crashed between 1925 and 1935). For most of the fifteen years that he was a member and chair of NACA's Main and Executive Committees, he was also head of both the Departments of Aeronautical Engineering and Mechanical Engineering at MIT. He held a lot of power, and for a long time.[53]

Hunsaker's contradictory policies also created an inherent tension within NACA, between tradition and vision, between subsidiary basic research and vanguard R&D. In line with his return to research, he disbanded NACA's Special Committee on Guided Missiles at the end of the war, transferring its authority to the RDB-GMC and to the military services. NACA was to play a supportive rather than a lead role. Yet he also invested NACA in rocket and guided-missile research, committed to the "scientific and engineering research" for "the design of guided missiles and their means of propulsion and control."[54] NACA Langley engineers took up the R&D charge for supersonic aerodynamics, culminating early on in the X-1 and D-558-2 Skystreak rocket planes. These tests of supersonic aircraft "brought some of NACA's researchers out of their laboratories and into contact with the development process and the excitement generated by these projects."[55] NACA Lewis engineers took up the visionary R&D charge for liquid-propellant rocketry. Walter Olson, like Zucrow a combustion and fuels specialist, made the first initiatives. He brought liquid-propellant rockets to NACA Lewis in November of 1945, creating the Rocket Section and the Rocket Laboratory in his Combustion Branch with the support of his boss, Ben Pinkel.[56]

John Sloop took direction of the NACA Rocket Laboratory by 1948, then comprising "four temporary static-thrust test cells for the study of rocket fuels and cooling, located behind earthen protective mounds." Given the toxic propellants and risk of explosion, along with the noise, this laboratory was built at a forested edge of campus, perched at the cliffs leading into the Rocky River ravine. It predated the Purdue Rocket Lab by several years, though they eventually were both devoted to research of "high specific impulse propellants" and higher chamber pressures, along with such various related issues of heat transfer, film cooling, ignition delay, and instrument research (as in new "temperature-measuring probes" for rocket thermodynamics). These were all keys to improving the rocket engine for "ultra-high-speed flight," in service to the armed forces, who had "already visualized future wars being fought primarily with long range automatically guided rocket missiles."[57]

Zucrow played a vital role in the propulsion work of NACA Lewis, as both teacher and colleague. He had taught the first course and written the book on jet propulsion, after all. He had defined its initial national research agenda and was already conducting vanguard experiments at Purdue. He had built a network of connections and vast knowledge about the nation's efforts in liquid-propellant rocketry through Aerojet, Project Squid, the MPF, and the ARS. NACA Lewis often sought out his advice and active cooperation. From Purdue he offered the technical expertise, innovation, experience, and vision that MIT's Hunsaker could not. When Sloop taught a college course, Aircraft Propulsion Principles (1949) through the Case Institute of Technology, his main survey source for lecture no. 8, on rocket engine and performance, was none other than Zucrow's *Principles of Jet Propulsion*.[58] In time, the Purdue Rocket Lab and the NACA Lewis Rocket Lab settled into a fine division of labor. Zucrow focused on his high chamber pressures, Sloop on high-energy propellants. Both labs studied the urgent need to cool rocket motors from their intense heats. Sloop and his teams soon even contracted with Purdue for a "special" apparatus to record heat transfer and frictional pressure, and for its experimental data and insights on internal film cooling, which included both the "porous metal" and parallel "disks" approaches.[59]

As Virginia Dawson and Robert Arrighi have already documented, the NACA Lewis rocket researchers, under Olson's and Pinkel's lead, and in the work of Sloop and his teams, made rockets a small if significant part of normal research science at NACA Lewis. They were only about 3 percent of the total personnel, receiving at most 2.4

percent ($1.2 million) of the center budget, but the center featured their work in public forums, as for example in the brochures and exhibits, movies and tours of the Annual Inspections of 1947 and 1949.[60] At the Institute of the Aeronautical Sciences' Second Annual Flight Propulsion Meeting in Cleveland (28 March 1947), NACA Lewis premiered the full range of its rocket research. Olson and Pinkel also offered regular lectures and discussions on their expertise at Purdue University on topics like combustion and fuel problems, heat transfer, and cooling of turbojet, ramjet, nuclear, and rocket engines.[61] The NACA Lewis newsletter showcased its rocket power plants, designed for "high velocities, high acceleration, compactness, or flight outside the earth's atmosphere." The newsletter also featured the softball "battle of the century" at the center's Labor Day picnic in 1948. It pitted the test cell technicians, aka the Rocket Mechanics (including "Flash" Gordon Kelsey), against the "brains," aka the Rocket Engineers (including Bill Rowe, George Kinney, and Riley Miller). The Rocket Mechanics won, accompanied by a corn roast, horseshoes, and a sing-along.[62]

As one of the "brains," Sloop was an American visionary. He was a popular lecturer at NACA Lewis and in the Cleveland metropolitan area. The achievement of the X-1 rocket plane, along with the urge for "higher and higher altitudes," even beyond Earth, were some of his favorite themes by 1948. He accepted the rocket as a guided missile for defense and deterrence, but also as a vehicle for the exploration of "the upper atmosphere and interstellar space," including space stations, these in the years before von Braun's celebrity. Sloop lobbied for a "large-scale, vigorous research and development program" at NACA. In a talk at Purdue in March of 1952, or as he called it, the "University of Purdue," he discussed these platforms and their power plants of "future design." As he said, "It is within engineering possibility to build long-range missiles that would span a continent or travel into outer space."[63] Sloop eventually framed these imperatives in an elegant metaphor of America's ascending curve of small "steps" into space. Our great leap was awaiting. "The growth of speed from the Wright brothers until World War II was from refinements in the piston engine. The turbojet blossomed after the war and gave a rapid spurt to the curve. But it was the rocket-powered aircraft that bent the curve skyward," he wrote. "In a couple of years from now the NACA X-15 airplane will be far out on the curve. Its pilot will go higher and faster than any man yet and for a few minutes he will be beyond the earth's atmosphere—coasting through space."[64]

Thanks to all these interfaces, Zucrow became a valuable consultant at NACA Lewis. Between 1947 and 1952, Zucrow served as one of twelve members of NACA's Subcommittee on Propulsion Systems Analysis (of the Committee on Power Plants for Aircraft), devoted to such issues as piston and jet engine designs (including new carburetor types), thrust augmentation (via improved components like nozzles and tail pipes), thermodynamics and heat transfer, and the new measurement and instrumentation techniques.[65] Zucrow also worked closely at times with Walter Olson on the

NACA Subcommittee on Combustion and with Ben Pinkel on its Subcommittee on Heat Resisting and its Subcommittee on Propulsion Systems Analysis. NACA Lewis recognized the new maturity of rocket engine research, centered at Caltech's Jet Propulsion Laboratory and at Zucrow's Rocket Lab, resolving to follow their leads.[66] These connections were more than just local. Zucrow had already drawn Olson into his MPF work. He also lobbied his new colleagues at NACA to become active in the ARS. NACA, so he was urging, belonged more at the center of the US missile and spaceflight effort. It was a sentiment shared by Frederick Hovde, chair of the GMC and overseer of the Earth Satellite Vehicle Program, who saw NACA as the only truly creative force in the federal government.[67]

## WHEN NACA FAILED TO BECOME NASA

By the early 1950s, all signs showed promise of a rising NACA Lewis rocket agenda. Though Hunsaker and Victory had limited NACA's range of initiatives for civilian rocketry projects, other leading administrators gave it solid support. Hugh Dryden, as NACA director of aeronautical research, and director of NACA after 1949, was a proponent of guided missile and civilian rocket research. Abe Silverstein, chief of research at NACA Lewis (1949–1952), later associate director, invested in aircraft and rocket propulsion, especially high-energy alternative propellants. He renamed the High-Pressure Combustion Section the Rocket Research Branch, raising it a notch in the NACA hierarchy. Silverstein also funded its improvements, including a "central control room," including four new test cells, this to reduce "total instrumentation," and its expansion into the fuller range of rocket research fundamentals. Bruce Lundin, with a BSME from the University of California, Berkeley (1942), rose to become chief of the Engine Research Division at this time. Lundin, a mechanical engineer who, like Zucrow, had begun his career designing power and utility plants, was mostly occupied with research on turbojet and ramjet engines, but he was also a vocal and unflinching advocate for advanced research in propulsion for spaceflight.[68]

Zucrow played his own role at NACA Lewis as a catalyst to transform NACA into more of a NASA. In January of 1951, NACA Lewis appointed him as chair of the Special Subcommittee on Rocket Engines, within the Committee on Power Plants for Aircraft. He remained chair until 1953 and a member until 1958. The special subcommittee had its origins at a NACA Lewis conference in early 1950. Military officers, contractors, and staff met to discuss new priority "development projects" for the RDB-GMC. Two of Zucrow's close colleagues were there among the "rocket experts": Lt. Col. Marvin Demler, chief of the Power Plant Laboratory at Wright-Patterson Air Force Base, and Opie Chenoweth, its chief scientist and Zucrow's old friend from the

Purdue Engineering Experiment Station in 1923. NACA's Olson and Sloop were also there. The plan was to transfer some of the work of Zucrow's MPF to NACA Lewis through the new Subcommittee on Rocket Engines in order for its engineers "to consider all forms of rocket engines," both solid- and liquid-propellant, "used for flight propulsion of guided missiles, aircraft, and artillery rockets."[69] That summer under Silverstein's lead, NACA Lewis, in coordination with the MPF and RAND studies, also formed the Long-Range Missile Panel for research into ramjet and rocket systems with three thousand miles of range and beyond.[70]

The outbreak of the Korean War somewhat delayed these initiatives, but it also spurred NACA engineers to face the next possible crisis and accelerate research for "radically new developments."[71] By the summer of 1953, NACA's Committee on Aerodynamics lobbied to build upon the X plane achievements and advance to the next "step in studying the problems of flight in outer space." It had some powerful backers: Robert J. Woods (one of the designers for the Bell Aircraft X-1, X-2, and X-5), along with advisers from the Air Force. The committee resolved "to provide adequate planning and basic research on the problems of upper atmosphere, ionosphere, and space flight" and to build on RAND reports for development of "satellite orbits in outer space" for "military operations." This meant a dramatic turn to study "problems associated with unmanned and manned flights at altitudes from 50 miles to infinity and at speeds from Mach number 10 to the velocity of escape from the earth's gravity."[72] Escape velocity and infinity meant spaceflight.

At this same time, Zucrow advanced his own liquid-propellant rocketry agenda, partnering with NACA Lewis and the Wright Air Research Development Center to organize the NACA Conference on Supersonic Missile Performance (March of 1952). It was a culmination of all of Zucrow's technical writings and government service since the end of the war. Nearly three hundred of the country's top missile experts and managers were there. Many were his friends and colleagues. At least two were his former students: Lt. Col. Langdon Ayres (MSAE 1947) from the ARDC and Cdr. Richard Duncan (PhD AE 1950) from the Office of Naval Research. Topics included the "optimum design characteristics" of all ranges of engines, with a focus on the coming ICBM, the long-range missile reaching to a six-thousand-mile range.[73] As chair of the MPF, Zucrow also called for NACA to raise its profile in the work of long-range rocketry and spaceflight. A crucial "specific deficiency" in the MPF's national survey was in the research and testing of "multi-step liquid rockets" and in "reliable engineering data for scaling-up thrusts of liquid-propellant rockets." This meant the work of high-performance fuels and higher chamber pressures ("from 500 to 900 psi or above"), the very two specialties at NACA Lewis and at Zucrow's Rocket Lab.[74]

Sharing these visions for both the militarization and exploration of space, Zucrow's Subcommittee on Rocket Engines at NACA Lewis was a formidable group of rocketry pioneers and spaceflight enthusiasts. Zucrow was at the top of a pyramid of America's best minds in rocket propulsion. There were sixteen representatives on it at any one time, chosen from throughout the long-range rocketry laboratories and test stands around the country, the very network that Zucrow had traversed and advised in his years at the MPF. Many of the members were his old friends from Aerojet, the M.W. Kellogg Company, Reaction Motors Inc., RAND, and North American Aviation. His former student Langdon Ayres was also on it.[75] Wernher von Braun and the members of the Peenemünde team were conspicuously absent. As John Sloop remembered, the subcommittee was "a significant milestone in NACA recognition of the importance of rocket research. In addition to its great value for coordinating and exchanging information on rocket research and development, the subcommittee was a political force for assuring a fair share of attention to rocket propulsion." It was "a unique body of rocket experts from government, industry, and universities" dedicated to the "national planning" of R&D.[76] Sloop's superiors at the laboratory recalled the moment, too, when Zucrow and a handful of leading engineers encouraged NACA Lewis to start "pushing rockets generally." Among them were William Bollay at North American Aviation (Rocketdyne), Richard Porter at General Electric, Homer Newell of Project Viking, Langdon Ayres at the ARDC, and several directors at the ARS (Andrew Haley, Milt Rosen, and Robert Truax). They sought out NACA to ally with them "in achieving orbital flight with rockets."[77]

Most of the initial work of the special subcommittee was oversight, reviewing the technical details of the NACA Rocket Laboratory in such areas as propellant chemistry, chamber and nozzle design, injection methods, instrumentation, and high-speed photography. The members were testing NACA Lewis against their own high standards, successes, and failures. Several subcommittee recommendations bore the indelible stamp of Maurice Zucrow, as when it requested more experiments with white fuming nitric acid ignition and combustion processes of the very kind at the Purdue Rocket Lab, or when it recommended a whole new approach, reminiscent of Zucrow's Aerojet days, to search for more exact "causes and remedies," not general effects, of combustion vibration. By the time of its 26–27 June 1952 meetings, the subcommittee also made a stunning priority recommendation: for NACA Lewis to launch the "investigation, with advanced high-energy propellants, of scaling factors up to 100 times present scale, to provide information necessary for the successful design of rocket engines of high thrust levels." The subcommittee was basically asking for NACA to bridge the work of its Committee on Aerodynamics with its Committee on Power Plants in order to create an architecture of spaceflight machinery. The special subcommittee's powerful statement in June reads like a proclamation of independence for NACA. It had all the right talent, but too few facilities and funds. It deserved to fulfill its destiny as an "advanced research agency" all its own. Zucrow defended the resolution before the NACA Power Plant Committee. The NACA Executive Committee approved it as well.[78]

Zucrow put his name and reputation on the special subcommittee's work. He was the dynamo, coalescing a band of main line American rocket engineers to stake their power and reputations for spaceflight. He was, in essence, helping to formulate a shakeup of NACA Lewis, endorsing its move into heavy-lift rockets as long-range ballistic missiles and launchers for outer space. Just finishing his tenure of five years on the MPF of the RDB-GMC, so dominated by military interests, he was also attempting to raise the profile of NACA as a counterforce. Zucrow's subcommittee was a bold gambit, an attempt to redress the imbalance and seize the initiative for NACA. It was all a potential turning point in America's guided missile and spaceflight investments.

Thus inspired, NACA Lewis advanced it plans to design a new "rocket facility" in or around Cleveland, one to accommodate the testing of engines up to twenty thousand pounds of thrust and with "future development" of a "full scale" test stand for one hundred thousand pounds. NACA Lewis was slated to progress into the research *and* development of heavy-lift rockets for outer space. Financing was paramount, with an initial cost of $4.5 million.[79] The work of the special subcommittee peaked in November of 1952, with Zucrow's personal endorsement. He and Walter Olson discussed a Bureau of the Budget proposal for an even more advanced $8.5 million "rocket facility," perhaps even out west. Zucrow offered his "hearty endorsement," a valuable one by NACA's standards, given how he was "fully cognizant" of rocketry R&D in the US. He confirmed that it was "exceedingly important" for NACA to research large-thrust engines, not just at one hundred thousand but even at three hundred thousand pounds of thrust. In other words, "engines of practical size." The subcommittee backed the plan too, even calling for a more sophisticated testing facility modeled on industrial sites.[80] This Zucrow endorsement was actually a capstone career moment, culminating all his rocketry work to date. He was the lynchpin. He had a unique perspective on the whole. No one else in the country had his breadth and depth of knowledge and networks. As to the companies and laboratories involved in rocketry and missile work, "I know them all," he once boasted. This also helps us understand some of his reticence in promoting spaceflight ventures in the ARS, or in expressly endorsing the von Braun satellite plan. He placed his confidence, instead, in the engineering and technical expertise of the people he knew from the defense and aerospace communities. He trusted his own spaceflight "vision of what ought to be."[81]

All of this high promise ended in defeat. The defeat came from within. Hunsaker, the pioneer of aerospace engineering at MIT, the conservative guardian of NACA's traditional role of basic research and service to the military, threw his weight against the plan. As Sloop commented on one copy of the proposal for an expanded NACA Lewis rocket facility, in a quick handwritten note: "Caused quite a stir. . . . Hunsaker violently opposed." To be fair to Hunsaker, he was under intense pressure at the time, from Congress and a series of General Accounting Office audits, to discipline his spending and any presumably errant research projects. So it was easier and smarter for him to make a tactical retreat and ensure that NACA remained beholden to military patronage, not equal to it. NACA would stay in the "race" to assist the military in creating missiles for "extreme altitudes across intercontinental distances." Hunsaker wanted missiles to hit the Soviet Union, not fly in outer space.[82] The defeat ultimately also came from without. Sloop remembers the Bureau of the Budget waging a war of "attrition" against the financial plans of the NACA rocketeers, as "the bubbles burst one by one." Furthermore, at a meeting of the MPF to discuss the new "rocket research facility," the issue was simply removed from the agenda, even though its members spoke in favor of NACA conducting research on larger thrust rockets, beyond the thousand-pound thrusts in current use. The crucial moment came when the Air Force representative criticized these plans. Researching rockets with thrusts of one hundred thousand pounds was far beyond NACA's capabilities and talents, he remarked. It was better off doing "the simpler research" of the "hows and whys" of smaller engines.[83]

Given these kinds of odds, Zucrow had clearly overestimated his own authority. The research triangle that he cultivated had turned into a research trap. Neither Purdue's Rocket Lab nor NACA Lewis could displace the US Air Force. NACA Lewis researchers had no choice but to return to their earlier priorities of simpler research contracts. In one mostly positive development, a consolation prize, NACA Lewis engineers did take on the work of Zucrow's MPF after the Department of Defense reorganization of 1953. They now became the technical experts who advised.[84] NACA engineers filled the new technical panels in the Department of Defense ranks. Abe Silverstein became chair of its Ad Hoc Group on Liquid Propellant Rocket Engines, eager to offer his laboratories for the defense cause. "We desire to learn of the problems facing the military services and to assist them in every possible way," he wrote in a memorandum.[85] Sloop's teams at the NACA Rocket Laboratory continued their full array liquid-propellant rocket engine studies (including ramjets) but remained most focused on the "theoretical and experimental studies of high-energy propellants." Not for civilian rockets, but for "long-range missiles."[86]

Such were the immediate outcomes of the 1952 events, a case study in how the US fell behind the USSR five years before Sputnik, a lost chapter in the Cold War politics of R&D. Ten years later, testifying before Congress, Zucrow and his colleagues remembered them as a moment of loss for the US civilian spaceflight program. It all rested on NACA Lewis, of course, but the Air Force intervened to defeat it. James G. Fulton, Republican congressman from Pennsylvania, summarized the discussion in these terms: "The scientific program was cut to the size made for the military weapon system." Zucrow agreed. George P. Miller, chair of the Committees on Science and Astronautics and Democrat from California, put it even more succinctly: "We should have had NASA a little sooner."[87] He really meant could have.

IN AN IRONIC TURN ON THESE EVENTS, AS NACA'S STATUS FADED, WERNHER von Braun's celebrity rose. His media tours during 1952 culminated with his appearances on television, sponsored by none other than Walt Disney, in a series of spaceflight shows between 1955 and 1957. Von Braun had the good fortune to raise his media profile along with his team's Redstone missile, the pride of Huntsville, Alabama, their new home. Huntsville gave von Braun and his team a legitimacy and spotlight few Americans could resist. His title: director of the Development Operations Division of the Army Ballistic Missile Agency of the Redstone Arsenal. In the context of the International Geophysical Year, the Army and von Braun Jupiter explorer mission soon competed with the US Navy's Viking–Vanguard project for the president's and the public's favor.

Von Braun rose to star power with his V-2 terror weapon, his Redstone guided missile, and his own mad dash for publicity. The ARS welcomed him in this role when, as a member of the board in 1956, he claimed the mantle of national spokesman, receiving loads of fan mail from young people interested in "rockets and jet propulsion." The guided missile industry also celebrated him as "the world's greatest authority on ballistic missiles," one of the founding fathers of American rocket science. He and Walter Dornberger broke into the limelight with books and awards by 1956: Dornberger (now of Bell Aircraft) winning the ARS Pendray Award for Literature, von Braun the ARS Space Flight Award.[88]

The media and publishing houses were primed to redeem von Braun of his Nazi past. Take Beryl and Sam Epstein's celebratory tale. As writers of children's literature, they were also biographers of Paul Revere and Winston Churchill. Now, with the support of the ARS, they turned to rocketry, applauding the Germans for all their years of "monotonous designing, building, testing, redesigning, rebuilding, retesting" of the V-2 missile. Von Braun and his team were paragons of the scientific method and unstoppable technological innovation who also just so happened to be building a weapon of mass destruction and "annihilation" effect. For "the Führer." As a weapon it was but a footnote to the bigger story of the "birth" of the "long-range rocket" for outer space. Scholars eventually joined this chorus, historians who became publicists, dazzled by the very publicity spotlight that von Braun focused on himself, as if he alone were the driving force and inspiration for the US spaceflight movement.[89]

One small chapter in von Braun's American debut was his visit to Purdue on 6 May 1953, when he gave a talk titled "Space Travel: How Soon?" for the fourth annual banquet of the ARS Indiana Section at Zucrow's invitation. The talk was a success; 150

people came to hear von Braun speak. At only forty-one years old, he was in his prime, America's rocket expert, there to talk about his space plane and wheeled space station. He called it an "artificial planet" that might even launch atomic rockets against enemy cities around the world, or enable a five-day trip to the Moon within a decade. Rocket Lab graduate student Elliott Katz was at the talk. Ten years before, in July–August of 1944, Katz had served as a navigator for one of the B-24s sent to bomb von Braun's Peenemünde missile site, and ideally kill him. "My only regret is that we didn't flatten the whole place," he later remembered. Katz did not mention this when they met and shook hands that evening. One of his classmates, J. P. Sellers, recalling the Nazi atrocities of the war, refused to do even that. Fellow graduate student Bruce Reese simply remembered the two-hour talk as going on way too long. He had a young family waiting for him back home.[90]

Zucrow and his close colleagues were also the wiser. They did not exaggerate the prestige and influence of von Braun and the Nazi rocketeers. For them, "the Germans" were always a group apart. Not at the center of the US aerospace program, but at its margins.[91] This was not a simple matter of Zucrow disliking the Germans as Nazis. It was a matter of him rejecting their authoritarian (Nazi) loyalties and their failed R&D methods. He and his colleagues valued instead the treasury of American invention and experience, as represented in the rocketry contributions of places like Caltech and Aerojet, Reaction Motors Inc., North American Aviation, and the Kellogg Company. Purdue memorialized the von Braun visit in glowing terms and in a much-reproduced photograph of Zucrow and von Braun at one of the Rocket Lab test cells. But the original caption was most revealing, which had Zucrow "explain a rocket motor experiment" to von Braun. Never a propulsion or power-plant man, von Braun probably felt somewhat out of place at the test cell.[92]

As Zucrow later remarked, in deference to his own legacy and those of his close colleagues in propulsion engineering, "Von Braun and his associates did not develop rocket engines in this country. They purchased them, for the missile or launching rockets they built; they're systems people."[93] Rocket pioneers like Grayson Merrill (US Navy) and Homer Stewart (Caltech) agreed. Von Braun was not a "highly educated engineer," but simply a "good manager." Abe Silverstein and John Sloop of NACA extrapolated this to the whole German team: as engineers, "they were the most conservative type of people. They built in I-beam bridge type of thinking." The Germans were not part of the American tribe. They were more like a cult, with von Braun fixated on "image building" and his people fiercely devoted to him.[94]

Zucrow and his friends could be even more candid about all of this. He and Werner von Kirchner from Aerojet liked to joke about the von Braun group in fake German accents as the "Oberkomando der Huntsvillen Wehrmacht." Or as von Kirchner parodied in an imaginary dialogue and in a pun on the playing of a flute: "It is better to be ein nazion of flautist zen floutist." Translation: It is better to be a team player than a rogue. Von Braun was the rogue. Von Kirchner, whom Zucrow coined as "the other Wernher," had unimpeachable credentials. He also had a real German accent. Rather than join the Nazi regime and German war effort like von Braun, he dedicated his youth to fighting against it, fleeing the Nazis for the West, joining the Polish Fighter Squadron and flying for the British Royal Air Force.[95] He had made the right moral choices. Yes, he certainly was the "other" Wernher.

The Zucrow family (*at left*), Maurice, wife Lillian, and daughter Barbara, at a Passover Seder in 1931 with their extended Lafayette family, the Feinsteins, Zovods, and Freedmans. (REPRODUCED BY PERMISSION FROM CRAWFORDSVILLE DISTRICT PUBLIC LIBRARY)

DISSERTATION

"DISCHARGE CHARACTERISTICS OF SUBMERGED JETS"

OUTLINE

Flow of Viscous Fluids.

Hydrodynamic Equations for the Flow of Fluids.

Dimensional Treatment of Flow Through a Jet.

Flow Through Jets with the Same Ratio of Length to Diameter. C

Influence of Chamfering Upstream Edge.

Types of Jets.

Characteristics of Jets with Different Ratios of Length to Diameter.

Other Experiments.

Application of Results to Carbureters.

Conclusions.

Bibliography.

PURDUE UNIVERSITY

The Graduate Committee

announces

THE FINAL EXAMINATION

of

Maurice Joseph Zucrow

for the Degree of

DOCTOR OF PHILOSOPHY

Tuesday, May 22, 1928, 4 P. M.

Trustees' Room, Eliza Fowler Hall.

M. J. Zucrow's final examination flier as Purdue's first "earned" PhD in 1928, front and back. (FROM MAURICE J. ZUCROW PAPERS; COURTESY OF PURDUE UNIVERSITY LIBRARIES, KARNES ARCHIVES AND SPECIAL COLLECTIONS)

EXAMINING COMMITTEE

A. A. Potter, Chairman

W. H. Bair
R. G. Dukes
W. E. Edington
F. W. Greve
H. M. Jacklin
G. A. Young

MAURICE JOSEPH ZUCROW

B. S. in M. E., Harvard University 1922
M. S. in M. E., Harvard University 1923

Research Assistant, Engineering Experiment Station, Purdue University 1923 - 1928.
Research Associate, Engineering Experiment Station, Purdue University 1928 -

PLAN OF STUDY

| | |
|---|---|
| MAJOR | Automotive Engineering<br>Internal Combustion Engines<br>Kinematics of Machinery<br>Mathematical Physics<br>Power Engineering |
| FIRST MINOR | - Hydraulic Engineering |
| SECOND MINOR | - Industrial Engineering |

M. J. Zucrow's final examination flier, as Purdue's first "earned" PhD in 1928, inside. (FROM MAURICE J. ZUCROW PAPERS; COURTESY OF PURDUE UNIVERSITY LIBRARIES, KARNES ARCHIVES AND SPECIAL COLLECTIONS)

Student cartoon titled "Celestial Research," lampooning Purdue's dean of engineering, A. A. Potter (*right center*); C. Francis Harding, head of electrical engineering (*top right*); William K. Hatt head of civil engineering (*bottom right*); Harry C. Peffer, head of chemical engineering (*center left*); and G. A. Young, head of mechanical engineering (*far left*). "Into the realm of Einstein's fourth dimension, the land of the cosmic rays, or the strata where the gases are really rare—where couldn't this crew take our ship?" (FROM *PURDUE ENGINEERING REVIEW* [JANUARY 1928]; COURTESY OF PURDUE UNIVERSITY LIBRARIES, KARNES ARCHIVES AND SPECIAL COLLECTIONS)

Aerojet team for the solid-propellant JATO. *Left to right, front row:* M. J. Zucrow, the 12-AS-1000 JATO unit, and Glenn Miller; *back:* Norton Moore, A. I. Antonio, Brooks Morris, Bernie Dorman, Martin Summerfield, and Robert D. Young. (FROM CHARLES M. EHRESMAN PAPERS; COURTESY OF PURDUE UNIVERSITY LIBRARIES, KARNES ARCHIVES AND SPECIAL COLLECTIONS)

The Northrop MX-324, America's first rocket plane, with a Zucrow-engineered power plant. At Harper Dry Lake, Mojave Desert, California (1944). (FROM CHARLES M. EHRESMAN PAPERS; COURTESY OF PURDUE UNIVERSITY LIBRARIES, KARNES ARCHIVES AND SPECIAL COLLECTIONS)

The Northrop MX-324 in flight. (FROM CHARLES M. EHRESMAN PAPERS; COURTESY OF PURDUE UNIVERSITY LIBRARIES, KARNES ARCHIVES AND SPECIAL COLLECTIONS)

M. J. Zucrow, professor of jet propulsion and gas turbines (1946–1966). (FROM MAURICE J. ZUCROW PAPERS; COURTESY OF PURDUE UNIVERSITY LIBRARIES, KARNES ARCHIVES AND SPECIAL COLLECTIONS)

**THE UNIVERSITY OF CALIFORNIA**

**LOS ANGELES**

**WAR TRAINING CENTER**

***Starting June 16, 1944, a Tuition-Free ESMWT Course in***

# PRINCIPLES OF JET PROPULSION

**DR. M. J. ZUCROW, Instructor**

**7:00-9:30 p.m., Fridays for 16 weeks**

**Room 695, 714 SO. HILL STREET, LOS ANGELES**

**(Course 949A2)**

This course will be restricted to a discussion of the basic principles of jet propulsion and will not include a discussion of actual contructions, materials, or operating conditions. In effect the work will review available information in the technical press and apply the principles of thermodynamics and fluid flow to determine generalized operating characteristics. The material will include performance and limitations of present engine propeller combinations; review of momentum principle, thermodynamics and study of gas turbine cycles; high speed compressors; theoretical Whittle type machine; rocket propulsion principles; thermal jet systems.

PREREQUISITE: B.S. Degree in Science or Engineering including a knowledge of thermodynamic and fluid flow; empolyed in aircraft engineering design work.

**Offered as part of the nation-wide Engineering, Science, and Management War Training Program authorized and financed by the U. S. Office of Education. Application for additional gasoline, if needed to attend classes, may be made at time of enrollment.**

**PLEASE POST** **For Additional Information See Reverse Side**

A1-121

Flier for M. J. Zucrow's college-level course Principles of Jet Propulsion (1943–1944), an American first, offered under the Engineering, Science, Management and War Training program. (REPRODUCED BY PERMISSION FROM UNIVERSITY ARCHIVES, LIBRARY SPECIAL COLLECTIONS, CHARLES E. YOUNG RESEARCH LIBRARY, UCLA)

Members of the Panel on Propulsion and Fuels, part of the Guided Missiles Committee, Research and Development Board (1947–1952). (FROM MAURICE J. ZUCROW PAPERS; COURTESY OF PURDUE UNIVERSITY LIBRARIES, KARNES ARCHIVES AND SPECIAL COLLECTIONS)

M. J. Zucrow teaching an advanced jet propulsion course at the Purdue Airport (fall 1948). (FROM *PURDUE ALUMNUS* [FEBRUARY 1949]; COURTESY OF PURDUE UNIVERSITY LIBRARIES, KARNES ARCHIVES AND SPECIAL COLLECTIONS)

The S curve career trajectory for the new aerospace engineer, reaching to ever higher altitudes. (FROM *PURDUE ENGINEER* [APRIL 1951]; COURTESY OF PURDUE UNIVERSITY LIBRARIES, KARNES ARCHIVES AND SPECIAL COLLECTIONS)

Faculty–student team on the completion of the Rocket Lab (1948). *Left to right, front row:* George Palmer, M. J. Zucrow, Cecil Warner; *back:* Walter Hesse, Charles Trent. (FROM CHARLES M. EHRESMAN PAPERS; COURTESY OF PURDUE UNIVERSITY LIBRARIES, KARNES ARCHIVES AND SPECIAL COLLECTIONS)

Rear view of the original Purdue Rocket Lab with test firing cells (1949). (FROM JERRY ROSS PAPERS; COURTESY OF PURDUE UNIVERSITY LIBRARIES, KARNES ARCHIVES AND SPECIAL COLLECTIONS)

Front view of the Purdue Rocket Lab (*center*) with the Headquarters Building (*right*) and Gas Turbine Laboratory (*left*) in 1954. (FROM CHARLES M. EHRESMAN PAPERS; COURTESY OF PURDUE UNIVERSITY LIBRARIES, KARNES ARCHIVES AND SPECIAL COLLECTIONS)

Rocket motor control panel with periscopic window (1951). (FROM CHARLES M. EHRESMAN PAPERS; COURTESY OF PURDUE UNIVERSITY LIBRARIES, KARNES ARCHIVES AND SPECIAL COLLECTIONS)

Graduate student Clair Beighley at the periscopic window (1951). (FROM CHARLES M. EHRESMAN PAPERS; COURTESY OF PURDUE UNIVERSITY LIBRARIES, KARNES ARCHIVES AND SPECIAL COLLECTIONS)

Rocket motor test firing, with downward stream of cooling water (1951).
(FROM CHARLES M. EHRESMAN PAPERS; COURTESY OF PURDUE UNIVERSITY LIBRARIES, KARNES ARCHIVES AND SPECIAL COLLECTIONS)

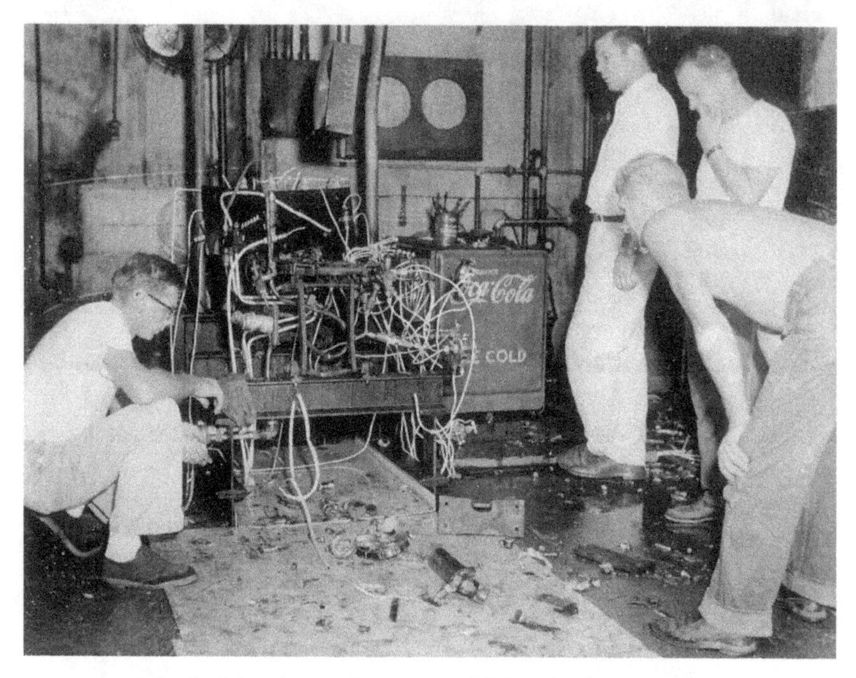


Rocket Lab graduate students inspect a failed test (1956). *Left to right:* J. P. Sellers, J. R. Osborne, and two unidentified students. (FROM J. P. SELLERS III PERSONAL COLLECTION; REPRODUCED BY PERMISSION)

Elliott Katz with his PhD dissertation and Wayne Colahan with his MS thesis (1954), both off to California, Katz to the Convair Division of General Dynamics and Colahan to the Jet Propulsion Lab. The Zucrow slogan "Beware of Optimism" is in the foreground; the student response in Latin (an adaptation of a Second World War slogan, here referring to Zucrow), "Don't let the bastard get you down," is in the background. (FROM CHARLES M. EHRESMAN PAPERS; COURTESY OF PURDUE UNIVERSITY LIBRARIES, KARNES ARCHIVES AND SPECIAL COLLECTIONS)

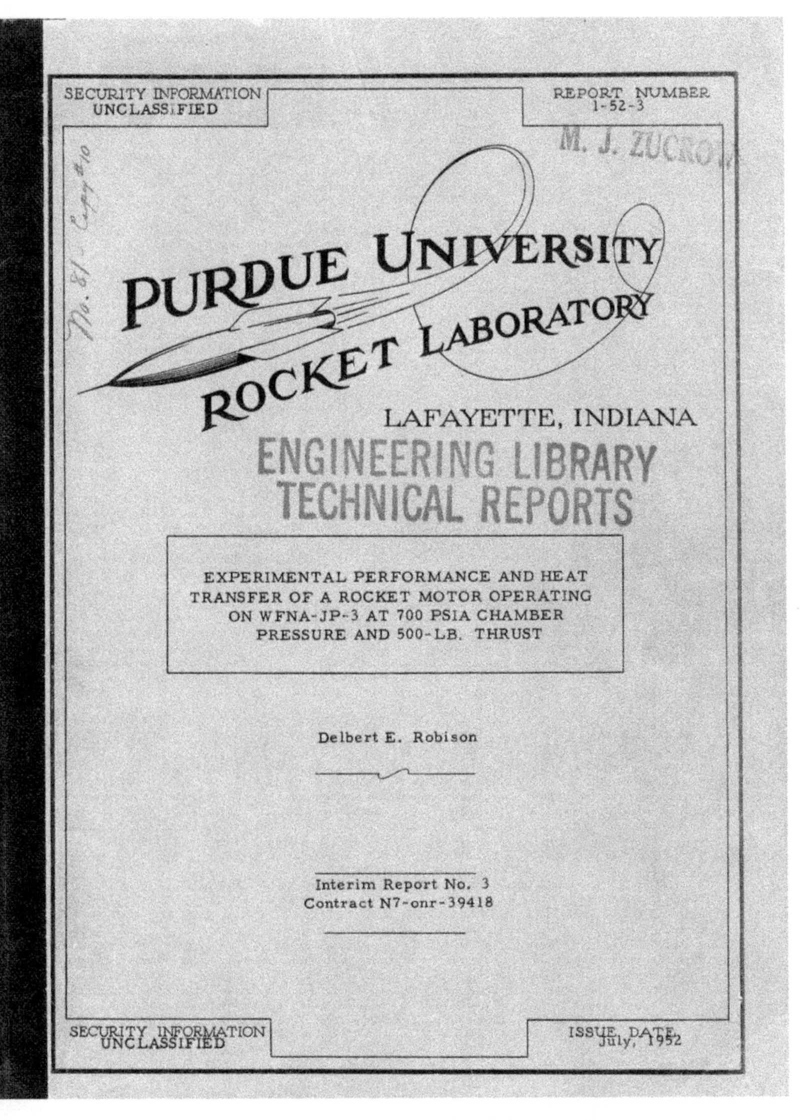


Front cover of Rocket Lab Interim Report No. 3 (1952). (COURTESY OF PURDUE UNIVERSITY LIBRARIES, KARNES ARCHIVES AND SPECIAL COLLECTIONS)

Launching banquet flier for the Indiana Section of the American Rocket Society (1950), front. (FROM CHARLES M. EHRESMAN PAPERS; COURTESY OF PURDUE UNIVERSITY LIBRARIES, KARNES ARCHIVES AND SPECIAL COLLECTIONS)

Purdue Memorial Union—1950

Speakers

Toastmaster................Clair M. Beighley, Purdue University

William L. Gore............................................*National President, American Rocket Society, Washington, D. C.*

Dr. M. J. Zucrow................................*Professor of Gas Turbines And Jet Propulsion, Purdue University*

Colonel M. Demler....................*Chief, Power Plant Laboratory, Wright-Patterson Air Force Base, Dayton, Ohio*

Menu

Breaded Pork Loin

Creamed Potatoes — Asparagus

Fruit Salad — Rolls and Butter

Coconut Cream Tarts

Coffee

Guests

Dr. F. L. Hovde, *President of Purdue University*

Mr. G. S. Meikle, *Director Purdue Research Foundation*

Prof. H. L. Solberg, *Head of the School of Mechanical Engineering*

Prof. E. F. Bruhn, *Head of the School of Aeronautical Engineering*

Launching banquet flier for the Indiana Section of the American Rocket Society (1950), inside. (FROM CHARLES M. EHRESMAN PAPERS; COURTESY OF PURDUE UNIVERSITY LIBRARIES, KARNES ARCHIVES AND SPECIAL COLLECTIONS)

Purdue University Rocket Lab demonstration test stand (1956) for public exhibitions. (FROM *PURDUE ENGINEER* [FEBRUARY 1956]; COURTESY OF PURDUE UNIVERSITY LIBRARIES, KARNES ARCHIVES AND SPECIAL COLLECTIONS)

*Guided Missile Engineers*

Mascot Purdue Pete as a guided missile engineer (*left*) and a space cadet (*below*). (FROM *PURDUE ENGINEER* [APRIL 1957]; COURTESY OF PURDUE UNIVERSITY LIBRARIES, KARNES ARCHIVES AND SPECIAL COLLECTIONS)

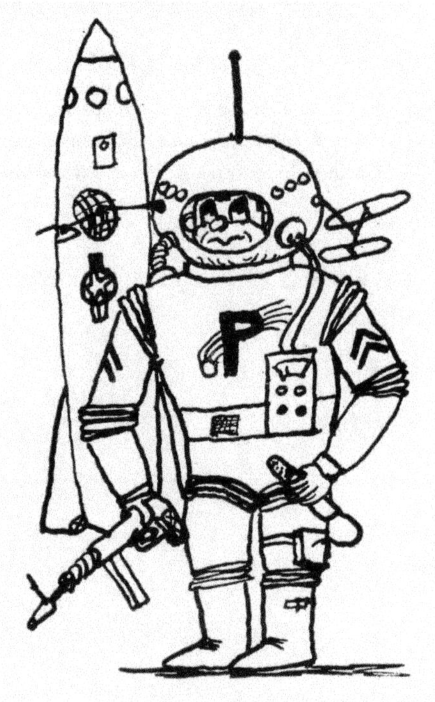

*Space Cadet ROTC*

The NACA-Lewis Rocket Research group inspects a model of the new Rocket Engine Test Facility (1957). *Left to right:* unidentified student intern, former Zucrow students Robert Graham and Elinore Costilow. (REPRODUCED BY PERMISSION FROM NASA GLENN RESEARCH CENTER)

Und Alzo das OBERKOMANDO
DER HUNTSVILLEN WEHRMACHT
GIBT BEKANT:
ZAT

Der azzumption zat this Bloke from "Pneenemunder" (der other Wernher) made mit dem Handschtand und Somersaulten to land on dem Feet der U.S.A.

FESTENRAKETENSTOFFENPROGRAM

is ein bunch of baloney. He even freeked ous when der Thixotropisher stoffen were drippen on him.

But alas und also acha, -- everybody knows that, eef the Hercules, und der Nike (mit dem Zussen umlaut) und der Schprint (to booten) would not get Herr Doctor Zucrows peachen massagen, all kind of raindropfen would be fallen on our head.

- Zo, as Dr. Zucrow always was expounden, -- "It is better to be ein nazion of flautist zen floutist"; we conzekwently kwick as ein wink, got all ze lead out of dem A. B. M., so he can have ein jolly 75 birthday.

Werner von Kirchner

You can see mein English is Okie Dokie, since through mine conshpicuous persevierence combined with mine Samuraien dedication I lost mine Sauerbratten accent.

Werner von Kirchner's parody of the Wernher von Braun German Huntsville group for M. J. Zucrow's seventy-fifth birthday (1974) (FROM MAURICE J. ZUCROW PAPERS; COURTESY OF PURDUE UNIVERSITY LIBRARIES, KARNES ARCHIVES AND SPECIAL COLLECTIONS)

Aerial view of the Jet Propulsion Center (1966). Chaffee Hall is at the top left. The Rocket Lab complex is at the center. The new High Pressure Combustion Lab is at the bottom. (FROM CHARLES M. EHRESMAN PAPERS; COURTESY OF PURDUE UNIVERSITY LIBRARIES, KARNES ARCHIVES AND SPECIAL COLLECTIONS)

Professor Cecil Warner and graduate student Michael Byrd prepare a staged combustion rocket motor test (1968). (FROM THE *BACKGROUNDER* [MAY 1968, PURDUE UNIVERSITY SCHOOL OF ENGINEERING AND MATHEMATICAL SCIENCES]; COURTESY OF PURDUE UNIVERSITY LIBRARIES, KARNES ARCHIVES AND SPECIAL COLLECTIONS)

**"Hope it doesn't blow up." Prof. Reese and graduate student James R. Hoffman have modified a rocket motor for simulation of the combustion in the supersonic combustion ramjet (scramjet).**

Graduate student Joe Hoffman and Professor Bruce Reese inspect a modified rocket motor for supersonic testing. (FROM THE *BACKGROUNDER* [MAY 1968, PURDUE UNIVERSITY SCHOOL OF ENGINEERING AND MATHEMATICAL SCIENCES];COURTESY OF PURDUE UNIVERSITY LIBRARIES, KARNES ARCHIVES AND SPECIAL COLLECTIONS)

Graduate student George Schneiter at work in his office at Chaffee Hall (1967). (FROM CHARLES M. EHRESMAN PAPERS; COURTESY OF PURDUE UNIVERSITY LIBRARIES, KARNES ARCHIVES AND SPECIAL COLLECTIONS)

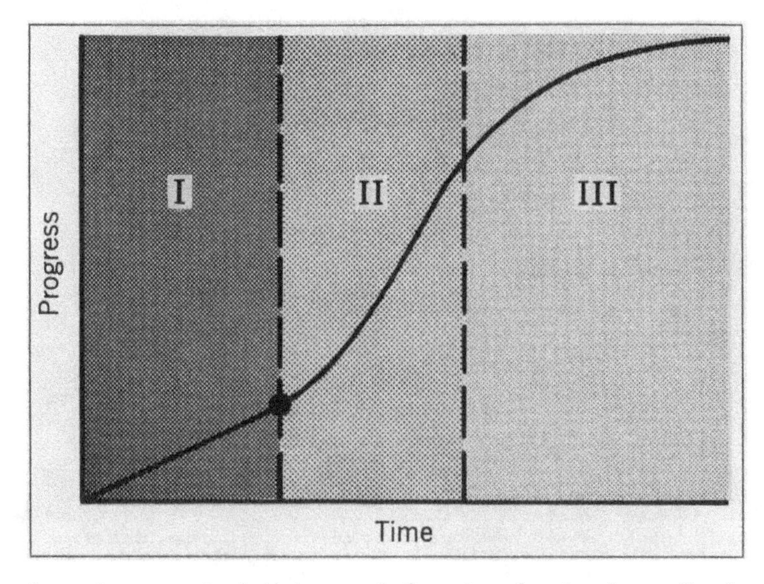


S curve chart representing the ideal career path of an engineer: from the early career "launching pad" (*left*), to mid-career "blast-off and acceleration" (*center*), to late career "orbit" (*right*). "Vital to the success of the flight is the efficiency of the propellant fuel by way of motivation, discipline, participation, and enthusiasm." (FROM *PURDUE ENGINEER* [MAY 1964]; COURTESY OF PURDUE UNIVERSITY LIBRARIES, KARNES ARCHIVES AND SPECIAL COLLECTIONS)

# 7

# ROCKET LAB

## Doc Zucrow and His Boys

MAURICE ZUCROW'S CAREER DREAM WAS THE ROCKET LAB, OR THE R-Lab for those who studied there. In this naming process, Zucrow took a principled stand on the rocket as the power plant for the future. As he wrote at the start of his Purdue career, "Today invention has become a scientific pursuit. Large, well-equipped laboratories have been built where research workers endeavor to uncover the secrets of nature. Invention has become a logical, orderly, scientific occupation, which has for its aim the discovery of new scientific principles and their applications to the benefits and use of man." Zucrow understood his lab as a pioneering site of rocket science: "To operate these laboratories requires competent staffs of men thoroughly trained in the basic and applied sciences. . . . These laboratories are the keys for opening those new frontiers so necessary to the healthy development of our society."[1]

The Rocket Lab was a bridge architecture. It bridged people, in the teamwork between the mentors and students, and between the students themselves. Zucrow called them his "boys," hearkening back to the strict but fun-loving educational regime of the Central Foundation School in his own youth. He and his staff also bridged research science and development engineering, both scientific "imagination" and workplace "capability." As Elliott Katz remembered, "We did theory, and then we did an experiment. Zucrow did not tolerate pure theory. He was a very pragmatic individual. You could babble the theory all you want, you could learn all you want about the theory and fill your head with it, but he was a meat and potatoes person." For Zucrow, the Rocket Lab was a place where "the scientist gets the knowledge (for missiles and spaceflight) and the engineer uses it." Or as he translated into his metaphor of the dry goods store, we must "stock our shelves with trained scientific teams and basic research knowledge."[2] He took Vannevar Bush's "intellectual banks" and Andrey Potter's "stock pile of basic scientific research" one step beyond. This was research *for* development. The Rocket Lab was in the business of "the training of engineering students for the benefit of industry and the national defense departments."[3]

## PURDUE'S ROCKET LAB

Zucrow's travels on the Panel on Propulsion and Fuels (MPF) may have taken him far and wide throughout the US, but he always returned to higher education. He built and expanded the Rocket Lab at a constant pace. In 1951, Zucrow enlarged the original Rocket Laboratory with two more rocket motor test stands and control rooms, a larger machine shop, and more storage and office space. By 1953, given continued "overcrowding" and growing numbers of students, Zucrow built the "Headquarters Building," finally outfitted with normal telephone service and electricity from the main campus. It offered a full suite of administrative offices, a classroom, a drafting room, a classified library, and three test cells and control rooms for the study of "heat transfer, fluid flow and combustion." In 1954, Zucrow added the Gas Turbine Laboratory, including three test cells and control rooms, for experimentation with gas turbines and "associated fundamental studies in fluid flow." It also had a larger machine shop, a welding shop, and an "experimental equipment assembly room." The Rocket Lab was self-sustaining. In 1955, he remodeled and expanded the Headquarters Building with a new wing of tests cells and control rooms for "liquid and gas flow investigations." The full Jet Propulsion Center became, he recalled, a "small but efficient rocket laboratory, and many bright, enthusiastic students did their research therein for the MSME and PhD degrees."[4]

Zucrow did all of this on the small and on the cheap. Purdue and the PRF gave Zucrow only building materials, what he called "bricks and mortar; i.e. buildings of very cheap construction." By keeping the costs of any one job under $55,000, the Physical Plant took on the actual building project without any outside bids or complications. Zucrow creatively negotiated these and other costs with R. B. Stewart based on the lab's many contract overheads. Over the years, Zucrow charged his sponsors overhead rates as low as 20 percent and as high as 47 percent "upon that portion of those salaries paid from research funds." The rate depended on the length and overall nature of the project contract, as well as on the demands of his patrons and on his own bargaining power. With the overhead, Zucrow built the Rocket Lab. But not all of it. He only used about $200,000 of overhead receipts to build the lab's physical plant between 1947 and 1958. That was but one-third of the total overhead proceeds. Purdue kept most of the rest. For operational funding, he scraped together small annual research contracts, on average between $20,000 and $60,000. This was by design, as he believed that "a lot of small contracts give stability." As of 1962, "the dollar value of the research contracts at the Jet Propulsion Center were approximately $300,000 per annum." Zucrow knew the hidden costs in these numbers: it was in "the enormous amount of time in writing proposals, progress reports, and sales visit to the customers."[5] As for outfitting the Rocket Lab, that came cheap too. Harry Solberg called him "the best scrounger of war surplus material Purdue University ever had." Frederick Hovde referred to him as the "foremost university master of that art of levitation called 'bootstrapping.'" He "begged, borrowed, stole from, cajoled or browbeat everybody who had something you needed."[6]

Zucrow's creative accounting and economy measures were rare in this age of Big Science. His lab was a relatively small part of the growing defense establishment. As he scraped together funds for his Jet Propulsion Center, Princeton University moved its premier "jet and rocket propulsion projects" to the massive James Forrestal Research Center near campus, out on Route 1. The university paid $1.5 million for it, with the assistance of the Rockefeller fortune, and opened a campaign to raise $2 million more to rebuild and fund it. Princeton was pricey.[7] Stanford's military contracts tripled in these years. MIT doubled its funding for defense electronics research, also adding a Heat Transfer Laboratory and Fluid Mechanics Laboratory to its massive portfolio. The Jet Propulsion Lab (Caltech) defense funding (for guided missiles) doubled as well between 1950 and 1953, to $11 million. Compared to these sums, Zucrow's annual $300,000 was not a lot of money.[8]

By 1958, one writer described the new Rocket Lab as an "unimposing cluster" accessed by a "gravel road." But it had also graduated "over 100 superlatively-trained engineers, many of whom now direct rocket research projects all over the nation." Zucrow's approach to sponsored research and contracting, and to the very organization and functioning of the lab, made it unique among the few jet propulsion labs around the country, even among the many university labs doing defense work. His approach was educational. The lab was meant to educate young people in jet propulsion. Zucrow accepted no contract that did not include at least one graduate student at its center, either MS or PhD. In this, he was following Purdue traditions, which held that faculty alone governed the research project. The administration was expressly forbidden to "actively pursue research contracts for the purpose of assigning staff members to carry out contract objectives." The PRF prime directive also held fast: "The engineering school's policy is to accept only those Government programs that can be incorporated into a thesis and every experiment is performed by at least one graduate student. That makes Purdue different from any other university in the field."[9] The administration was steadfast in teaching the "active research" of "fundamental problems" and avoiding any academic work or preparations devoted to weapons systems. Zucrow rejected any talk of "guided missile engineering." He educated mechanical engineers, period.[10] This meant that "engineers trained in breadth and depth," he wrote, "are the prime product of our work." He added, "They don't just go to school here; they get their hands dirty in the rocket business." Compared to Princeton and Caltech, "ours is the only one where the students do the work," he said. This meant that most of his graduate students were "on the Government payroll when not in class." Their jobs were "vital to the national security." This also meant trial and error. "Oh, my boys make mistakes—I let them—for they learn," he wrote. One visiting general to the Rocket Lab remembered how when he "pushed the button—the whole thing blew up." Just another lesson in rocket science.[11]

Generals mattered. Guided missiles were a part of everyday life on Purdue's campus through the 1950s. Air Force ROTC cadets featured the Matador and Lark on the front cover of their newsletter, the *Air Force Missile*. They put models of the Matador on display in the Student Union. Purdue was also invested in the defense establishment. At the retirement tribute for Dean Potter in 1953, Maj. Gen. Leslie Simon, chief of Army Ordnance Research and Development, spoke on the role of the engineer as the "sine qua non" of national defense.[12] The Office of Naval Research had its own office in Purdue's Executive Building to administer its twenty-five research project contracts focused on national defense and the buildup of "scientific manpower."[13]

Zucrow was committed to serving the "retaliatory" defense posture of the US throughout his career. He understood the USSR as a technological and existential threat: "We face armament danger. Only if our retaliatory powers are so great no one dares to touch us, will we be safe." Or as he warned, "We're in a scientific and engineering war with the USSR. The country who wins will be the one with the greatest technological ability in both breadth and depth."[14] In this, he saw no contradiction with his progressive political views. He was a Cold War democrat in the Roosevelt–Truman mold. Two Zucrow books joined a growing library of works on guided missiles as centerpieces of Cold War strategy. *Aerodynamics, Propulsion, Structures and Design Practice* (1956) was a volume in Greyson Merrill's Principles of Guided Missile Design series. Zucrow wrote the central section on propulsion. *Propulsion and Propellants* (1959) instructed Army Ordnance engineers on the basic science and practice of guided-missile power plants. No small thing this. The handbook was in use at the Guided Missile School, Fort Bliss, Texas, and at the Ordnance Guided Missile School (Army Ballistic Missile Agency), at Huntsville, Alabama, reaching over 3,550 students. These were the very old haunts of von Braun and the German teams, now transferred to NASA. Zucrow had taken their place.[15]

Zucrow also proved his Cold War credentials in service. He continued to advise the US Army on improvements to the Nike propulsion system in its variations and applications throughout 1950s, not just as an antiaircraft but also as an antiballistic missile (ABM) platform. ABM was controversial. Critics saw it as disrupting the delicate balance of nuclear deterrence. Zucrow valued it as a potentially decisive American strategic advantage. He also passed this responsibility on to Bruce Reese, who spent two years at Redstone Arsenal, rising by 1962 to become chief (civilian) technical director of the Army Nike Zeus program. He had the "responsibility for all systems engineering work, planning, directing, and executing all technical aspects" of the wider antimissile missile and space defense system. He kept "track of the technical aspects of a world-wide missile program stretching from Ascension Island in the South Atlantic to Kwajalein Atoll in the Pacific, and the expenditure of approximately one million dollars per day." He reported directly to defense secretary Robert McNamara and the President's Scientific Advisory Board. The system was challenging: how to identify and target the incoming enemy missile (reentry vehicle, or RV) in time to intercept and destroy it, requiring sophisticated ground equipment for control and guidance, and the speediest and most reliable of propulsion systems. Reese was the technical director when the US made the first so-called "intercept of an RV going into Kwajalein from the west coast," as he remembered. "We got within a kill distance where our nuclear weapon would have destroyed their nuclear weapon and that was quite an achievement."[16]

The military was the primary sponsor at the Rocket Lab. Most of its research contracts were with the Office of Naval Research, the Air Force Office of Scientific Research, and the Army Ballistic Research Laboratory, with a few NACA and Aerojet contracts. The Navy remained the largest financier, funding 86 percent of the lab's contracts between 1953 and 1958.[17] Zucrow enjoyed a cooperative relationship with the military services. They were "farsighted and sympathetic" to the Rocket Lab's independence, he remembered. He had complete freedom of action for research projects, from start to end, with all their subsequent revisions and new pathways. "All of our proposals for research have been unsolicited proposals," he reported back to the US Congress. "I wish to complement every agency with which I have dealt. I never had them dictate to me what they thought I ought to do."[18]

How did the actual Rocket Lab operate? The problem assignments of Project Squid's Phase 7 set the framework for its teaching and research missions. Chuck Ehresman, chronicler of the lab's history, remembered that they became a fascinating series of cascade effects. As he put it, the "ground breaking fundamental research" in high-pressure chamber studies and heat transfer led to several more "technological ground breakers" in rocket propulsion between 1947 and 1955. Ehresman made it all sound so easy and seamless. PRF publicists were less optimistic, more on the mark. Rocket science was hard. The Problem Assignments actually turned into a series of "frustrating vicious circles: as one factor related to ultimate performance shoots toward the theoretical maximum, other factors crop up to decrease efficiency." Each advance in performance in one area created new problems in others. The work of the Rocket Lab was a delicate balancing act between applying high combustion chamber pressure and confronting all of its aggravating effects. The teams had to investigate propellant choices and chamber and nozzle designs; study heat transfer and cooling methods; and solve problems with droplet formations, combustion instability, ignition lag, and rocket engine oscillations.[19]

The Rocket Lab's first experiments centered on Dr. Zucrow's hypothesis that high chamber pressure would improve the performance (thrust) of America's rocket engines. He and Cecil Warner set the stage with their foundational work on the relationship between high combustion chamber pressure and heat transfer, on how the "theoretical rates of heat transfer" would "increase linearly with chamber pressure," based on a five-hundred-pound thrust motor at chamber pressures of 1,500 psi and higher.[20] Graduate student Charles Trent took up this charge, proving that higher chamber pressures meant up to a 20 percent rise in thrust performance. In the approach of the lab, one of inheritance and innovation, a new class of graduate students (D. L. Dynes, Clair Beighley, and Del Robison) assumed this research track, often working in pairs, testing white fuming nitric acid and JP-3 jet fuel in chamber pressures ranging from 300 to 700 psi, and on occasion even up to 2,250 psi.[21]

These technical contributions were small but significant: small improvements in specific impulse and reaction power that improved the throw weight and range of nuclear weapons, or the launch weight of a spaceflight mission. As the PRF boasted, a "performance increase of 10% can mean a range increase of 30%." This meant that "an increase of as much as 30% in performance may be realized by operating at 1,500 to 2,000 psi as compared with operation at 300–500 psi." The Navy circulated these results via classified lists throughout the industry and military, with the effect that by 1955 "all of the large missile motors" were using a "much higher chamber pressure." Zucrow's results built thrust and built confidence. In the case of the Viking sounding rocket, its early motors operated at 400 psi. The improved Viking 11 rocket motor operated at nearly 1,000 psi.[22]

Building on this work, the Rocket Lab also diversified, adding the Air Force and NACA to its Project Squid portfolios. Several other initial tracks studied the chemical reactions, heat transfer values, "fluid friction," and optimal performances of a variety rocket engine fuels and oxidizers at different chamber pressures. The lab created enthalpy–entropy diagrams and manuals on experimental procedures and calculations for national distribution. Students conducted fundamental studies of rocket propellant thermochemistry and thermodynamics. They researched the "operating characteristics of various rocket motor propellant combinations when used in gas generators." They made detailed studies of injection methods and back pressures on the ignition of rocket propellants. They also created altogether novel apparatuses and devices to register precise readings for these experiments. All of these findings were "disseminated throughout the rocket industry."[23]

The higher chamber pressures and higher temperatures (up to 7,000°F) created extreme heat loads against the rocket motor's inner walls. Traditional methods of regenerative cooling simply did not work against such heat loads, requiring experimentation with new methods, as for example internal film cooling, defined as a film "between the wall and the reacting hot gases," with various platforms, thicknesses, and uniformities. The film was at times prone to form droplets that escaped into the propellant stream, requiring perforated grids to prevent droplet formation. One early experiment involved the "cooling of rocket-motor nozzles by transpiration cooling through parallel disks." The Rocket Lab's mechanics built a rocket "constructed of a pile of circular ring-disks between which is injected the coolant. The pile of disks is being held together mechanically and serves as the combustion chamber of the rocket." Cecil Warner took up the challenge of film cooling, guided by the high-pressure studies defining it, leading to the work of Eldon "Al" Knuth, Alfred "Al" Graham, Donald "Don" Emmons, and J. P. "Jack" Sellers on the role of liquid film cooling in reducing heat.[24]

A small disaster opened up another research track, as Ehresman remembered: "An explosion resulted from an ignition delay, thought to be caused by the low temperature of the hypergolic fluid used to start the rocket motor combustion process." Ignition delay was an old problem that acquired a new urgency. Hence a new research track for the Investigation of Ignition Lag of Spontaneously Ignitable Propellants, initiated by Stanley Gunn. Project Squid was offered liberal contracts for studies of the "variables which affect the time lag, short as it is, between the entry, evaporation and ultimate reaction of various propellants." Or as Gunn said, "the ignition lags ranged from about 10 to about 400 milliseconds, depending upon such variables as temperature, acid composition, fuel compositions, and metallic additives." Precision was the key: he had to design and build a test apparatus to learn "the minimum quantity" of RFNA and aniline propellant necessary to avoid "affecting the ignition time." He crafted an "entire electronic ignition lag timer" and even "a device for calibrating the ignition lag timer." As Ehresman recalled, Gunn's "open cup work and finally the development of the impingement of injected streams of propellants became the universal method to determine ignition delay with self-igniting propellants."[25]

These ignition delays were one possible cause of another propulsion dilemma in rocket engines: oscillations or "pressure fluctuations," sometimes leading to "structural damage," "mechanical failure," and resulting explosions. The Rocket Lab set out to search for "a design technique to avoid rapid changes in combustion pressure" brought on by the mixing of the fuels and the rise in temperatures. To avoid "trial and error" dead ends and establish surer and safer general parameters, the teams tested and calculated a storehouse of variables: kinds of fuels and injectors, sizes of combustion chamber, and the presence of carbon deposits. James Bottorff built his own test cell to isolate correct procedures. In 1957, the Rocket Lab began a sustained research program to study the related and recurring problem of "'screaming' high frequency sounds" being emitted from the rocket engines. Dr. Zucrow assigned teams to examine its possible causes in a variety of propellant combinations. John Robert "Bob" Osborn, like Reese a graduate student soon hired on as an assistant professor, took over this track with his own new corps of graduate students, including J. Michael Murphy. Osborn also helped build a new program in solid-propellant studies in close cooperation with the Thiokol Chemical Corporation in Huntsville, Alabama, where he had interned as a junior engineer (June 1950 to September 1951) and senior engineer (the summers of 1957, 1958, and 1959). Osborn later won the AIAA Wyld Propulsion Award (1995) for his role in the development of the solid rocket boosters for the space shuttle.[26]

## DOC ZUCROW'S LEGACIES

The Rocket Lab reproduced these many studies in several formats. As scripted briefings and updates for current sponsors. As local "interim and quarterly reports" of the Rocket Lab's technical findings. As Project Squid reports (quarterly and annually) via classified distribution lists. As project proposals for future sponsored research. And most importantly, as master's theses, PhD dissertations, and scholarly articles. The whole point was not only to capture data about jet propulsion but to record and archive it, interpret and confirm it, write and revise it, present and debate it, document and share it. The Rocket Lab created what Bruno Latour has called the great "inscriptions" of normal laboratory science.[27] Most of this work was not classified. Some of it was restricted or confidential due to the sources that the students used. Only a few were secret. All were eventually declassified, some within a few years, by Zucrow himself. The Rocket Lab archived these studies on site until 1973, when Purdue transferred most of the lab's official reports, along with most of Zucrow's library, to the basement library stacks of the new Potter Hall. There, for the next forty years, curious engineering students with a little time on their hands could leaf through the legacies of the Rocket Lab.

Zucrow had some fun with his first interim reports of the lab, their covers graced by the playful swoosh of a rocket in flight, propulsion contrail swirling below. It was none other than the Caltech WAC Corporal, of which Zucrow was so obviously proud, as one of the designers of its Aerojet power plant. As a second stage to a V-2, now called Project Bumper, the WAC Corporal was the first rocket to reach outer space, speeding 5,150 mph to 244 miles altitude, on 24 February 1949. If we add the other booster rocket for the WAC Corporal, the Tiny Tim solid rocket, built under Hovde's guidance at OSRD, the Corporal was a historic achievement of the Zucrow–Hovde rocket team now at Purdue University. The students at the Rocket Lab were part of something big.

Zucrow and Warner often wrote and published their findings as coauthors with their students. From the start, their accent was on collegiality and team effort. The students themselves began their project assignments as individuals or teams, which they then passed on to new generations of students, who "followed the predecessor's path," as if in a research relay race. This was not unlike the Jewish *havruta* partnership and companionship method that Zucrow's father, Solomon, had taught in his home. Older and established students became "phase leaders" for the new classes joining the work of the lab and the phase. Each class advanced the state of the art with new research dimensions. In time, each also cited the earlier achievements of their ongoing research projects. The lab was building upon its own precedents.[28]

Zucrow was thereby preparing his students for the project management system in their coming professions in the military or industry. This meant how to choose a topic, study its history and literature, prepare the elements of experimentation and research, design the apparatuses and methods, work with veteran and novice classmates on the project, organize and write up the results, and prepare reports on them. Robert Glick remembered learning all of this from an older PhD student, Helmut Wolf, who told him in no uncertain terms: "What you should do is sit down and seriously think about your project, what the problems are, what the challenges are, how you can get at them. . . . I mean really sit down and think about it and analyze it." For Glick, this was the stepwise technique in engineering research, "Step 1. Step 2. . . . I consider that moment, the absolutely crucial moment in my whole education at the Rocket Lab." J. Michael Murphy turned this Purdue stepwise method into something more ambitious, what he called a "*straight-ahead* approach to hardware projects," by which he meant to "initiate a success-oriented approach to an important system, define the technology that nature will allow, and not wait for approval before taking the critical initial steps."[29]

Zucrow's students had each other in terms of the collegiality of the phases. They also depended on the staff of the Rocket Lab, especially the technicians, initially Gordon Hurst and Dallas West, to build their test platforms and apparatuses. The secretaries typed their reports and provided office support. But they were otherwise on their own. Zucrow and Warner did not supervise them. "You didn't work with him. You produced results or not. That's the way it was," said Robert Strickler. Each graduate also owed something back to Zucrow. Each "had to make some contribution to the lab." Walter Hesse and Edward Govignon had set this trend from the start in their master's theses, when they helped build and outfit the lab. These contributions took many forms, as for example new test stands or equipment, or published articles. Murphy built a storage magazine for the new solid propellants of the early 1960s. Del Robison and George Schneiter wrote detailed surveys of the Rocket Lab's work for national release.[30]

What kind of a teacher was Doc Zucrow? Demanding. Paul Petty remembered him as a strict disciplinarian. In the classroom, he could fill three walls of a blackboard with equations by the end of an hour of class time. Students even competed with each other to find his rare mistakes. Others who dozed off or lost attention bore his wrath, as in the case of one young man who gathered his belongings to leave a gas dynamics course just a bit too early. "Hold on young man," said Zucrow, "I have one more pearl of wisdom to throw out at you." As for the rest of the class, on seeing this, "nobody else moved."[31] His graduate students remembered one evening exam in Steam and Gas Turbines (ME 139) in his and their first semester, fall of 1946. It lasted three hours, until they all agreed, teacher and students, that seven out of the impossible nine problems would equal 100 percent. Students remembered an assignment he gave them over Christmas break that ruined their vacation, only to discover on the first day back at class that he meant it as a thought piece, nothing more. These were ordeals, recalled Eldon Knuth in a fun Zucrow accent, well before Apollo, that took these students "240,000 miles to the moon as Zu Crow flies."[32]

Zucrow set a constant pace of edits and rewrites for student work. Accuracy was the key. Not just in measurements and calculations, to replicate and confirm data for the records, but also in "the art of writing clearly and forcibly," as he preached, so as to allow the reader "the minimum expenditure of time and effort." He read and reread students' theses and dissertations and their reports to research sponsors, filling them with "red ink marks and pungent commentary," correcting their grammar with "the King's English," as one student remembered. "When you got through working under Maury Zucrow," said Elliott Katz, "you knew how to write, you knew how to present, and you knew how to perform, and you knew how to report." Zucrow also scheduled multiple dry runs for their upcoming presentations to military and industrial sponsors, or for the formal "Research Reviews" for visiting guests and the Purdue administration, including the deans of Engineering and President Hovde. These could last up to an hour, filled with elaborate "poster-board graphics" and even movies, and were invaluable experiences in how to sell an actual research proposal or its results. They taught the graduates how to become efficient and smart managers, how to communicate in a style of writing, speaking, and presenting that made sense. Mel L'Ecuyer remembered it all as "terrifying at the time."[33]

Always demanding of the students' "best effort," Zucrow also taught compassion. His kind and honest advice usually encompassed just a few words, as in one of his favorite flight expressions: "There's no lift without drag." He and his wife, Lillian, considered his students as "part of our family," worthy of life's "intimate associations." As he once wrote, "In my humble opinion real happiness is in large part based on a sincere interest in people, on a desire to help them when necessary, to share their successes, and to share their sorrows." It all meant, by the testimonies of the graduates, that the "boys" were bound by a "pact," a deep sense of loyalty to their mentor and to each other.[34] Many of his students, men and women both, spoke openly of their love for Zucrow, a love given in return, or of his sense of humor and "humanism," both in ideas and spirit. These sentiments were inseparable from the engineering fundamentals, from the very content of his jet propulsion lectures. His students called the Rocket Lab "Zucrow University" and themselves the "Zucrowites." This meant that even some of the wives considered themselves as his "boys." He and Lillian invited them for suppers and conversations, took time to know their children and shower them with attention and toys. He arranged summer internships in order to help along courtships and romances. He gave Walter Hesse a week off from studies for his honeymoon but made him take a scheduled quiz on his return.[35]

Zucrow had power to dispense at the Rocket Lab. In 1956, a doctoral student preparing for the PhD, as a full-time research assistant, earned $350 a month, with six free credit hours a semester, and the opportunity to teach once they earned a master's degree.[36] Graduate student salaries were competitive with the best jet propulsion university programs in the country. Zucrow also recommended students for postgraduate internships and jobs. He got them deferments from military service, or made sure they fulfilled their commitments doing jet propulsion internships instead. Bruce Reese, a lieutenant junior grade in the Naval Reserve, did his two weeks of training in 1953 at the Research and Development Board, Naval Research Reserve Unit, Washington, DC. George Schneiter completed his obligations to the US Army at Aberdeen Proving Grounds and Redstone Arsenal.

The Rocket Lab was an interesting place to visit. Cecil Warner's son, James, remembered it as a wonderland. As a youngster, he learned to drive the family car down the long gravel road and played there on weekends while his father was working. For him, it was also the only place in town where he could buy the exotic soda, Dr. Pepper, in a vending machine. He also used the machine shop to build a go-cart and grind out a bowie knife. It was also an interesting place to work. Zucrow was somewhat eccentric. Students remember him hurrying back and forth, oblivious to the world around him. "Dr. Zucrow is not happy unless he is busy," wrote one, "and he is happy most of the time." He genuinely enjoyed fishing. As he once joked, "I don't think about anything else but the fish." A doctor once agreed, prescribing a fishing trip to Canada for rest and relaxation. Zucrow tried to take two briefcases along with the fishing rod and tackle box.[37]

As to Doc's "boys," the graduate students enjoyed a rare sense of camaraderie and fun. Several Zucrow graduate students and faculty drove fast cars. Tom Carpenter had a 1955 MG TF roadster. George Schneiter first drove a Triumph TR4 and then a 1966 Corvette. Stanley Gunn had one too. Bob Osborn owned a series of Corvettes, competing in autocrosses and active in the National Council of Corvette Clubs. Some of them also "used alcohol for other things besides rocket propellants," recalled J. Michael Murphy. Graduate students tapped into the reserves of 100-proof alcohol, provisioned in gallon jugs from Purdue's chemical stores. "Therefore," remembered Robert Glick, "'cokes' at the Gas Turbine Lab were often energized." One group also "fabricated a 'bar' for their apartment from junk materials at the Rocket Lab." To lighten the pressures of their courses and lab experiments, they also engaged in a variety of pranks and tricks. They sent condolence cards to fellow students about to get the PhD dissertation back from Dr. Zucrow, filled with corrections in red ink, as if the work was dead on arrival. They locked their professors and classmates in the darkroom by screwing the door shut. They set out vats of smoking liquid nitrogen to scare the staff that the building was on fire. Even Doc Zucrow was not immune. His cigar once exploded from a tiny charge planted by some pranksters, only to be surprised by another one as he finished smoking it.[38] As with any small and intense professional setting, there were also cliques and personality clashes that became contentious, disagreements that turned into grudges, not soon forgiven.

Social affairs were the rule of the day. Mechanical Engineering staged an annual Christmas ball in the Purdue Memorial Union. For their celebrations, the Rocket Lab teams frequented Sarge Biltz's restaurant and bar at the Lafayette bypass intersecting

Routes 52 and 25. Against Dr. Zucrow's wishes, his staff and students often visited the Skyroom restaurant at Purdue Airport for coffee and meals. It was just a short drive away, or walk by hopping the fence. Zucrow may have been worried about Soviet spies, or more ominously his rivals and enemies in Aeronautical Engineering. But the food and companionship were irresistible. There his staff and students could rest and energize, eat a donut and have a smoke. As Warner put it: "Many ideas upon ways of improving research techniques were exchanged. Numerous baffling problems were solved." The Skyroom was also a place for rocket science.[39]

Zucrow became a model of success for his students, following his principles of hard work and fundamentals, and the admonition "Go thou and do likewise."[40] He did it by living the values of honesty and loyalty. As Chuck Ehresman said, "He was able to inspire action in people, often by example." His teaching and mentorship helped them get the best jobs in their fields. This he did with resounding success. Hovde offered high praise: You "gave more trained men to and did more to advance the art and science of jet propulsion than any other university laboratory in the entire world."[41] Zucrow's students enjoyed some remarkable careers, as confirmed in appendices 1 and 2. They graduated to either direct or staff the propulsion labs at Wright-Patterson and Edwards Air Force Bases. They ran major defense and spacecraft programs for the US military. They went to work for NASA, for the aerospace industries, for the government. They taught in universities. Each of these careers has a longer and bigger untold story all its own. Yet each also reveals the transformative power of a Rocket Lab education, Purdue as a crossroads of careers.

These kinds of job placements were not unique to Purdue, although Zucrow was quite good at facilitating them. Any of the major engineering programs around the country could claim them too. In the twenty years between 1946 and 1966, aerospace careers in either missiles or rockets were abundant, alluring, and upbeat. Science and engineering were prospective, like the S curve of progress, and even mirroring the countdown, launch, and orbit of rockets. Zucrow confirmed it: "Because science knows no bounds, the field of the professional engineer is a changing but unlimited one." The engineer was, by definition, one of the "forward-looking members of society."[42] The aerospace industry offered America's graduating engineers the best of both worlds: outer space and upward mobility. Advertising metaphors proved it. Bendix Aviation Corporation asked engineering graduates to "measure your future" by the rocket's parabola trajectory through space, fueled by their own "talent and ambition." McDonnell Aircraft at St. Louis built a whole multi-ad campaign around this peculiar career trajectory: from launch to acceleration, and on to orbit and recovery: "Your career, like a missile, must first get off the ground," and be sure to choose McDonnell so as not to self-"DESTRUCT." The Bell, Boeing, and Douglas companies all framed their advertisements with parabolic arcs into the heavens, a symbol of the vast potential of a well-chosen career path, promises to come. Or take the Martin Company ad to future engineers, a play on the "X" (unknown) factor: "You are 'Project X,'" it teased. You are your own "beginning engineer's dream," just like the "first man-made satellite," presumably to "be launched by Martin" within a few months. Join us for "the greatest engineering project of all time—the conquest of space."[43]

For the most part, Zucrow's Rocket Lab, like these other sites, was a preserve of masculine privilege. Rocketry and spaceflight enthusiasts were men, as a rule. Their shared interests and talents both bonded and isolated them. "Stag" rules often applied. Women were not allowed, so the prejudice went, because they distracted the men from demanding mathematical equations, sophisticated thinking, and safety routines. America's engineers, numbering just over half a million strong, were classic organization men, white collar commuters enjoying the suburban lifestyle and the challenge of nine-to-five careers.[44] Women were present in these worlds, if in subordinate roles. They reappeared in rocketry publications as idols, in quick cartoons and sketches as attractive women on the minds of busy engineers. Purdue administrators tended to see women as "girls," as for example the wives of GI Bill veterans, best suited for secretarial use in the office. Many Rocket Lab theses and dissertations thanked either the female secretaries or the wives for typing the manuscripts, and sometimes even for their sacrifices in enabling the husband to complete the degree. Women were wives, but also secretaries, historians, and engineers. They ran the offices and archives of the OSRD and of the Purdue Rocket Lab. In the case of Ethel De Haven and Mary Self, they wrote superb, interpretive histories for the US Air Force.[45]

Women engineers broke through these constraints and ranks, rising to academic achievements and career success. The barriers were formidable. Between 1897 and 1946, only thirty-two women graduated with engineering degrees from Purdue. By 1966, 99 percent of engineering alumni were still males (65 percent from the Midwest). One break came in 1947, with the foundation of the Purdue section of the Society of Women Engineers, the Gamma chapter of Pi Omicron.[46] In Zucrow's orbit, there were a few women, including Mary Lou Nave, a secretary in the School of Aeronautics, whose husband was studying for his electrical engineering degree, and staff member Eileen Dimmick. Betty Jane Yost (BSEE 1949) and Florine Klatt Cain (BSAE 1951) were active students and researchers at the Rocket Lab. Yost moved to Dayton, Ohio, for a career with the General Motors Component Division as an electrical engineer. Cain's career included positions at Lockheed's Missile Division and Rockwell's Strategic Systems Division, working on propulsion and design elements of the X-7 test ramjet and the Minuteman missile.

Eleanor Louise Costilow (MSAE 1949), known as Ricky, was an active member of the Institute of the Aeronautical Sciences and American Rocket Society at Purdue. At the Rocket Lab, she specialized in airfoils and exhaust nozzle designs, building a career at the NACA Lewis Flight Propulsion Laboratory, where she became a rocket power plant engineer by 1958. She graduated from and entered into a "man's world." But she was also joining a NACA that offered pioneering women engineers a place for research, engagement, and professional advancement. Virginia E. Morrell was at Lewis, a specialist on mathematical models to define and evaluate a rocket's thrust, as well as on combustion-chamber pressure and nozzle designs (1950–1951). So was Dorothy Simon, an expert in the properties and speeds of hydrocarbon flames, as well as the phenomenon of "quenching" in aircraft and rocket engines. Simon eventually went on to a long career at the Avco Corporation, where she helped to design new materials to withstand the heat and flames for the reentry shields of spacecraft. At NACA, all of these women engineers had secret or top secret clearances.

## THE RISE OF ENGINEERING SCIENCE

Into the early 1950s, Purdue Engineering remained in the Andrey Potter mold, dedicated to his one-third rule: one-third of coursework for the fundamentals in science and mathematics, one-third of courses for specialized engineering, and one-third for the liberal arts. Potter institutionalized this with a brief "liberalization" of the engineering curriculum in 1950, requiring eight to ten courses in either language, literature, history, government, economics, or psychology. He named engineering as a "liberal education," part of transforming Purdue from a technical university into more of a modern university. Engineers had to be specialists, to be sure, but also needed to learn the "best" of art and music, literature, and the social sciences. They needed to cultivate the "active life" of the mind and express a "sensibility to human relationships and good citizenship." Their vocation was not to injure but to "elevate humanity."[47] These efforts were part of a national trend, what the American Society for Engineering Education called the creation of "a social-humanistic stem" within "engineering education." This was not the monolithic STEM (science, technology, engineering, and mathematics) framework that we know today, but a more dynamic and open-ended engineering education that included the liberal arts as part of its very existence. Not as a branch but as a stem, coequal to the "scientific technological stem."[48]

During the Second World War, Potter and Warren Howland fused these liberalization efforts for engineers with a unique program for women students. They called it "Liberal Science" (LS), largely because, by an informal arrangement, Purdue only offered the bachelor of science (BS) degree, while the other major public campus in the state, Indiana University, offered the bachelor of arts (BA). Purdue's liberal science offered women, and eventually men students, a series of science and liberal arts courses along the pathway to the BS degree. The 190 women who graduated from the program (1939–1949) were mostly preparing for teaching or business careers. They took cloistered science and technology survey courses together in their first two years, along with courses in English, history, economics, government, journalism, business, philosophy, and the social sciences. They also shared a course called Science and Technology in Present-Day Life (LS 50), which included a section on the history and philosophy of science. The leading physicist (and feminist) on campus, Karl Lark-Horowitz, taught it to great acclaim. Purdue was a true innovator here. Liberal science included one of the first university offerings in science, technology, and society (STS) studies.[49]

As juniors and seniors, the women (the "co-eds") of liberal science eventually joined the male students (the "eds") in open upper-level courses. By 1949, Purdue opened all liberal science courses to the men. The veterans on the GI Bill lobbied for this, partly to mix with the opposite sex, partly to broaden their education with new courses in international relations, world economics, and the Russian language. Howland established an inter-curriculum colloquium for better communication between the humanities and engineering. In fact, he was anticipating how to close C. P. Snow's famous gap between the two cultures: the diffuse and disparate fields of the humanities, as opposed to the exact and unitary culture of science and engineering. Purdue was already well-versed in how they could talk with each other.[50] President Hovde went even further, promoting new units and resources, consolidating them by 1964 in the School of Humanities, Social Science, and Education. Purdue's "deeply scientific community" demanded such vibrant and challenging "intellectual pursuits in the humanities."[51]

Zucrow followed these new approaches with an even more spacious view. At the start of his Purdue teaching career, he argued that the successful engineering graduate required a "broad" education in the sciences and engineering, meaning all of "research, development, experimentation, test, design, production, management, sales, services." But the engineer was also an "applied scientist," and to become a truly "useful member of society," required "more of the humanities" than ever before. "For one lives neither by bread alone or by technical competence alone," he said. This required self-education and a whole lifetime of reading dedicated to the "general knowledge so necessary to cooperative effort." The engineer needed to understand "those forces, scientific and social, that are continually exerting a modifying influence upon his function."[52] Without the humanities, the engineer lived simply by "self-liquidating" project goals. One project followed upon another, defined mostly by information. Only the liberal arts and social sciences could teach the broader capacity for judgment. According to Zucrow, "The primary ideal of education is to waken, within the individual, the spark that encourages him to be a student throughout his or her life. . . . The spark must be aroused that makes him alert and sensitive to human values, to his relationship to other people and their relationship to him."[53]

Zucrow's first great professional crisis came not from the humanities but from within engineering. His successes with the Rocket Lab put him in an awkward position by 1952. At his initial hire, his time was split 70 percent Mechanical Engineering

(ME), 30 percent Aeronautical Engineering (AE). It was a balance of forces that made sense right after the war, when gas turbines and jet propulsion were brand new. It also made sense in that ME was the larger and more established school, AE smaller and only a few years old. It especially made sense as a kind of gentleman's agreement, solidified between Harry Solberg, Elmer Bruhn, and Maurice Zucrow, old-timers and Purdue men who understood and worked with each other well. The trouble started in 1950, when Bruhn stepped down as dean of AE and returned to teaching. Purdue replaced him with a rising star in the field: Milton U. Clauser, an "alert" and "dynamic" thirty-seven-year-old. He was the twin brother of Francis, who was head of AE at Johns Hopkins University. Both had originally planned to attend Purdue but ended up at Caltech instead (with doctorates in 1937), and both then went on to work jointly at Douglas Aircraft Corporation. Milton Clauser was like a triple play for Purdue: a specialist in stress analysis, power plant design, and electronics and instrumentation. He was an energetic trendsetter, a proponent of systems engineering and engineering science. At a time when Purdue engineering students were still required to take summer workshops in sand casting, metalworking, and welding, he advocated an end to this shop culture campus. The scientific/technical courses mattered more. "The day of the handbook engineer is past" and the dawn of a new appreciation of the "fundamentals" is here, he proclaimed.[54]

Clauser came in the fall of 1950 as head of the School of Aeronautics. He was nominally Zucrow's boss, or at least 30 percent of him. Purdue press releases framed them as equals. Clauser was hired to "head instructional work in aerodynamics and mechanical equipment." Bruhn was to direct work in airplane structures. Zucrow to "head instructional work in jet propulsion." Herein lay one source of the dispute: a new personality in the mix, a change-maker. Clauser had come to Purdue, so he once said, with big dreams. Once he got there, he was "amazed" at the potential. "It is certainly the great land of unexploited opportunities," so he wrote to his friend and fellow Caltech graduate Harold de Groff, who soon joined the faculty. Purdue was a place of wasted resources and so much unused talent, wrote Clauser. Students were "as good as any anywhere" but not challenged by their AE professors. Thus neglected, they often left Purdue to do their graduate work elsewhere. All "except Zucrow who has built up quite a group of researchers around him. (He is a very nice fellow incidentally)." Clauser significantly mentioned the Gas Turbine Laboratory in the Quonset hut at the airport, just southwest of Hangar 3, superbly outfitted with six Allison engines "driving twelve primary stage compressors." But it was in disrepair; along with dormant "diesel engines, superchargers, air compressors, tanks, electric motors."[55]

Zucrow was mostly to blame for this neglect. He had focused on his Rocket Lab at the expense of his turbojet responsibilities. He admitted this in one of his formal reports to the university. The Office of Naval Research (ONR) was funding "radial-flow turbine" studies ($56,700 worth) in what he called his "Air-Burning Engine Laboratory" at the Purdue Airport. But as his progress with rocketry sped up, his work on "gas turbines, turbojet and ramjet engines" slowed. This was in part because such research was "extremely costly." Zucrow also needed a road to connect the Rocket Lab to the airport itself. They were isolated from each other by a lack of direct access. In sum, he reported, "Unfortunately, the writer has been unable to devote the requisite time to the planning of a sound and logical program for developing such a laboratory."[56]

Clauser must have offended Zucrow at some point when he attempted, with good reason, to assume authority over jet propulsion, or at least the Gas Turbine Laboratory at the airport. This dispute was worsened by a controversy over clearance authority, stemming from Zucrow's secret work for the Guided Missiles Committee (GMC). Clauser had demanded the right to see all official and classified correspondence of the Rocket Lab. He shared a top secret and Q clearance with Zucrow. They were both embedded in the GMC. But now Clauser was leveraging power over Zucrow's small "empire" at Purdue, as one student recalled. On this Zucrow did not yield. He was "a man who controlled everything." He also had the full backing of President Hovde, which empowered him in ME, but also made him "quite a few enemies" on campus.[57]

The controversy peaked on 4 April 1952 after Zucrow decided to devote himself to ME full time, 100 percent. Clauser addressed the move at a formal staff meeting of the faculty of the School of Aeronautics by discussing what he called "the problem of the transfer of the Professor of Gas Turbines and Jet Propulsion to the School of Mechanical Engineering." He never mentioned Zucrow by name. Clauser's faculty were disturbed by the loss of Zucrow and his teams, and by the damage to "the future reputation of the School of Aeronautics." He assured them that the transfer was purely for "administrative coordination." But it clearly was not. Clauser admitted that he and his teams now faced an immense "challenge" in meeting the "rapidly increasing complexity of the problems in aeronautics."[58]

Both sides very quickly formed ranks. Thirty-two of Zucrow's colleagues and students signed a proclamation on 4 April, honoring his "scholarly attitude, his friendliness and his high ideals," along with his "perceptive guidance and inspired teaching in the art of engineering and scientific research." His graduate students threw a "surprise dinner," replete with spoof "progress reports," an honorary scroll, and the gift of a large briefcase.[59] Zucrow's full-time turn to ME made sense from his own point of view. It had been his academic home since his graduate work in the 1920s. He thought of himself unequivocally as a mechanical engineer. ME also remained (followed by Chemical and Civil Engineering) the strongest, richest, and most influential of all the schools of engineering at Purdue, the one with the most students and faculty, the one with the most active and diverse research.[60] Gas turbines might have belonged to it, yet jet propulsion without a doubt belonged in the School of Aeronautics. Its loss of Zucrow meant that Purdue lost the opportunity to build a stronger and more reputable national program in AE. For the next twenty years, Zucrow and ME monopolized graduate education in jet propulsion.[61] This became a very public breach. The transfer was also personal and political, a sore point that deeply divided the two schools for many years to come. AE did not soon forgive Zucrow. Its faculty wrote him and the lab out of their departmental histories.[62]

Clauser stayed on at Purdue for another academic year, but the break with Zucrow had diminished his authority, and he resigned abruptly in September of 1954. One further decisive event might have been the fire in March of 1953 at the Gas Turbine Laboratory in the airport Quonset hut. The fire started at the "fuel research project," when gasoline spilled from a fuel line and ignited under a nearby refrigerator. Two hundred onlookers spent part of the sunny spring day watching the fire.[63] Clauser's main laboratory was lost. He went on to build an impressive aerospace career, first as director of the Aeronautical Research Laboratory at the Ramo-Wooldridge Corporation, then working on Air Force systems engineering for ballistic missiles, later directing the Lincoln Laboratory at MIT, and holding other leading positions in research and academia. Clauser may have lost Purdue, but he gained a leading national reputation. Purdue, in turn, atoned to Clauser at the spring commencement of 1967, when it awarded him an honorary doctorate in AE, the year after Zucrow retired.

Purdue replaced Clauser with Harold de Groff, who proceeded to complete a series of radical reforms for engineering science, what the Purdue Research Foundation soon called a "happy medium between basic science and applied engineering." These began when the Faculty Committee on Freshman Engineering established a new "theoretical" curriculum for first-year engineering students to be introduced in 1954–1955. There were new requirements for college preparatory courses in high school: in algebra, plane geometry, trigonometry, physics, and chemistry. At Purdue, machine shop and drawing courses were waylaid in exchange for basic science and math courses for first-year students. The new program placed a stress on the study of the "fundamental" principles in mathematics, science, and physics, all in preparation for a series of more advanced upper-level technical courses. In exchange for the added courses, the new curriculum dispensed with broad required studies in the liberal arts, focusing instead on the simple sequence of one course in English composition (English 101) and one in speech communication (Speech 114), with no other required nontechnical electives in literature, history, and government. Purdue Engineering wanted its students to acquire only the technical skills of good writing and speaking. The rest of a "general education" (Gen Ed) was up to them. This was quite a turn away from Potter's liberalization from just a few years before.[64]

De Groff also helped establish a whole new Division of Engineering Sciences at Purdue, centered on courses to help prepare engineers for the advanced R&D fields:

Advanced Calculus, Fluid Mechanics, Thermodynamics, Heat Transfer, Properties of Materials, Mechanics of Solids, and Electrical Theory. Purdue was the first major engineering school to be accredited with such a new such division, part of a rising trend, and well-financed by industrial monies, focused on the "basic physical laws and their combinations," with intense studies of the "pure sciences," numerical analysis, and theory. This was to prepare students to "solve the engineering problems of the future."[65] One of Zucrow's students, H. Doyle Thompson, was in the new Division of Engineering Sciences before he joined the Rocket Lab. Though a math whiz, he struggled in Advanced Calculus. "I almost failed out the first summer." As a small elite on campus, the division had its detractors. The other engineering students liked to make fun of it. One lampoon claimed to have found a midterm on "the steps of the Engineering Sciences Building." It was a fifty-minute exam to answer seven out of seven questions. One was to "give a brief critical analysis of Einstein's Unified Field Theory." Another asked, "Using the score cards provided, derive a valid expression for President Eisenhower's golf scores as functions of sunspot activity. You may neglect second-order differentials." The final question: "Describe the universe and give three examples."[66]

The retooling toward engineering science did not come easy. Zucrow mounted a critique, blaming it for dropping enrollments in the late 1950s. In his view, engineering science was simply too demanding and small-minded, "teaching things in little classified packages." He advocated a return to the Potter way, joining science and engineering, the humanities and practical workshop skills, for the "broader point of view." He wanted less theory, more practice. This meant teaching advanced mathematics not to "be misled by mathematical treatments of unreal problems," but teaching it for "engineering problems," for "design and development."[67]

As chair of the Mechanical Engineering Design Committee in 1959, Zucrow helped to modernize the curriculum with a design and lab experience. The first two years were engineering basics (fluid mechanics, thermodynamics, heat flow, automatic controls, electronics and circuitry, higher math, and computer technology). The third and fourth years were laboratory experiences in "electronics and electromagnetic fields, high speed instrumentation, computation, automatic controls, and systems design." Zucrow modeled these labs on the "project" or "systems approach." They joined students with interdisciplinary teams of professors to confront "new problems to be solved" without handbooks or textbooks, but with real-world experience. This meant learning "not just one correct solution but perhaps many acceptable solutions from which one is chosen as optimum under given technical and economic constraints." After a ten-year low of 590 enrollments in 1960, Purdue ME rebounded to 767 by 1962. Engineering programs around the country also took notice. Over the next few years, the Zucrow approach of joining the fundamentals of engineering science with problem-solving and practical learning became a national norm.[68]

AE also had its own star, professor Joseph "Piston Joe" Liston (ME 1930, PhD ME 1935), who also applied creative laboratory methods to challenge his students. In his Advanced Design course, he broke them into teams to build optimal power plants for aircraft or missiles. They presented formal defenses in front of visiting engineers and faculty. Liston challenged his students to "examine and improve the engines of today." They learned by inventing, or as one AE senior remembered, "old scraps of wood, spare pieces of plastic, and cardboard were transformed into the products of imagination." Liston also taught Creative Engineering Synthesis (ES 473), a whole semester devoted to breaking bad engineering "habits" and inspiring "deliberate creativeness." One student tried to invent a whole new kind of sounding rocket but ended up devising a unique propellant mixture instead. Another invented an automatic feeding system for soldering. A third crafted an automobile brake light that also signaled danger.[69]

The dean of engineering, George Hawkins (1953–1967), was the sponsor of these turns to engineering science and design experiences, what he called a hybrid "systems" approach to learning. It was not an easy system to keep balanced, as he was also an advocate for a humanistic education for his engineering students. To better understand their skills and expectations, he commissioned a landmark series of social status and opinion polls beginning in 1957, designed and conducted by the Purdue Department of Sociology and Professor William K. LeBold. They found a student body mostly from the Midwest, mostly middle class, and still mostly male. They found students disappointed in the new theoretical approaches, eager to do practical engineering instead. Hawkins fine-tuned his course requirements in response.[70] He was also the author of the undergraduate portion of the national *Goals of Engineering Education* report (1968). In it, Hawkins and his team made some wild promises: desalinization plants and fully automated libraries and workplaces by 1984; ocean-based farming, synthetic foods, and colonies on the Moon and Mars by 2000. Engineers would build it all. Hawkins also advocated a controversial plan to create a nationwide five-year study program for engineers, this to accommodate the various curricular demands and to graduate young people ready to master "complex engineering systems." He wanted to credential them as truer professionals, like doctors or lawyers. It did not take. Nonetheless, Purdue's demanding curriculums, pairing the hard sciences and advanced engineering, did become a national model under his tenure.[71]

These trends somewhat diminished student appreciation for the Potter ideal and the liberal arts. In the results of one engineering survey, including alumni, faculty, and students, literature and history received the lowest scores in terms of their perceived "importance" as educational "goals." They just did not matter so much anymore. The sciences mattered more. But these engineers, both young and old, still valued smart writing and speaking skills. They considered them technical skills, built into their prejudice for "technical-scientific goals" and good "engineering judgement." Also scoring low were such life goals as the "development of moral, ethical and social concepts" necessary for good citizenship, and the "attainment of an interest and pleasure in social-humanistic pursuits." Students had high expectations of a "humanistic" experience in their first year at Purdue, but they soon became disappointed by the routines of their general education courses and the poverty of cultural events in Lafayette.[72] Dean Potter, now retired, must have been disappointed.

One intrepid student warned about these contradictions, what he called the dilemma of the "vertical" versus the "horizontal." Both the US and USSR were racing for the ultimate heights, for conquest of outer space, by way of powerful machines of our own making. This exaggerated sense of the vertical was technology-obsessed and "research-extruded." With such a fixation on "technological verticalness," we were losing our humanity, the values and varieties that encompassed fuller livelihoods on Earth, best represented by a Leonardo da Vinci or a Benjamin Franklin or an Albert Einstein. These inventors represented the beauty of the horizontal: technical expertise liberated by creative expression. In the tradition of Potter and Zucrow, it all sounded so perfectly old Purdue.[73]

## THE SPUTNIK CHALLENGE

The rise of engineering science brings us, ironically, to a crisis in science. A crisis that came to the US from afar and from above. Sputnik, by surprise. We were waiting for a Soviet nuclear attack. Instead we got a 128-pound aluminum sphere beeping from Earth orbit. Actually, it was not just one victorious Sputnik, but three. Sputnik came in multiples: first, the satellite on 4 October 1957; second, the dog Laika on 3 November; and third, the scientific satellite *Sputnik 3* on 15 May 1958. And not just three, but three minus one, because the US matched the series of Soviet successes with a losing *Vanguard*, its own dramatic failure on 6 December 1957. These were a traumatic few months in the US media.

The "shock" of Sputnik was not really technical or technological. It was cultural and political. Engineers were not surprised. They were expecting a satellite, either one of ours or one of theirs. The shock was in the media and government reaction, the shock not so much that the Soviets had done it, but that the Americans had not. Part of the problem was the drumbeat of signals in the popular media and scientific/technical journals in the months before, that the US was destined to get there first, in a literal "race for the nation's survival." The US had mastered a new "era of technical achievement" beginning with Robert Goddard and culminating with North American Aviation's new rockets headed on a "road to the stars."[74] In the wake of such high expectations and confidence, Sputnik was a perfect storm of self-interests. The media saw advantage in the crisis to sell the news, and the Democratic Party to field advantages in the upcoming congressional and presidential elections.

The Sputnik events also prompted a wave of elite and popular anguish that America had fallen behind in basic science. The common perception at the time held that science always predated and informed technology, that science informed the engineering solutions and technologies that shaped our lives.[75] Thus, to overcome its failures in technology, America needed to reinvest in science. Publicists argued that the dominant model of development had not worked: basic science received a paltry share of the national research budget (9 percent), most of it went to applied and developmental funds (91 percent). The media and political outcry focused on basic science as the solution to the country's decline. Shamed by failure, the Eisenhower administration sought to regain the advantage through its new President's Science Advisory Committee.[76]

The outcry came from all quarters. Gerard Piel, editor of *Scientific American*, blamed America's "one-note siren song of utility," its obsession with things, from "the space ships that are being delivered this year by our automobile factories, to cancer cures, to bigger bombers and faster jets." Sputnik justified his claims; we needed fewer things, more science. The Purdue Research Foundation approved. Americans were obsessed with "material wealth and privilege" as "status symbols." We needed less "know-how," more "imagination."[77] James Killian, special assistant to the president for science and technology, decried America's "almost endless list of 'things,' such as jet planes, missiles, nuclear reactors, satellites, insecticides, vaccines, drugs, rocket fuels, computers, and space suits." The country was "technology happy," had an "undue attention to those things which are glamorous." The American Association for the Advancement of Science and the National Science Foundation agreed. President Eisenhower also gave his imprimatur, targeting the rise of the R&D project system for scorn, how the universities had sold out to "regimented research" and vast "government financial support" instead of following democratic values and "creative genius."[78]

Engineering and development lobbies joined this chorus that America needed to follow a purported Soviet model and reinvest in basic science. The US Navy, through its Naval Research Advisory Committee, and recalling its earlier investment in "fundamental science" with Project Squid, called on raising support for basic research in the universities by 40 percent.[79] The American Rocket Society (ARS) finally completed several initiatives, led by Andrew Haley and the Policy Committee, to navigate the group away from its engineering ethos and toward what they called the science of "space flight and astronautics." Haley found a ready ally in Wernher von Braun. Both were eager to subdue the dominant "propulsion" wing in the ARS, which was obsessed with "straightforward engineering," in favor of new issues like "guidance, telemetry, re-entry." Von Braun repeated these arguments several years later, bemoaning how the ARS had become little more than a "National Association of Liquid and Solid Fuel Powerplant

Manufacturers," built around the "pure inertia" of a "$2 billion a year business." It had forsaken its role as "a rallying point for the young generation" and needed a "rejuvenation" for true astronautics.[80] America had made too many guided missiles for bombs, too few rockets for satellites. This was quite a rebuke of the propulsion engineers who had helped build the ARS. Sputnik resolved the dilemma, empowering the idealists in the ARS to launch their own manifesto for the conquest of space: the "Space Flight Program" of 1957. It offered details on the short- and long-term possibilities for missions into space, along with a new American philosophy of spaceflight. The ARS, and the country, were ready to leave behind the government–industry market for guided missiles, ready to aim higher.[81]

Industry followed the ARS trend. It discovered the peace dividend. Why not turn the guided missile into a space rocket, just as we transformed the jet bomber into a commercial jet airliner? Why not teach the missile engineer to become an aeronautical engineer? The aerospace industries, if still bound to the guided missile, adapted. General Electric renamed its Missile and Ordnance Systems Department as the Missile and Space Vehicles Department. Lockheed created a Missiles and Space Division; the Ramo-Woolridge Corporation formed its Space Technology Laboratories. The "Missile Age" was becoming a "Space Age."[82]

This new excitement created a dilemma for the ARS in the winter of 1957–1958. It was a rocket society, after all, and decided to mobilize America's youth on a national level. The ARS set out to create a national network of "Missile and Rocket Clubs," beginning with the high schools of the Washington, DC, area. It was to be funded by the US government and aerospace industry to train "tomorrow's scientists" with educational courses, including "static test firing" and "live firings." Here was a program for education in the "research, development, engineering and production of guided missiles, particularly ballistic missiles."[83] These plans backfired with news reports about young people suffering injuries and deaths from rocket mishaps. To avoid legal liabilities, the ARS bylaws soon prohibited any and all such "experimental work." Its leadership even had to admonish von Braun at least twice, with formal letters requesting that he cease and desist from promoting youth amateur rocketry experimentation.[84]

Although he had encouraged the practice, even giving live model rockets to the children of his students and colleagues, Zucrow soon joined the conservatives and came out against amateur rocketry as "extremely dangerous." In an interesting turn of events, he was memorialized by one of the young experimenters, Homer Hickam of Coalwood, West Virginia, author of the memoir *Rocket Boys*. He and his boyhood friends, otherwise destined for the coal mines, were inspired by the launch of Sputnik to experiment with homemade rockets. Their first attempt, built from part of a flashlight and the flash powder of thirty cherry bombs, ended in a "huge ball of fire." Success only came after the young men learned something about rocket science. In their senior year, their chemistry and physics teacher gave them a book in a bright red binding from the series called Principles of Guided Missile Design. Homer called Zucrow's chapter "Fundamentals of Rocket Engines," the "most wonderful title of a chapter in any book I had ever held." With help from their calculus teacher, the boys eventually learned the key Zucrow lesson on the "characteristics of flow passage for subsonic and supersonic flow expansion and compression." That taught them how to build the right kind of chamber and nozzle. The feature film about the memoir, *October Sky*, showcased Zucrow's book in one dramatic scene, sitting on Homer's desk, only to reveal an autographed photo of Wernher von Braun underneath. Zucrow taught him rocket propulsion by the book, but the young man remained obsessed with his rocket star, von Braun.[85]

Purdue was quite well prepared for Sputnik, still boasting the country's largest School of Engineering in terms of enrollment (5,800 students in 1958), graduating more engineers than MIT, Georgia Tech, or UCLA.[86] Some detractors ridiculed this claim, insinuating that Purdue had size not smarts, that it had "brass tacks" but not a "purely scientific" spirit. Yet bigness had consequences. By 1960, Purdue had "placed into industry approximately one out of every 25 engineers practicing in the engineering profession in the United States." These were also the better jobs, the higher income jobs. Purdue also enjoyed "the largest instructional program" in mathematics, chemistry, physics, and the biological sciences than any other institution in the world. It was also the fourth-ranked university in the US in terms of doctorates in the natural sciences.[87] The director of the university libraries even boasted that over 831,321 people had crossed the threshold of the main campus library in the last fiscal year (1962), more than "the combined total of all those who attended convocations, football, and basketball games." Purdue students were living the life of the mind. This was a surprising statistic, almost heresy for sports-crazy Indiana. The librarian might have gone too far.[88]

As to Sputnik, Hovde did not overreact. The US Senate wrote to him about the rising fears, expressed by such leading voices as Vannevar Bush and Jimmy Doolittle, that "American scientific education lags behind that of Russia's," that "we are being far outstripped." Hovde recognized that there was a "truly serious shortage" of science teachers and laboratory equipment in American elementary and high schools but rejected any forced Soviet-style measures or government subsidies. He remained in favor of local responsibility and free-market forces.[89] In private, to the secretary of the Army in one case, he rejected the calls for "hysterical action or over-hasty reorganization." The Sputnik failure was not one of money, or resources, or intelligence. He spoke from experience, of course, from his work for the OSRD and GMC. Sputnik was the result of bad management and "delays in decision-making," along with all the chaotic interservice rivalries. The US simply needed "to support 'projects' at optimum rates," and speed their progress from "conception to use." He also put in a plug for higher education, especially for the education of science and engineering graduate students.[90]

Purdue scholars helped to make sense of the Sputnik crisis with a series of pathbreaking public opinion polls. They found the culprit in ourselves, our American selves, namely in our misguided perceptions about science and scientists. The polls discovered that significant proportions of teenagers, those from the cities and suburbs, both boys and girls, had faith in science for its discoveries and products, but rather little faith in scientists. These teenagers, at the peak of the baby boom, were a major statistical cohort. There certainly were a lot of them. The image of the "egghead" scientist was "burned into the picture tube of many teenagers' minds." It was "the Buck Rogers stereotype of the oddball, impractical absent-minded scientist, high in the ivory tower, far removed from the rest of us, sacrificing people right and left in his dangerous experiments and caring little about the consequences of his work." This posed quite a problem, so the authors cautioned, if the US ever wanted to match the Soviets on the battlefield of science.[91] Yet another Purdue study, by sociologist Walter Hirsch, this time of science fiction magazines between 1926 and 1950, read by some six million mostly young men, found that conspiratorial scientists were the favorite villains, madmen intent not on saving the world but destroying it.[92]

The poll and study matched well with at least one of the strange letters that President Hovde received in the aftermath of Sputnik. "All scientists and militarists are partly nuts," wrote one local Hoosier, who called for the US to get out of the UN and end its drive for global conquest by way of the ICBM. "I say all our leaders should read the bible and try to obey it and there would be more happy people without ulcers."[93] These anti-science stereotypes stung. Two electrical engineering undergraduates answered them and Sputnik's challenges with winning campus essays. Bill O'Neel wrote that math and science still had social value, that engineers were not the cartoonish "weird species of homo sapiens" buried in their books and experiments. Wallace Flueckiger warned campus about "the self-contented unconcern" of the average American consumer. "High-powered rocket and space research" were the way to win again. A new Purdue opinion poll confirmed their points. It asked teenagers about their attitudes toward engineers. The results were surprisingly favorable. The engineer was seen as "a practical, substantial citizen closely associated with business and industry and a respected and admired member of the community."[94]

In response to Sputnik, and to student demand, Purdue also fortified its Russian language program. By spring of 1961, enrollments in Russian classes were at an all-time high of 281 students. That fall, the Department of Modern Languages offered its first Russian major and minor.[95] For the first time in his life, Dr. Zucrow also began to learn the language, and his Rocket Lab now required technical Russian as one of its graduate requirements. His students were expected to take a few semesters of basic grammar and pass the examinations in order to read the Soviet academic press on aerospace and propulsion topics. They also had machine translations from Wright-Patterson Air

Force Base, whose "print reader" and computer could translate Russian to English at thirty-five words per second. With a handy technical dictionary, the Rocket Lab students did the rest, although it was not always easy. J. Michael Murphy's very first course as a PhD student was Introductory Russian. After three and a half years, he was still in it. Only the kind act of his language teacher, with an easy take-home final exam, allowed him to pass.[96]

The Soviets took notice of Zucrow and his Rocket Lab, publishing a prestigious Russian-language volume of Zucrow's contributions in *Aerodynamics, Propulsion, Structures and Design Practice* and translating and publishing his students in a variety of technical journals. These citations became hallmarks of their scholarly résumés along the path to tenure and promotion. In the world of unclassified research, recognition by the enemy was now an honor. When the Armed Services Technical Information Agency translated Ia. M. Paushkin's *The Chemistry of Reaction Fuels* (1962) from Russian into English, it found a series of references to Zucrow's pioneering works, along with those of his students Stanley Gunn, Charles Trent, Clair Beighley, Henry Wood, and David Charvonia on propellant combinations, heat transfer, and film cooling. Both Soviet engineers and American intelligence officers agreed that the work of the Rocket Lab was "unique and advanced in the world." It was "state of the art." Purdue had actually helped to teach the Soviets how to launch better rockets.[97] Werner von Kirchner, the "other Wernher," parodied all this on Zucrow's seventy-fifth birthday, with a decree from the Politburo of the USSR for a "reasonably happy birthday" to "citizen" Zucrow "for his conspicuous contribution to the imperialistic and capitalistic United States rocket propulsion development program."[98]

For the 1958–1959 school year, Zucrow added two new dual undergraduate and graduate courses to his curriculum: Guided Missile Systems (ME 550) and Rocket Propulsion Systems (ME 552). He also revised and enlarged his classic textbook, now published in two volumes as *Aircraft and Missile Propulsion*, dedicated to the thermodynamics and fluid mechanics of jet propulsion and its variety of power plants. It was a telling title. The market for guided missiles still informed the terminology. The publishers marketed it as America's answer to Sputnik. "Launched!" said one advertisement. The books were focused on teaching college seniors and first-year graduate students the fundamentals across a wide array of historical, theoretical, scientific, and technological topics, once again introducing gas turbines as used in locomotives, marine propulsion, and electric power, along with the turboprop and turbojet, ramjets, and rockets. It was also a superb reference set for aerospace and propulsion engineers, selling 6,005 copies from 1958 to 1966, with use in college engineering courses at MIT, Caltech, UCLA, Berkeley, Michigan, and Notre Dame, among others. One reviewer called it "magnificent." It was "technically thorough and beautifully organized," with precise studies of engine cycles and performance parameters, be they in the design, static testing, or in-flight stages. Each chapter stood independent, like a book all its own.[99] Another reviewer, Allen Fuhs, praised it from personal experience. He taught it to his engineering students at the Naval Postgraduate School (Monterrey). It was the perfect union of theory and practice, transposing the "ideal device" and its "real effects."[100]

It is difficult to weigh the influence of any one textbook or author, of course. Yet Zucrow's works had a multiplying effect, both on the professors and students who used them at Purdue and the thousands more beyond. This included the nearby NACA Lewis in Cleveland, where Zucrow's textbooks were the primary references for some of America's premier propulsion engineers. Thanks in part to Zucrow's mentoring, NACA Lewis became a matrix for American civilian spaceflight endeavors. There, his acolytes and colleagues continued to promote the country's greatest challenge: creating the foundations of a civilian spaceflight program. Walter Olson and John Sloop and their teams maintained their cutting-edge work in the "theoretical and experimental studies of high-energy propellants."[101] On 6 May 1955, Olson made a remarkable bid, building on the initiatives from 1952. He "proposed to the senior managers at Lewis" to get NACA "into the rocket business in a big way, including spaceflight, including flights into outer space." Buoyed by the propulsion achievements of his own researchers, and still backed by the Rocket Subcommittee (Zucrow was now a member), Olson called for NACA to take leadership over space exploration: including propulsion to get there (chemical and nuclear), along with the physics and biology of surviving there (for satellites, probes, and piloted spacecraft), and all the "logistics" of communication and orbits to return to Earth. It was a blueprint for the Mercury, Gemini, and Apollo eras.[102]

Olson and NACA Lewis were also encouraged by their new Rocket Engine Test Facility (RETF) to test higher-thrust rocket engines, focused on high-energy propellants (including liquid hydrogen). Their "grand plans" for "a huge site in a remote area of the West" were scaled back to this smaller one in Cleveland, not worth $8.5 million as originally planned but a more modest $2.5. It was also limited to tests of engines at twenty thousand pounds of thrust.[103] To celebrate the opening of the RETF and spotlight its continuing investment in spaceflight, Lewis offered the Seminar on Rocket Propulsion in February and March of 1957. Sloop organized a comprehensive twenty-six lectures on such topics as flight equations, chamber design, pumping systems, and instrumentation. Two Zucrow alumni helped to lead the proceedings. Robert Graham, head of the Fluid Dynamics Section, lectured on gas dynamics and heat transfer; Eleanor L. Costilow on engine and nozzle designs. Buoyed by the work of NACA Lewis, Abe Silverstein called a meeting of the Research Planning Council in September of 1957, a month before Sputnik, to answer the calls of the "Propulsion Systems Division" for a "reorientation" of NACA Lewis to spaceflight. Bruce Lundin led the charge for coordinated propulsion research and initiatives in "satellite vehicles" and "space flight missions."[104]

Sputnik was the tipping point at NACA Lewis for a momentum long in play, ever since Zucrow's initiatives in 1952. As Olson put it, "We were ready to go with the space mission" back then, but only Sputnik became the "magic" moment when "almost the whole center converted to rockets," when Lewis got "ready to go with the space mission."[105] On 9 December 1957, Lundin formalized the plan for radical reforms. Stung by years of subservience to the military, he noted that "some line should be drawn." NACA deserved the "leadership" role as the centerpiece "single independent Government agency" for long-range rocket propulsion to space. He called for doubling the staff by 1962, switching Lewis personnel to 75 percent dedicated to "advanced research" and only 25 percent to "air-breathing systems." He called for a new "space laboratory," centered on the rocket research facility first proposed with Zucrow's support in 1952: providing for "ground test stands for large high-energy-fueled rockets," flight tests for sounding rockets and satellites, and provisions for nuclear rockets and nuclear electric power, along with an architecture for human spaceflight, including "re-entry, rendezvous, and recovery." Lundin admitted his plan was "bold and visionary." Yet unlike Wernher von Braun's space station dreams from 1952, or President John F. Kennedy's coming "moonshot" initiative in 1961, Lundin cast a wise eye to the sustainable. He manifestly argued against any single "dramatic project" simply to "fire the imagination of the staff and be a focal point for our research." It was wrong to "dangerously limit our goals, restrict the range of our thinking, and give us nothing to grow on." He wanted more: "long-range research" in "space technology" for the "exploration of our universe." For "research is a dynamic and ever-expanding activity and needs broader and wider objectives than any end-point article."[106]

More than any other single document, Lundin's plan transformed NACA into NASA, not by political action from above but by technical expertise and confidence from below. NACA Lewis committees discussed his proposal over the late fall and winter of 1957, confirming such details as mass ratios, high-energy propellants, velocity requirements, and optimum propellants for the "intensive research and development" of "moon satellites and landing on the moon."[107] Abe Silverstein adapted the plan into a blueprint for NASA. His edits reveal a wonderful sleight of hand. He turned Lundin's "I" into Lewis's "we." He added a table of Walter Olson's "15 Points" at the center. He took Lundin's signed "Some Remarks" and remade it into the "Lewis Laboratory's Opinion."[108] It was the leverage he needed in the coming months to convince NACA headquarters, and its chair James Doolittle, that NACA Lewis was ready to assume the lead. At the time, NACA Langley and Ames were skeptical. NACA Lewis, said Silverstein, was the true "seedbed" of space enthusiasts. The rest is NASA history.[109]

HOW DID THE EISENHOWER ADMINISTRATION RESPOND TO SPUTNIK? THE ONLY way America knew how. The project way. Its initial response was the Advanced Research Projects Agency (ARPA), organized in February of 1958 to develop surveillance and deployment technologies against a surprise nuclear attack, targeted to deter and match any coming Soviet threat. ARPA was a return to the project contract approach of Vannevar Bush's Office of Scientific Research and Development (OSRD) during the Second World War. It was a government office even more focused and agile than the OSRD. It was also remarkably unlike the bulky Research and Development Board (RDB), which had attempted to govern all federally funded military and civilian R&D. ARPA was svelte and fast, out to speed ahead of the Soviets for strategic American advantage. Zucrow advised its second director, Austin Betts, on propulsion and antiballistic missile (Nike Zeus) issues.[110]

The new NASA, legislated into action by July of 1958, inherited many of ARPA's early "advanced" propulsion projects: in power generation for both launchers and spacecraft, in energy conversion by way of fuel cells and batteries, and in various kinds of novel propulsion (nuclear, electric, and plasma). NASA also took on the primary charge to manage civilian spaceflight, just as NACA's Walter Olsen and Bruce Lundin had earlier proposed, with Zucrow's blessing. This meant the forthcoming Projects Mercury, Gemini, and Apollo. Quite an array, these. Silverstein remembered how NASA, building from the momentum of NACA Lewis, suddenly became a "procurement" agency, joining the other "big government agencies" as a maker of "research and development contracts," initially reaching above $125 million, soon into the billions, like the contracts to Rocketdyne for one of the Saturn engines, and to McDonnell for the Mercury capsule. NASA's contracting procedures came straight out of the "armed services procurement regulations," remembered Silverstein. They were also known as R. B. Stewart's Blue Book.[111]

Back at Purdue, the media responses to the Sputnik furor were varied. One student editorial warned Americans not to fall for the media hysteria. The Russians, after all, had invested in Sputnik with a single-minded obsession. "We did not take away researchers from other fields to concentrate on this project. We did not insist that the people in the United States sacrifice their time and their standard of living to contribute to the work." Americans may have lost one superpower duel, but not their native prosperity and promise. Dr. Ralph Morgen, research director of the PRF, along with Purdue professor A. G. Guy, School of Metallurgical Engineering, confirmed the judgment. During a short window of detente in 1958, they took a five-thousand-mile "inspection tour" of Soviet engineering institutes, sponsored by the National Science Foundation and American Society for Engineering Education. They found an impressive educational system, but one obsessed with state procurement, not geared to the freedom and creativity of the individual. The enterprising Guy stayed on, driving his wife and two teenage daughters in a Volkswagen Microbus Camper from Stuttgart to Moscow, where they attempted to live and work. Alas, the authorities isolated them from colleagues and social contacts and did not even allow Guy to show home movies of "daily life" in Lafayette, Indiana, and at Purdue University.[112] That would have been truly subversive.

The Sputnik media wave turned Zucrow and Hovde into minor celebrities, two of America's premier "rocket scientists." Both of them advised that the US answer Sputnik with a robotic Moon landing, or even lunar soil capture and return.[113] The *Indianapolis Times* soon featured Zucrow's mysterious Rocket Lab, where "blackboards and notebooks look like foreign code"; where "in the low-slung buildings of the rocket lab they peer through bulletproof glass inside cells with walls a foot and a half thick," toying with "the dangerous valves and wires and cylinders that make up rocket motors." Zucrow responded with flair: "We'll go farther—we'll go to Jupiter and Mars—and beyond." It was a line he borrowed from the popular tune "Fly Me to the Moon," soon a Frank Sinatra hit. "There is no end," he said, "for when we get there, we'll look for another planet. But to blast off, we need two things. People and money." The *New York Times* went a bit further, writing about Zucrow's "elite corps of engineering graduates," who "weigh baffling outer space problems," manipulating their "missile enigmas" from their headquarters in "three squat white buildings set in a cornfield." The article also introduced him as the "Russian-born Dr. Maurice J. Zucrow." That was all the Sputnik-inspired cachet American readers needed.[114]

Purdue eventually picked up these metaphors, raising the prominence of its airport and Rocket Lab as emblems of the university's achievements. It was no longer a "cow college," no more the high-grade "vocational school." People like Amelia Earhart and Maurice Zucrow had helped to transform the university into something more. "In the heart of the cornfields of Indiana, Purdue has become the greatest technical school in America—and possibly the world."[115]

# 8

# MOON RACE

## Purdue's Pathways to Outer Space

THE SPUTNIK INCIDENT IN THE FALL OF 1957 FRAMED THE POLITICS OF the 1960 presidential campaign, fueled by the media tumult. News events like Lyndon Johnson's Senate investigations added to the controversy. For its own advantage, the Democratic Party played up the presumed "missile gap," that the US had fallen desperately behind in the race for ICBMs, and the "space gap," that the USSR had achieved an overwhelming lead in spaceflight. Candidate John F. Kennedy retooled these critiques into the optimism of his New Frontier platform, with calls to double America's R&D commitments. As president, Kennedy made good on these promises with his political and financial support for Project Mercury and new ventures into Projects Apollo and Gemini.

Public-minded engineers shared this enthusiasm. Some in colorful ways. The imaginative Krafft Ehricke, formerly of Peenemünde, now the director of advanced propulsion studies at General Dynamics, linked spaceflight propulsion with the metaphor of the egg. Rockets were pressure vessels like eggs, he wrote, as if propulsion was the source of all energy and life. "The prehistoric prototype of our space vehicle—the egg—has a mass ratio (percentage of expendable contents) somewhat better than that of the best current rocket systems—about 93 per cent." Besides the shell, the egg was all thrust. We humans needed to become more like the egg and surge out of the envelope of Earth's gravity into life beyond it.[1] This ideal of bursting forth had its origins in one of the earliest known conceptions of rocket propulsion (from AD 1280): the "self-moving and combustible egg" of the Syrian polymath Al-Hassan er-Rammah. The Aerojet-General Corporation paid homage to him, complete with a cartoon, in an ad hiring engineers dedicated to "the creation of tomorrow's realities from yesterday's dreams."[2]

This evolutionary vision was quite popular. Dorothy Simon, a lead engineer at the Avco Aerospace Corporation, took it a step beyond. She compared Earth's biosphere to a "spherical shell" and "circular line" whose "upper limit" was constantly rising thanks to human aviation and invention. Now was the moment for humanity to break free into the new spherical dimension of outer space. It was as if "an ameba swimming in a stagnant pool" had suddenly become conscious, now "free to leave the surface and explore at will." This was the "great evolutionary step for man."[3] Humanity was rising above and beyond its timeworn horizons. Maurice Zucrow contributed to these spaceflight metaphors, in his own indomitable way, as one of America's elder statesmen for rocket propulsion, advancing new confidence about interplanetary exploration through his public appearances on a national stage, as well as through the quieter educational mission of the Purdue Rocket Lab.

## THE MILITARY–INDUSTRIAL COMPLEX

America's spaceflight progress was not without its many twists and turns, breakthroughs from abroad on the part of the USSR, and critiques at home by the Democratic Party. President Dwight Eisenhower contributed his own complaints in his famous Farewell Address on 17 January 1961, a symbolic threshold for America in the Cold War. He was not known for high drama, at least outside of the Sputnik fervor, and his very public and embarrassing lie about the U-2 spy plane incident. The address created plenty of controversy for years to come with its spotlight on a "military–industrial complex," along with its "scientific–technological elite" that was coming to dominate American science and engineering.[4]

The origins of this infamous term, "military–industrial complex," so often used as both a mode of analysis and term of abuse, are in dispute. Some historians have argued that the president himself crafted it after seeing all the advertisements for guided missiles and military weapons in professional journals and popular magazines; others that the president's speechwriters originally framed it to include the universities in a rising "military–industrial–academic" nexus. In either case, President Eisenhower was expressing his conservative distaste for big government and heavy spending, along with his firsthand knowledge of a "complex" of military and industrial interests invested in the guided missiles and weapons systems of the early Cold War. He was in part responsible for this troubling phenomenon, having mobilized new justifications and resources for national security and defense, though now warning of their effects.[5] Eisenhower took special aim at government-sponsored research projects: "the prospect of domination of the nation's scholars by federal employment, project allocations, and the power of money." The complex was corrupting higher education and basic science: where "research has become central, it also becomes more formalized, complex, and costly," and "a government contract becomes virtually a substitute for intellectual curiosity."[6]

Eisenhower was targeting the project system for reproach. It certainly had grown by the time of his address. Public and private spending for R&D had risen from $5 billion in 1954 to $15 billion in 1962, although most of that came after the Sputnik crisis and its fierce partisan battles. Most of this money was dedicated to applied research and engineering studies, "work related to large development efforts." Researchers in the basic sciences received but a fraction of the overall funds, just over $1 billion in 1962. In compiling these statistics, the National Academy of Sciences joined the president's critique, arguing for a fairer distribution of funds to raise the profile of basic research in the sciences. Most scientists and engineers considered the project system as sound. It had created a viable research apparatus in the universities, in a "government–university alliance." Fiscal administrators and principal investigators became partners by promissory contract. The system was based on a plural balance of power, along with consultative peer review, a system rewarding expertise. Most important was the "individual investigator," the creative researcher "at the farthest edge of the frontier."[7] The trouble was that the project system was dominated by "large scale research and development projects" that corrupted the universities. Sponsored research was not an equal partnership between the government and academia. It was a patron–client relationship and the universities were the clients, often subject to the terms and dictates of the sponsoring government agency or military service. Harvard University agreed, its president sounding the alarm that the federal government was playing an oversized and corrosive role in his university's affairs. The project system, as a result, was eroding the university's social function as a "creative force," especially in the mentor's relationships with undergraduates. Students really received no benefit at all.[8]

Archie Palmer's 1962 definitive study found that universities like Harvard were the lucky ones. Out of the country's more than 2,000 institutions of higher learning, only 349 were worth reviewing for their formal research policies and procedures, and only 147 of the top schools made it into his work. Only they had the sophisticated administrative apparatuses to meet the challenges of intense, segregated, classified research. They did this, moreover, "for the performance of short-term or hardware type research and development," at the expense of the basic sciences, social sciences, and humanities. All this meant that the project system, aka sponsored research, was the "source of deep satisfaction and also of serious concern."[9]

For all of R. B. Stewart's earlier interventions, the rules for overhead were a recurrent problem in R&D, at least from the perspective of the military and government. The universities were exploiting overhead in their own favor. It was one of the reasons that there was widespread approval of the project system among researchers, some of whom now received their very own salaries as overhead. They tended to worry not about their academic freedoms but about their overhead profits.[10] This was a "serious fiscal problem," a by-product of the project system's success. Stewart's principles were certainly long-lived, culminating in the Office of Management and Budget Circular A-21 (1958), defining overhead as "a negotiated rate based on an audit of the expenditures of the institution." It required that universities maintain robust accounting and business staffs for self-monitoring, as well as spend federal funds with "prudence, economy, and probity." The government and military, in turn, were duty bound to respect the "total function of the university," its freedoms and traditions, its dedication to the "advancement of knowledge" as a "community of scholars." The trouble was that these Stewart principles, however fair and flexible, were not universally or consistently applied. So many different government and military agencies, aligned with so many different and complex university settings, translated into multiple and conflicting rules and accounting practices.[11]

The issue of academic overhead is a puzzling one for most historians. Aside from a few general reports, the truths about it remain hidden in confidential military documents or in proprietary files. The published reports suggest that overhead benefits to universities were decreasing by the 1960s. Military and government offices began to better police the practice and replace it with the grant system, wherein universities figured overhead as part of their own cost sharing. Yet the practice of padding contracts with excessive overhead continued. It was a privilege of those clients with special talents that were in high government demand. It was a flexible point of leverage and reward in the negotiations. After a formal evaluation of Project Squid's various contracts in April of 1969, Bruce Reese found that of the seventeen universities participating, the average was 47 percent overhead. Purdue was one of the highest chargers, along with the University of Denver, Rensselaer Polytechnic Institute, and MIT, charging 50–61 percent overhead rates. Reese warned Purdue not to price itself out of the market.[12]

Zucrow understood the perils of the project system, how it threatened the independence of the university. He saw an equal peril in the monetization of education, all the new business tasks that too often corrupted the creative spirit of inquiry and discovery. His inspiration was not President Eisenhower's complex but economist Peter Drucker's competence, which advised that the individual scientist and engineer be free from organizational and monetary demands, free to pursue their technical vocations for social betterment. Zucrow saw something of himself in Drucker's humanism, "the new power to organize men of knowledge and high skill for joint effort and performance through the exercise of responsible judgement."[13]

By the early 1960s, Zucrow complained, the research professor had become a "salesman," writing sponsored research proposals, making visit and sales pitches, building networks, negotiating contracts, even making "fence-mending visits." He took these demands seriously, even though they were a drain on his own time and energy, in order to shield his staff and students from all the business distractions, what he powerfully called "the uncreative areas of certainty." The integrity of the research project was in peril, he warned. "Its technical merit is secondary to its salability." To stimulate creativity, "one must be concerned with the new, the novel, the untried, and the unpredictable."

Instead, he said, "the present system is so bad, that some of us—and I am getting to that age—wonder whether it is worth the battle."[14] His philosophy matched closely with President Eisenhower's farewell address, as when Zucrow warned against big government and bureaucracy, focused merely on "the appropriation, administration, and disbursal of funds." What mattered instead was the "Principle of Self Improvement." His institutional ideal was a "National Institute of Science and Technology," devoted not just to defense but to the "socio-technical problems" of pollution, crime, poverty, transportation, and health, and answering not to the president but to the House of Representatives. "I am against czars of any kind," he said.[15]

Some of Zucrow's engineering colleagues agreed. Market forces had intervened in academic research the way they already worked in industry, whereby firms prepared bids for military contracts, sometimes at a cost of several million dollars, as well as "millions of scarce engineering manhours."[16] The researcher was turning into a "businessman," someone in mortal combat with other scholars, wasting time with proposal writing, greedy for overhead. "Sponsored research" was corrupting higher education just like "semi-professional intercollegiate sports," or akin to putting "slot machines in the halls" of academia.[17]

President Kennedy and his advisers chose to ignore Eisenhower's counsel, along with these various warnings. In fact, the new administration fortified the project system through the "Bell Report" (1961), written by David Bell, director of the Bureau of the Budget, with the concurrence of the heads of the Atomic Energy Commission, the National Science Foundation (NSF), NASA, and the secretary of defense. Their review confirmed the dynamism of government and military project contracts with "private institutions and enterprises," along with the traditional roles of government and military labs around the country. In conjunction with them, projects with industry and academia would provide the conduit for the Kennedy administration's significant expansion of R&D. Subsequent reports fine-tuned the approach, recommending closer government technical supervision of R&D contracts, if ideally in balance with independent academic research.[18]

These kinds of initiatives meant that the universities now had to escalate their competition for federal monies. Purdue joined with purpose, what the PRF called the "race for knowledge" at the "frontiers of the human intellect." Hovde reorganized sponsored research, under the lead of R. B. Stewart, to prove Purdue's credentials for the new federal project contracts. The reforms were significant: a new University Research Policy Committee, along with an Industrial Communication Research Council. Both now partnered with a fortified PRF Office of Research Grants and Contracts. The PRF was also responsible for a streamlined system of "receiving and distributing classified material." The results were impressive: Purdue's federal funding rose from $3.9 million in 1961 to $6.8 million in 1963. This included funds from the NSF, the Department of Defense (DOD), the Navy and Army, along with an Advanced Research Projects Agency (ARPA) material sciences research grant.[19] By the 1965–1966 fiscal year, the PRF boasted that it was managing one thousand research projects, including "all classified research projects." With $9.3 million in expenditures and $26 million in assets, it was one of the top ten businesses in Tippecanoe County.[20]

McClure Park Inc. was one of the PRF's new ventures in these years, the university's first industrial and research park, named for the land donated by alumnus Frank McClure, a vast plot of initially one hundred acres, and up to a thousand. It was located north of campus, just off the US Route 52 bypass, with ease of access to Chicago and Indianapolis, adjoining the new West Lafayette suburban home developments that the PRF had helped to build. It set the stage for President Hovde's grand plan for 1964: a new electronics research center, something akin to Boston's Route 128, near MIT and Harvard; or like Palo Alto's Silicon Valley, neighboring Stanford University. The push came in December, as a Purdue delegation, joining university and PRF representatives and local politicians, lobbied NASA for its proposed $50 million electronics laboratory. Hovde made a strong case for the city of Lafayette. It was, he argued, "not a soot-ridden, over-built and over-populated area" like other applicants. He touted the campus for its critical mass of scientists, engineers, and students, especially the Department of Computer Science, the School of Electrical Engineering, the Department of Physics, and the "Dr. Zucrow group" in Jet Propulsion. Purdue, ranking eleventh among universities producing doctorates (1961), was "one of the most eminent scientific and technological institutions in the nation."[21] In the end, Boston won the contract.

The results of the Kennedy initiatives were all-encompassing. In 1965, the DOD instituted mandatory "project management" techniques for the military. A small world of acronyms now governed weapons systems and technologies, subject to surveillance and systems engineering: the principal investigator (PI), by way of the request for proposal (RFP), under the direction of the project [program] evaluation and review technique (PERT). The US Navy created PERT in order to speedily and safely develop the Polaris submarine-launched missiles. The US Air Force and NASA adapted it is as their own. PERT was a culmination of the project system. It translated the sponsored research project into an algorithm of formulas and flow charts that the average engineer could understand and manipulate. It quantified and computerized the project into a time and cost schedule of discrete tasks and events, configured as circles and arrows, along a critical pathway to completion. It allowed for the vertical integration of projects into the overarching and autonomous "program" for the timely management of "resource allocation" (people, money, and things). In a revealing front cover to one of its manuals, the Air Force projected a PERT diagram of circles and arrows right onto outer space as a new spiraled architecture for cosmic conquest.[22]

Government and military offices now disseminated these approaches, first refined in the 1950s, into their new network of "large-scale projects," all during the Kennedy "age of massive engineering."[23] With these imprimaturs, the policies of the Kennedy and Johnson administrations spurred a redoubling of the military–industrial complex, giving a new currency to the very term. Under Eisenhower, it had been a rather "loose collaboration" of the government, military, industry, and academia, based largely on free-market mechanisms. During the Kennedy–Johnson years, it became a veritable "para-state" of centralized managerial control of industry and academia. By 1968, Secretary of Defense Robert McNamara ruled over $44 billion in defense subsidies. Several of these billions were earmarked for academia, whose beneficiaries including some familiar names at the top of the funding list: MIT and Johns Hopkins, California and Columbia, Stanford and Michigan.[24] The excitement of the Apollo missions made much of this possible, now joined by the modest figure of Maurice Zucrow, Purdue's professor of jet propulsion.

## PROJECT APOLLO

With the advent of the presidential campaign in 1960, and a Democratic congress, Zucrow entered the realm of politics: deputized as a leading member and principal adviser on the Panel on Science and Technology within the Committee on Science and Astronautics of the US House of Representatives. The panel was a stellar culmination to his career. He was nominated, with his fellow ten panel members, by the heads of NASA, the NSF, and the National Bureau of Standards. He served for nearly ten years with a number of notable colleagues, including James Van Allen from the State University of Iowa (famous for the *Explorer* satellite), Fred Whipple of the Smithsonian Observatory, and several of his friends and colleagues from the aerospace world, especially Clifford Furnas, chancellor of the University of Buffalo, and H. Guyford Stever of MIT. He shared the stage with, among others, his former student Walter Hesse, with Frank Malina, formerly of the Jet Propulsion Laboratory (JPL), with Secretary of State Dean Rusk, and with Vice President Hubert Humphrey. They were invited to advise the US government on the pressing spaceflight issues of the day, from advanced propulsion to geodetic satellites, and broader issues like urban renewal and national transportation networks.[25]

Along with his Fowler Hall PhD defense in 1928 and his Los Angeles speech to the American Society of Mechanical Engineers in 1945, the panel offered Zucrow the third command performance of his career. He was a lively discussant, never afraid of a fight, adding feistiness and drama to the sessions. Many of the other testimonies were bland

and monotonous, framed as official science, punctuated by the grandstanding of some of the congressmen. Several panelists and guests either read or inserted long scientific reports for the committee's attention. Zucrow preferred to contribute extemporaneous comments. At times, the hearings devolved into stand-up comedy routines, speakers telling scripted stories and jokes. Zucrow's humor was more natural. He could get the whole room laughing with just a word. To a claim that engineering was a technical/vocational discipline, not a university discipline, he answered in perfect timing, "Well . . ." When someone mistakenly introduced him as coming from Lafayette, Louisiana, he answered with the deadpan "Indiana."[26] He often set the theme and pace of discussion with a sharp turn of a phrase, raising issues that other speakers then echoed and reinforced.[27] His committee style was contentious but respectful, even endearing. When the whole panel came out in chorus in favor of the metric system, Zucrow alone opposed it. All of industry had its "machine tools and engines in English units," or "my units" as he called them.[28] He brazenly faced down congressmen for their biases, or factual mistakes, or unwillingness to appropriate more funds to aerospace.[29]

The mandate of the panel was to discuss how scientific research might better promote the "national welfare and security." It sounded innocuous enough. But this term research was fraught with all the deep politics of the Sputnik event, tied up with the wave of opinion in favor of basic over applied science. Zucrow rejected the separation as "ridiculous and stupid." He argued for the unity of all research and for cooperation, not antagonism, between scientists and engineers.[30] He was really making an argument for the essential research component in R&D, for the applied scientist or engineer as basic researcher. Basic research scientists could make no such claim. They remained cloistered in their labs, apart from development. Only applied scientists could do both basic and applied. As Zucrow argued, "You just can't divorce basic and applied research as two separate streams. They really are one stream, interwoven, one is eddy currents within the other."[31] Spoken like a true jet propulsion man.

From the start, at the 4 May 1960 panel meeting, Zucrow mounted a deliberate and sustained defense of engineering as an applied science. "Now, I am not a scientist. I am an engineer. I am very proud of it, too. [Applause.]" He had to remind the audience that engineers built the Titan missile: "The scientists had nothing to do with it." Engineers were the ones "who fired it, designed it, made it. [Laughter.]" He summarized this defense of engineering with a metaphor of "the beans." It was a dry goods metaphor, that research without application amounted to "so many cans of beans" on the supermarket shelf, alone and unused. "Basic research without the proper effort of translating it into something is sterile." It needed the crucial "translation" into utility. What counted as the ultimate product was not the can of beans, but the can of beans on the plate, cooked and ready to eat. Science must become engineering. Knowledge must become hardware. "The rapid translation of the findings of basic research into technical devices and processes is an essential for achieving a dynamic society," he said.[32] Zucrow put his formula even more bluntly at a forthcoming panel meeting: "I think also with all due respect to scientists, scientists sometimes make engineering decisions. My observation over the last 20 years is that the number of times they are wrong exceeds the number of times they are right."[33]

The panel's meetings started out promising enough. Kennedy's campaign metaphor of the New Frontier captivated the proceedings, for Democrats and Republicans alike, both eager to fund the Space Race. Patriotism and national purpose demanded no less. Space was our new manifest destiny, ready for exploitation by American know-how and machines, the way that railroads and highways, the locomotive and the automobile, had already opened the West.[34] Zucrow supported Kennedy in the 1960 election, likely because he recognized a new young presence with reputable Cold War credentials, as well as a promising reform platform for civil rights and social welfare. He played up the New Frontier slogan, beginning in the spring of 1960, offering that "the next big engineering venture is outer space." Our "responsibilities as a world power require that we must be dedicated to the conquest of space." We must be "supreme." We must "win that race."[35]

This spirit also revived the spaceflight visionaries within the American Rocket Society (ARS). By February of 1960, the board of directors (Zucrow was on it) endorsed a long-term plan, based on the framework and achievements of its technical committees, to achieve true US interplanetary flight within twenty-five years. Technical experts in NASA agreed, setting the "Ten-Year Plan" to ensure the "lead time" to achieve a Moon landing, along with the necessary booster and spacecraft architectures, mission techniques, communication and control, and auxiliary space science networks. NASA also planned for advanced propulsion platforms to move from the Moon to Mars and the planets beyond.[36]

These initiatives culminated with the much publicized Space Flight Report to the Nation, an event between 9 and 16 October 1961, attended by some two thousand engineers and scientists at the New York Coliseum. Fourteen thousand people made their way through the exhibits and special events, dedicated to the "state of the art" in all the facets of spaceflight technologies, including the varieties of propulsion systems and interplanetary trajectories to which Zucrow and his corps of students were contributing. Zucrow was in charge of Space Education Day, the Saturday sessions devoted to high school and university students. It might not have been the most glamorous moment at the conference, not as alluring as the Redstone rocket framing the entrance to the Coliseum at Columbus Circle, or Wernher von Braun's stirring words as the keynote speaker. But it certainly was one of the most important, as confirmed by Theodore von Kármán, who advanced ARS appeals for more curriculums around the country in astronautics. That Saturday was also the only set of technical conference papers open to the public. Zucrow made sure that American democracy was on display.[37]

The ARS now prepared for an even more dramatic transformation: its fusion with the Institute of the Aeronautical Sciences (IAS), which had only recently renamed itself for the aerospace sciences. The ARS and IAS had memberships of about twenty thousand each in 1961, with about 16 percent of that shared. Both also saw an "overlapping of programs with duplication of effort," including a growing number of technical meetings every year. Leading professional bodies, like the Aerospace Industries Association, the American Society for Mechanical Engineering, the Society of Automobile Engineers, and the Institute of Radio Engineers, approved the unification. As a member of the ARS Board of Directors (1957–1961), Zucrow advocated the merger with the IAS. He then served on the board of the new American Institute for Aeronautics and Astronautics (AIAA) between 1961 and 1965. This was nothing radically new for him. He had been merging the two societies right from the start, promoting their joint meetings at Purdue and attending both of their annual conferences throughout the late 1940s and the 1950s.[38]

The proposed merger set Zucrow and his allies against some formidable opponents. As vice president of the ARS, Martin Summerfield lobbied against the merger, fearing that the "IAS 'Old Guard'" conservatives, representing big business and industry, would silence the visionary and "technically minded" engineers of the ARS, devoted to "rocket propulsion and space power."[39] Edward Pendray and Wernher von Braun did not take the merger well, seeing it as a sellout to the new industrial "aerospace" interests, a word Pendray counted as "faintly emetic." He saw the new "brobdingnagian" AIAA as a threat to true space exploration and all the ARS zealots with their "intelligent propagation of the faith."[40]

Von Braun's profile was still rising in the US at the time, feted as that rare "German" with the "vital extra-national drive . . . and intense desire to get into space." As the new director of the Marshall Space Flight Center in Huntsville, Alabama, he was "America's top missile expert," now sharing the media spotlight with celebrities like preacher Norman Vincent Peale and actress Donna Reed.[41] One of von Braun's speeches at this time revealed his powers of media persuasion, this at a conference on research and management sponsored by the Purdue Engineering Experimental Station, no less. It was a talk filled with his typical humor and lighthearted American clichés, references to rock and roll and the twist, crabgrass and cocktail parties. He mentioned how, as a youngster, he was "ready to hop in a rocket and blast off for the moon," how he and the Nazis "hit the big time" with the V-2, and how after "years of wandering in the wilderness" he had now become America's "prophet" of spaceflight. Most of the speech advised the audience how to become "ideal scientist-managers" and lead a "complex research and development project." He claimed that what he did in Nazi Germany at Peenemünde in the 1930s, a combination of innovative thinking and close-knit discipline, was a formula that still applied at NASA in the 1960s. "It works on the same fundamental rules in this country as it did in those years in Germany," he concluded, in a stunning admission of easy equivalence.[42] Von Braun, whose Peenemünde team had only recently transferred to NASA, his first civilian job ever in his career, was still thinking in the formal terms of closed military bureaucracies and arsenal systems.

Zucrow worked within a completely different life experience, one at the heart of the American project system. At the Panel on Science and Technology in April of 1961, he made his strongest claims on the issue of propulsion, his expertise. He warned against being "satisfied far too readily with accomplishments." Like the achievements of the Saturn booster, of all things. It was nothing more, he said, than "an assembly of available engines." The US, he advised, needed new engines with larger thrusts for true spaceflight within the solar system. Here was an echo of his NACA appeals from 1952, only now by orders of magnitude beyond, even up to fifteen million pounds of takeoff thrust. He called for advances in nuclear propulsion with Project Rover to advance beyond Earth orbit. "The key to space is in propulsion," he announced, and therefore the US needed "a spirit of urgency," a "crash program."[43] All of this came in his short speech before the congressional panel on 17 April 1961, one week after Yuri Gagarin's flight into orbit, one month before President Kennedy's famous "Moonshot" speech. Perhaps the president or his advisers were listening.

President Kennedy's forthcoming "Special Message to the Congress on Urgent National Needs" (25 May 1961) dramatically announced the Apollo goal: "I believe that this nation should commit itself to achieving the goal, before this decade is out, of landing a man on the moon and returning him safely to the earth." Here were famous words that relaunched NASA's Project Apollo, and its Saturn rocket, from technical to political imperatives. The Kennedy initiative had many sources and correlations, as historian John Logsdon has explored. The president was responding to the mounting succession of Soviet space firsts, peaking with Gagarin. He was answering his own campaign promises of extending a New Frontier. He was also deeply embarrassed by the failure of his attempt to overthrow Fidel Castro's communist regime at the Bay of Pigs (17 to 20 April).[44] Kennedy's young administration (only a few months old) needed a space spectacular to raise American pride and his own reelection prospects.

We also tend to forget the very next sentence after his "I believe" quote: "No single space project in this period will be more impressive to mankind, or more important for the long-range exploration of space; and none will be so difficult or expensive to accomplish." This reference to project was an old currency in a new coin. One minted especially for Kennedy, to represent his investment in the military–industrial complex, his commitment to a new American technocracy. His administration featured the "best and brightest," stars like Robert McNamara (secretary of defense) and James Webb (director of NASA), proven technical managers of vast government offices and industrial enterprises. Or as one of the president's close allies, Democratic senator from

Washington Warren Magnuson, put it: "Managerial competence is essential in order to translate growing scientific and technological achievements into practical projects, programs and applications. As a nation we recognize urgent forces accelerating our drive to create new complex systems to prevent or win a war with communism—or send a man to the moon."[45]

When the Panel on Science and Technology met next, Zucrow was excited by the president's moonshot speech, seeing it as the culmination of a "great revolution" happening in the world since 1957. He certainly agreed with the president's further comments on the need to better "develop alternate liquid and solid fuel boosters, much larger than any now being developed." This included a tripling of funds to "accelerate development of the Rover nuclear rocket. This gives promise of someday providing a means for even more exciting and ambitious exploration of space, perhaps beyond the moon, perhaps to the very end of the solar system itself." These Kennedy appeals were a threshold for Zucrow. He now became one of their national advocates, America's expert on propulsion and the spaceflight future, as expressed in two manifestos: his thirty-four-page "Statement on Propulsion" as presented to the US Congress (March of 1962) and his contributions as a prestigious Sigma Xi lecturer encapsulated in a major article for its academic journal, the *American Scientist* (September 1962). These works were summations of his twenty-year career at the forefront of American rocketry and spaceflight. In them, Zucrow cast his unique style as a cautious visionary, balancing both skepticism and optimism. He recognized the technical challenges of factoring in the "lowest" cost and the "highest" reliability, but also called for advanced chemical propulsion systems, now raised to twenty million pounds of thrust, in order to build the space stations and exploratory flights of the 1980s, including the development of a reusable "manned booster for placing loads in orbit." He offered a "reasonable assurance" that the US could "proceed with confidence." Above all, he made the case for "escape velocity," for true space exploration beyond Earth. His were classic studies in what most congressmen, and average Americans, considered rocket science, filled with incomprehensible equations, thrust-to-weight ratios, specific impulses, energy cycles, and performance parameters. Zucrow also offered simpler primers on combustion chambers, exhaust nozzles, molecular weights, heat transfer, film cooling, combustion pressure oscillations, and electrodes. In all these cases, he said, propulsion was the "*rate-controlling influence*" for all space exploration.[46]

Zucrow also called for advanced propulsion systems and the energy systems to survive in space—of the nuclear, electric, ion, plasma, and solar varieties. These included "nuclear heat transfer and electrothermal [thermal arc-jet] rocket engines" of increasingly larger sizes. He expressed special confidence in the recent successful nuclear reactor engine tests for the Project Rover Kiwi-A reactor, the Nuclear Engine for Rocket Vehicle Application (NERVA), and Systems for Nuclear Auxiliary Power

(SNAP-8).[47] Zucrow was not alone in this advocacy for advanced propulsion. NASA's Office of Advanced Research and Technology (OART) was setting the stage for these new modes of "space power generation," planning for the "foreseeable future" of human and robotic missions to the "Earth, Moon, Mars, Venus, Jupiter, Saturn."[48] One of the Rocket Lab graduates, David Elliott, now an engineering group supervisor in the Propulsion Research and Advanced Concepts Section at JPL, was already designing "electric-propulsion power plants" and a "magnetohydrodynamic space power system" for true interplanetary flight.[49] These new trends built Zucrow's confidence. He was more forthcoming at the next panel meetings. In January 1963, he even provoked a small "furor" in Congress with his advocacy of going beyond the Moon and on to Mars and the further "ocean" of space. The US, he said, to the consternation of some of the other experts and congressmen, needed to "break out of its current lull" and mere 1 percent of gross national product funding in order to "go unhindered through space."[50]

These appeals were not just words. They were based on actions, investments Zucrow had made in nuclear propulsion ever since he returned to Purdue in 1946. Throughout these years, Zucrow worked closely with NACA and NASA Lewis, especially at its Plum Brook Reactor Facility in Sandusky, Ohio, where he consulted on its gas nuclear core reactor and nuclear propulsion projects.[51] He encouraged his graduate students to specialize in the physics and engineering of nuclear propulsion, both for aircraft and rockets. They took minors, wrote visionary reports and studies, and made careers in the field. Among them, in diverse capacities, were Walter Hesse, Richard Duncan, Elliott Katz, and Howard Rodean. Cecil Warner joined them to ensure a nuclear future for the Rocket Lab. For his second sabbatical at Aerojet (1963–1964), Warner served as a technical specialist in the Nuclear Operation Department, "engaged in the development of the first flyable nuclear rocket for NASA." He was responsible for ensuring that the nuclear reactor, then being designed and built by Westinghouse, "produced the high temperature gas required by the rocket." He worked on heat transfer, two-phase flow, computer programs, and evaluations of all components.[52]

Two of Zucrow's PhD students were among the country's top nuclear propulsion experts, eventually working for the joint NASA–Atomic Energy Commission (AEC) Space Nuclear Propulsion Office (1959–1973). Charles Trent (PhD 1951) ran the primary contract for a nuclear rocket engine at Aerojet. In an interesting homage to Purdue's long expertise in steam engines, Trent helped direct the Aerojet test facility (worth "several million" dollars) that used "several 400-kilowatt circulatory loops, fed by steam boilers," to heat a sodium–potassium alloy for reactor tests. Trent remained true to the Boilermaker tradition, now for rocket engines. He also helped run the tests of the Nuclear Rocket Experimental (NRX) A5 to A7 series at the Jackass Flats, Nevada, site, an old AEC bomb area. His boss at Aerojet, Chan Ross, the senior vice president and head of the Nuclear Division, remembered the "extremely severe" working environments and technical issues, the "enormity of the task" of placing a functioning nuclear rocket engine into the vacuum of space. Government inertia and inexperience also intervened, creating a "continual state of confusion."[53]

Stanley Gunn (PhD 1953) ran a parallel and more intense project at Rocketdyne, pioneering the liquid oxygen and especially liquid hydrogen "flooring systems" for its ion and nuclear propulsion programs, including the feed and cooling systems and nozzle designs. He planned the Rocketdyne reactor test facility, helped to plan the Jackass Flats site, and made major contributions to the Kiwi, Phoebus, and NRX programs. Presidents Lyndon Johnson and Richard Nixon canceled all these high-end nuclear engine hopes and plans, even amid a series of successful tests, along with all of the plans for building space stations and lunar colonies. Yet not all was lost. Gunn's expertise with liquid hydrogen for nuclear engines translated into real results, as Rocketdyne also designed and built the superb liquid hydrogen J-2 engines for the Saturn rocket. His work on heat transfer, combustion instability, and oscillation issues for the prototype J-2 engine helped convince a worrisome NASA that its "design" features were "workable." Gunn also briefed a doubtful Wernher von Braun on all these possibilities during one of his visits to Rocketdyne.[54]

Back at Purdue, S. N. B. Murthy, a Zucrow protégé, assigned these kinds of futuristic technologies as the centerpieces of his graduate seminar, Space Propulsion and Spaceflight Systems (ME 654), offered through the Rocket Lab in the fall of 1964. Using the latest data, often just a few years or months old, on loan from the files of Douglas and Aerojet, General Electric and Lockheed, students wrote term papers on the performance characteristics and costs analyses of varied projects. One student wrote on the possibilities of a bold reusable launch system to get ten astronauts to orbit, and eventually on to the surfaces of the Moon and Mars, all by the 1990s. Another wrote on ablative reentry technologies, the means to shield ballistic missiles, and spacecraft, and even engine nozzles from the extreme heat of launches and reentries. Space planes like the X-15 needed it as well for "hypersonic airframe construction" to break through engineering's "thermal barrier." Several other students wrote about futuristic propulsion systems, like electric and ion propulsion, nuclear reactors and solar cells, either to get us into space or to power us once there.[55]

In the spring of 1969, anticipating the thrill of Apollo 11's Moon landing, the Junior ROTC class at Purdue dedicated the whole spring semester to creating a viable space station for Earth orbit, what they called Project Metro. It was a dramatic contrast with their ROTC classmates from fifteen years earlier, so fixed had they been on the guided missile as the wave of the future. Under Maj. David H. Clegg's supervision, cooperating with a team of professors from aeronautical, electrical, and nuclear engineering, the class divided up into committees that focused on propulsion systems (with liquid-propellant and nuclear options), orbital trajectories, guidance and control issues, life support, caloric requirements, energy systems, radiation hazards, reusable resources (the station aimed to be 50 percent self-sustaining), and structural designs for this "Space City." This was a course in engineering, framed in higher mathematics and technical details. But the students included a chapter on space sociology, how to choose the right personnel for the project and ensure they would survive and thrive in orbit, with attention to religious and recreational needs and self-government. It was not all optimism. Our planet was doomed, the students warned: "Situation-Earth is rapidly becoming over-populated." All the more reason to build upon Apollo and make way for an orbiting city of one thousand people by the 1980s. The exercise was not purely academic. Among the junior class was one Jerry Ross, soon to be a graduate of the Zucrow Labs and a record-breaking astronaut for NASA on the space shuttle and the International Space Station. For him, at least a small part of the seminar came true.[56]

Purdue's graduates helped to fill the ranks of spaceflight industries and institutions in these years, a number of them in leadership positions. Dr. Harold W. Ritchey (PhD 1938) was executive vice president for Rocket Operations at the Thiokol Chemical Corporation. George Mueller (MSEE 1940) was associate administrator of NASA and director of the Office of Manned Space Flight. Dr. Raemer Schreiber (PhD physics 1941) was head of the nuclear rocket propulsion program at Los Alamos Scientific Laboratory. Harry L. Markus (BSEE 1941) was responsible for the development of ten of the flight instruments for the Gemini spacecraft. Dr. Donald Sulken (PhD 1950) was head of the recovery operations branch, Manned Spacecraft Center, Houston. Beyond this, "hundreds of Purdue alumni connected with NASA and the Bell System" participated in the launch of the first communications satellites, *Telstar 1* and *2* (1962–1963). "Thousands of Purdue alumni participated indirectly in the teaching, research, and production" of Project Gemini. Large and active Purdue alumni clubs popped up in places like Huntsville, Alabama, and Cocoa Beach, Florida. Where the space age went, Purdue followed.

And then there were the astronauts. Purdue's recognition of them began modestly enough. Gus Grissom (BSME 1950) and Neil Armstrong (BSAE 1955) were honored in the Purdue alumni magazine, Grissom for *Liberty Bell 7* and Armstrong for flying the X-15. The university now spotlighted the two more and more, together with Eugene Cernan (BSEE 1956) and Roger Chaffee (BSAE 1957), given their achievements in the Gemini and Apollo programs. They returned to campus for Gala Week and Founder's Day. They flew Boilermaker flags in space as gift souvenirs for Boilermakers back on Earth. They were feted as Distinguished Engineering Alumni. Grissom, called the "first 'Buck Rogers' of the space age," received a standing ovation from 250,000 people at the Indy 500 in May 1965, where he "orbited" the track in the pace car. All four astronaut alumni went to the Rose Bowl game on 2 January 1967. Purdue's float in the parade featured their names, a Gemini capsule, and the motto Alma Mater of Astronauts.

Star quarterback Bob Griese even took Purdue to a close win over the University of Southern California (14–13). Only three weeks later, on 27 January 1967, the Apollo 1 disaster took the lives of Grissom, Chaffee, and Ed White. The campus grieved over the spring semester, "where curious college students began to grow so big that they could never be content to spend all the days of their lives on a little planet called Earth."[57]

## THE NASA HIGH PRESSURE COMBUSTION LAB

Zucrow took his first sabbatical in the spring semester of 1959, five years overdue, turning it into a working national tour of "educational institutions, governmental installations, and manufacturing establishments in connection with the use and manufacturing of propulsion systems for guided missiles and aircraft."[58] It was not much of a rest. He visited the University of California, Aerojet in Azusa and Sacramento, Cape Canaveral, the Martin Company in Orlando, and Redstone Arsenal. He met with fifteen of his graduates in California to discuss possible curricular changes to the Rocket Lab program. And he went to Washington, DC, with Warner and Reese to pitch research proposals to the Office of Naval Research, Navy BuAer, the Air Force Office of Scientific Research, and NASA.

The tour included a sweep through the South: Tulane University, Louisiana State University at Baton Rouge, Louisiana Polytechnic Institute at Ruston, and Mississippi State University. These were not among the country's premier engineering schools, but they aspired to join the new trends in aerospace engineering, specifically in spaceflight. Zucrow gave them advice and encouragement about creating graduate programs in jet propulsion. In this, he was a trailblazer. This is exactly what NASA did just a few years later, seeking to close the "gap" in the "project system" between the elite and well-financed aerospace science and engineering programs and the poorer, less developed ones.[59] Zucrow returned to campus to claim the prestigious Vincent Bendix Award for "outstanding contributions in engineering education" in June of 1960 at the national conference of the American Society for Engineering Education (ASEE), held at Purdue with some three thousand attending, including Theodore von Kármán, his old boss from Aerojet. Purdue had also recognized Zucrow with a Distinguished Professorship in 1959, the first in engineering. His friend Herbert C. Brown, professor of chemistry and soon a Nobel laureate, received one too.

These events prefaced some of Zucrow's busiest final years on campus. One reward came in the rebuilding of the Rocket Lab, thanks to NASA's new Sustaining University Program (SUP), a program shaped by President Kennedy's reinvestment in the project system and by his moonshot imperative. Starting in September of 1961, through NASA's Office of Space Sciences and Applications and its Office of Grants and Research Contracts, SUP cultivated a series of "vital" and "decisive" university relationships to help educate the three million college students awaiting the country's spaceflight future. The pilot program chose ten schools for the initial research and training grants, reaching its "peak" in 1966 with $45 million to seventy-five universities and eventually funding some three thousand research projects.[60] SUP was part of James Webb's wider initiative to draw the regional parts of American society and industry into NASA's aerospace missions. He set out to turn America's universities into "spinoffs" of a kind, sites for technology transfer, science education, and economic potential.[61] Purdue joined the third wave of winners in 1965 with a series of NASA multidisciplinary grants, initially $600,000 for three years. One of the justifications for these Purdue grants came from the Rocket Lab and the placement of its graduates in industry and the military, "more than any other university." The monies funded research in "applied mathematics and the basic sciences" to help solve the "foreseen and unforeseen" problems of spaceflight. The accent was on the new, in terms of advanced topics and younger scholars, including assistant professors and graduate students who, in a favorite trope of the 1960s, would now be able to start "launching their own research projects."[62]

There were two significant complements to these grants. One was an initial $225,000 NASA grant to Purdue in 1966 for satellite multispectral sensing in agriculture. This was at the very beginning of geodesic Earth-centered studies from space, and for which Purdue established its vanguard Laboratory for Applications of Remote Sensing (LARS). By 1973, serviced by the Earth Resources Technology Satellite (ERTS) and later Landsat satellites, with NASA funding of $2 million a year, Purdue's LARS was helping the Department of Agriculture to map crops and predict yields, identify natural features of forests and rivers, and survey the spread of crop blights and forest fires.[63] A second grant came from the US Air Force to establish a new master's of science (MS) degree for officer graduate training in engineering. The program began in 1963 with fourteen graduates from the Air Force Academy. It was initially named the Aero-Space Environic Engineering Program for a "far reaching" new Air Force curriculum to handle the "substantial revolution in its mission, its weapons, and its environment." Students had two years and eight months to complete any MS degree in engineering, with a focused minor, as for example aeronautical engineering with a specialty in thermodynamics and heat transfer. Over the coming years, the program helped graduate a small cadre of Purdue alumni for the future space shuttle: seven in all. Purdue secured its legacy as a maker of astronauts.[64]

By 1963, amid this accelerated pace of funding, Zucrow and the Rocket Lab reached a ponderous "crossroads," in Cecil Warner's terms. The choice was to either remain small and overcrowded or double the size and significance of the lab and meet the real needs of its forty-some graduate students. SUP decided the question, especially its special category of facilities grants to support advanced research, directed by NASA's OART. Purdue and its main rival, Princeton's Guggenheim Jet Propulsion Center, competed for the prize, with draft proposals readied by February of 1963. Warner called Princeton's application "in many respects a carbon copy of Purdue's proposal." On a visit to Princeton's campus, seeing its rising fortunes, he was impressed with how "Purdue's Jet Propulsion Center in a very real sense is in direct competition with Princeton for graduate students, as well as financial sponsorship for research."[65]

Princeton received the windfall first, for technical education in "scientific space exploration" and "practical space systems." The construction of its new facility began in November 1963 and was finished a year later, "ahead of schedule." NASA awarded $625,000 and Princeton added another $125,000 for premier research labs for chemical, electric, and nuclear propulsion. Princeton was flush with money in these years, having just concluded a campaign to raise $53 million to rebuild its campus, including $8 million for a central new Engineering complex.[66]

The terms of the Princeton dedication set the rivalry with Purdue into high gear, framed by all the stock hyperbole of the early Apollo years. President Robert Goheen positioned Princeton as the nation's undisputed leader in rocket propulsion research, heir to Robert Goddard, Charles Lindbergh, and the Guggenheim Foundation. Princeton represented the "surging, conjoined, forward thrust of scientific inquiry and engineering practice." Raymond Bisplinghoff, NASA associate administrator in OART, called Princeton's new center the best "anywhere" in the world, comparable to the great pyramids of old, though reaching ever higher, even into Earth orbit, by "three orders of magnitude." "By 1970," he boasted, "we are confident that we will command the space occupying a concentric sphere extending from the surface of the Earth to the orbit of the Moon, an additional three orders of magnitude." Only the planets and stars remained. "Each successive leap into the new space beyond demands a vast growth in technology—a geometric progression in growth." These leaps depended on places like Princeton.[67]

Purdue's final draft application for the facilities grant was in place by December 1963, although the initial NASA approval was not without controversy. Robert Caro, in his biography of the new president, Lyndon Baines Johnson, argued that the negotiations for the grant were the result of politics, not science; favoritism and a backroom deals, not merit. The story began with Charlie Halleck, Republican congressman from Purdue's Indiana district, and also the House Republican leader. Johnson needed Halleck's crucial support for the coming civil rights bill. So he injected some useful drama into the Purdue story. With Halleck sitting next to him in the Oval Office on 18 January 1964, President Johnson called NASA director James Webb on the telephone. "I need to do anything I can for Charlie Halleck," Johnson told Webb. Under this pressure, Webb agreed to keep Halleck "satisfied." At a subsequent 21 January meeting with

Halleck, Webb promised "that we could do some things at Purdue," including "a building that would run three-quarters of a million dollars, and we're talking about some research grants and contracts." The deal depended, of course, on Halleck's subservience. Johnson told Webb to ensure it with three-year grants "on an installment basis" for "as long as he cooperates with you."[68] These meetings were political theater of a high order, Webb knowing full well that the grants program was already in the works. Caro gave the event a false currency, as if Zucrow only received the funding by political accident. In time, the story made the rounds on the Purdue campus, a smear by some of his rivals to punish Zucrow for his earlier break from Aeronautical Engineering.

NASA and Purdue signed their memorandum of understanding (MOU) for Research Facilities Grant NsG(f)-21 in April of 1964. It was a foundational document, at least as far as Webb was concerned. He expected the MOUs in the SUP series to be a bridge between NASA's advanced technologies and the country's local economies, a means to embed a high-technology marketplace across the US, with active university participation. Most MOUs did not meet his high standards, so NASA analysts and historians have judged. Purdue's did. It was the centerpiece of the university's efforts to diversify its curriculums and the local economy in advanced research. Zucrow threaded the document with his congressional testimony and Sigma Xi lecture, and with a full review of his Rocket Lab's achievements on "propulsion science and technology" as "the key to space exploration . . . the rate-controlling influence."[69]

Purdue received an even larger package than Princeton: $829,000 from NASA for the laboratory itself and $410,000 a year (for three years) as start-up funds for liquid-propellant rocket research. The NSF shared one-third of the costs with Purdue (out of $650,000 total) for a new administrative building, named in honor of astronaut Roger Chaffee, with two stories of classrooms, offices, a classified and open library, and an electronics lab and drafting room. Zucrow's grant was a relatively modest one in the national scheme of things. It beat out Princeton's grant, though several schools received much larger facilities grants: MIT, $3 million; Case Western Reserve University, $2.2 million; the University of Minnesota, $2.5 million; and Stanford and UCLA, $2 million each. At about the same time (1959–1967), among NASA's "obligations to universities" were $87.5 million to MIT for the Apollo guidance system and $1.3 billion to JPL and Caltech, which included the Ranger and Surveyor lunar probes.[70]

The NASA monies enabled Zucrow to build the High Pressure Combustion Research Laboratory, the only one of its kind in the US. The new Purdue lab made sense. NASA Lewis, which had oversight for the project, had "little interest in high pressure engines" of its own. So this was a perfect fit. Zucrow also patterned the new lab's "control console layouts and data acquisition system" on the NASA Lewis Rocket Engine Test Facility and nuclear Plum Brook Station. The two NASA labs were his models.[71] Although nominally a NASA and NSF facility, the new High Pressure Lab was also a second Aerojet extension, its satellite in the heart of the Midwest. Zucrow built the first one with graduate student support beginning in 1946. Now, some twenty years later, another of his graduate students, Chuck Ehresman (BSME 1949, MSME 1951) returned to Purdue after ten busy years at Aerojet, where he had served as a leading test engineer. At Purdue, he was now responsible for the "design, implementation, and instrumentation" of the new High Pressure Lab, "on loan" from Aerojet, which paid half his salary during the project.

As with the Zucrow's original Rocket Lab, Ehresman planned and built it with a small team of current graduate students. H. Doyle Thompson and Donald Crabtree were in the mix. So were J. Michael Murphy and D. L. Carstens. But Robert Strickler, with a BSAE (1960) and an MSAES (1962) from Purdue, bore most of the responsibility. Zucrow brought him back from a stint at Aerojet, where Strickler had assisted in its design of a proposed Apollo lunar module. For him it was a transformative moment: "eye opening" and "career changing," his pathway to the PhD. Under Zucrow's and Ehresman's guidance, Strickler negotiated with contractors. He prepared budgets. He played a major role at a crucial NASA Lewis functional design review and approval meeting in Cleveland (12 November 1964). Zucrow was the team leader. Ehresman presented "the building plot plans," with a flowchart of tasks, test cell schematics, and the architectural design. Strickler was responsible for presenting on the "mechanical test system schematic," including the "pressurization gas supply," the "pressurization system," the "propellant supply system," the "water supply system," and the purge system, test stand, and noise attenuation system. That was quite a list of charges for a novice PhD student. NASA Lewis even questioned the very "feasibility of having graduate students operate the facility at Purdue University." This seemed incredulous given the inherent sophistication and dangers of rocket motor testing. But Zucrow defended the "caliber" of his students, also noting that he had a staff of long-standing "test mechanics" who ensured the safe operation of all the machinery.[72]

In perfect Zucrow style, the team prepared for that 12 November meeting at NASA Lewis with a series of "dry runs" the week before, including a Saturday 8:30 a.m. practice session. Ehresman demanded a "smooth-flowing verbal presentation," with a package of "clear eye-catching" visuals of slides, charts, and movies; the "ink penciled flip chart" for readability and comprehension"; and ready answers to the inevitable questions. The team brought all of these to Cleveland, along with Zucrow's sense of humor. He kept the meeting in stitches. "Humor properly used can often establish the perspective more effectively than technical facts," concluded Ehresman. Strickler also remembered Zucrow's flair for showmanship, as they flew to Cleveland's airport in a Purdue Aeronautics Corporation DC-3, taxiing over to the adjoining NASA-Lewis Center. "We flew from our airport over to NASA Lewis, and he had the pilot pull right up to the front door of the NASA administration building . . . the airplane just pulled up and blew their minds." Without airports of their own, that was a spectacle even Princeton and Caltech could not match. It was all part of the Zucrow style in "selling the package."[73]

To design and build the new High Pressure Lab, Ehresman wrote a minor classic on the role of the university laboratory within NASA's project system. It was a fitting tribute to his mentor, Zucrow. He broke down the mission of the "project engineer" into its experimental, developmental, and production elements. He celebrated basic research and all the "ingenuity" and hard work of the experimental scientist/engineer. He applied the PERT method to diagram a project's "logical sequences of events and activities," reducing the tasks into achievable and timely increments, gridded into the pattern of circles and arrows that defined the optimal "paths." Here was the classic "stepwise procedure" at the heart of the engineering approach, one by which to build the new lab and then operate it. At least for the next ten years. That was all he would give in terms of a life span before future technologies rendered it obsolete.[74]

Ehresman needed this project plan given the newness and immensity of the tasks, along with all of the technical challenges. The normal propulsion tests on rocket motors at chamber pressures of 5,000 psi required high-pressure storage tanks at 8,000 psi. No one had ever built such tanks and feeding systems for such a laboratory before. Ehresman and his team even calculated the TNT equivalents in the worst-case scenario of an accidental explosion. The Rocket Lab also planned "all major mechanical and instrumentation and control system components" on its own. These all required the prior approval and very stringent standards of NACA Lewis, via correspondence, telephone calls, reports, and official visits. Between the start of planning in September of 1964 and breaking ground in May of 1965, NASA Lewis administrators commanded precise attention to a series of further demands: risks to the nearby runways of Purdue Airport, the peril of "white noise" pollution to local residents and students, and dangers from the "release of toxic gases, contamination of ground water, and projectiles from an explosion."[75]

For Zucrow, building Chaffee Hall was a "difficult and unenjoyable task." He had to scrap plans to build a half-million-dollar "space laboratory," one to include the full array of nuclear and other novel propulsion systems, after Purdue failed to provide the full matching funds. President Hovde and the deans had approved $900,000 in spending, but the forceful new vice president and treasurer, L. J. Freehafer, delayed and hedged. "Well, the exercises I went through were thoroughly disgusting," remembered Zucrow. He had to scale back. Part of the problem was the nature of the funding. These were grants. There were no overhead costs to be siphoned. The earlier days of informal agreements had given way to more formal bureaucratic procedures. "The University is so organized that the emphasis is on business," remembered Zucrow. "It reminds me of a military organization rather than an educational organization."[76]

There were some rewards. The Indiana–Purdue football program for November 1964 featured Zucrow's portrait, calling his Rocket Lab "the nation's leading producer of engineers and scientists for the propulsion field."[77] There were also some interesting moments along the way. At one point Ehresman reached out to the Boiler Division of the Babcock and Wilcox Company for a quote on four propellant storage tanks with "a pressure rating of 150 to 250 psi." This may not have been the most sophisticated of the various storage systems. But they were boilers of a kind, made by the very outfit, one of the master builders of railroad locomotives, that had helped build Potter and Solberg's High Pressure Steam Generator Laboratory in 1929. Ehresman's work order brought Zucrow and Purdue full circle, back to their origins as Boilermakers. In another case, with the assistance of NASA Lewis, Ehresman and Cecil Warner procured a series of storage tanks from a decommissioned Atlas ICBM site near Cheyenne, Wyoming. The large ones were forty feet long and offered "high pressure storage capacity" at 8,000 psi. Here was one of the first peace dividends of the Cold War. Zucrow had helped to advance Atlas as a viable ICBM project in 1952. Now he turned the ground equipment for a nuclear-tipped missile into a laboratory storage tank.[78]

The High Pressure Lab concentrated its research on cooling methods for liquid-propellant engines, along with studies of combustion pressure oscillations, supersonic nozzle design, and solid propellant combustion. Two new test cells measured twenty by forty feet, with extra thick eighteen-inch concrete walls and with panes of observation glass. Blast walls caught the exhaust and cushioned the sound waves by angling them upward. Closed-circuit television covered the test cell firings. A twenty-four-channel oscillograph and nine-strip chart recorded the instruments for analog data, along with a seven-channel analog tape recording system. Computers recorded and relayed the data from the firings, fluid flows, and storage tanks. This included a sixteen-channel low-frequency digital data system with taped output and an IBM 7094 for data processing. What had once taken weeks to interpret now took only hours.[79]

Publicity featured the new lab with high hopes. The local newspaper called it Lafayette's own version of Cape Kennedy. Reporters promised that the Purdue's "jet set" lab "would thrust America into a fantasy-filled future of interplanetary travel," control of the weather, and even antigravity machines. Ehresman pledged to help build "earth orbital" and "inter-orbital space ferry vehicles (earth, earth-lunar, earth-interplanetary)" with "chemical, solid core nuclear, nuclear pulse, nuclear electric, gas core nuclear, [and] controlled thermonuclear propulsion."[80] At the new lab's April 1966 dedication, John Sloop, director of propulsion and power generation in OART, defined the present aerospace moment as an era of "geometric growth," just like the Renaissance and Industrial Revolution. Only this time humanity would break through to "advanced space flights for the 1980s," including missions to the Moon and Mars. Sloop named Zucrow's pioneering Rocket Lab as the "leading edge of propulsion technology" and the country's top producer of "knowledge and well-trained young engineers."[81]

The Purdue High Pressure Lab also represented a culmination of the project system for aerospace R&D, at least in terms of the American university. At this very moment, the vast majority of NASA funding for universities did not go to SUP multidisciplinary research grants ($50 million), or to its predoctoral training programs ($100 million), or to facilities grants ($43 million). Rather, two-thirds of SUP monies, $467 million in all, went to individual R&D projects. As one evaluation defined it, "The project system is characterized by grants or contracts to individual faculty members by a federal agency for investigations of a single problem or several closely related problems associated with some concept or phenomenon." In a remarkable recognition of Purdue's and Zucrow's own approach since the 1920s, the NASA task force admitted that these small and diversified grants, especially when centered on the education of graduate students, worked the best. NASA's investment drove a new peak in the production of engineering and science doctorates in the US, one to match the first peak at the time of Zucrow's own doctorate and one now achieved in the first years after Zucrow's retirement, and to which his career and writings contributed significantly.[82]

These busy years witnessed Zucrow's handover of the lab to his student successors. It actually began as early as 1958, when he "let these younger men take over the full responsibility for the operation of the Jet Propulsion activities, for the sake of the experience involved, and in order that we may watch them and judge their capabilities."[83] The initial plan did not proceed smoothly. Cecil Warner cautioned that the lab lacked an "effective organization" and was rent by division. Some on the staff were more single-minded than community-minded. These were years of occasional contention and dispute. Bruce Reese finally took over in September of 1965. Warner commended his ability to "evoke cooperation and loyalty." Zucrow called him the person most "technically competent" and someone "of high ethical principles."[84]

Meanwhile, Zucrow took a leave of absence without pay in the spring of 1963 "for personal reasons," mostly to "build up his health and do some technical writing" in a "warmer climate."[85] He returned the next academic year to experience the January 1964 blizzard, with 40 mph winds and 9 inches of snow, some drifts up to 8 feet high. Purdue closed the campus on the thirteenth, a rare event in that day. The fall of 1965 was Zucrow's last full-time semester at Purdue University. He took a "terminal leave" for spring semester and retired formally as of June 1966. Purdue organized a deserving send-off, with a banquet at the Purdue Memorial Union (in the West Faculty Lounge) for a capacity crowd of over one hundred people. Hovde praised his old friend for turning a $25,000 investment into a $6 million "complex," all thanks to "his longtime and stubborn insistence that the rocket jet was the propulsion plant of the future for both aircraft and missiles." His "state-of-the-art" innovations in higher pressures and cooling designs "squeezed more power from available fuels" than any other technique. Harry Solberg rightly noted that Zucrow's proudest achievements were "the men he developed in his engineering classes and laboratories," the rising leaders of American aerospace. The Rocket Lab was, after all, the "first in the country devoted exclusively to educational research."[86] At his retirement, Purdue's public relations office disseminated Dr. Zucrow's achievements in colorful terms, one of its writers turning him into the "Russian-born professor" and "Russian-born aero-space pioneer," though he only spent the first year of his life in Ukraine. It was as if, in this era of Sputniks and Gagarins, that small detail made him even more of an authority, a local boy done well, if from a country far away.[87]

Further honors poured in. In June of 1966, at a gathering of propulsions engineers in Colorado Springs, Zucrow received the ultimate approval, the James Wyld Propulsion Award "for outstanding contributions in the development or application of rocket propulsion systems." After the Pendray Award for literature and the Bendix Award for education, here was a rare trifecta of ARS and AIAA recognitions. The ceremonies took place on the new campus of the Air Force Academy, framed by the aerodynamic style of its Cadet Chapel and modernist buildings, set against the backdrop of the western edges of the Rocky Mountains. In what must have been a most gratifying moment at the conference, two of Zucrow's students, Joe D. Hoffman and H. Doyle Thompson, delivered a cutting-edge paper on computer models for rocket research. It was, appropriately, research cosponsored by Aerojet and NASA. Zucrow also received the Distinguished Civilian Service Award of the US Army in May of 1967, its first and "highest recognition" for a civilian, in a ceremony at Redstone Arsenal. The Army Missile Command praised Zucrow's capacity to "get things done" and to resolve seemingly "insurmountable" problems, especially with the Corporal, Sergeant, Lance, and Sprint missiles. At the time, Rocketdyne's William Bollay, one of America's premier rocket pioneers, gave Zucrow the highest praise, naming him the "new von Kármán."[88]

With Bruce Reese and of course Cecil Warner, five more of Zucrow's PhD students succeeded him at the Rocket Lab: John R. Osborn, Joe D. Hoffman, Mel R. L'Ecuyer, James G. Skifstad, and H. Doyle Thompson. Ehresman stayed on too. This was Zucrow forming his "own ball club" with Hovde's support, though other campus leaders feared the professional stigma of so many self-hires.[89] The NASA and NSF grants had set this team up for success, creating a vanguard research laboratory and new classroom building. Mechanical Engineering advertised all of this in smart new brochures for the Jet Propulsion Center and its studies in "Rockets, Jet Propulsion and Gas Turbines," with specialties in "film cooling of high-performance rocket motors, combustion pressure oscillations in chemical rocket motors, combustion of solid and hybrid rocket propellants, and a new field in energy transfer in electric arc plasma jets" for space travel. There was also a broad roster of courses by Zucrow's team. Mechanical Engineering's curriculum had undergone a remarkable transformation in twenty years. From Zucrow's revolutionary few courses in 1946, Purdue now offered over three dozen new undergraduate and graduate course listings.[90] The team also divided up the lab's graduate students as their advisees and thesis students, some seventeen PhD and twenty-one MS candidates.[91]

The transition was not seamless. Zucrow's students, now teachers, did not sufficiently plan how to integrate prospective MS and PhD students into the new high-pressure curriculums, which seemed flat and inflexible. This, at least, was the judgment of advisers in Mechanical Engineering, who saw a precipitous decline in the quality of students who accepted Purdue offers for 1967–1968. They advised that the Rocket Lab "propaganda" had to improve, and improve fast. It was "not getting our message through," they said.[92] Part of the trouble was Purdue's rather isolated location in north central Indiana. Undergraduates did not stay. Graduate applicants did not come. The West Coast was especially alluring, engineering graduates tempted by the opportunities for sailing and fishing, golfing and skiing, in places like Seattle and Santa Monica.[93] At the time, Purdue feared that Indiana was suffering a "brain drain," turning into a "drag on the nation's economy." The state had fine universities with excellent degrees, yet only small percentages of their graduates stayed local. One of the problems was that Midwestern economic markets focused on consumer manufacturing rather than the more glamorous military and aerospace R&D. The Midwest was in peril of becoming an "intellectual Appalachia." Zucrow agreed, noting that of all the Rocket Lab graduate students between 1946 and 1960, less than 10 percent were Indiana residents, and less than 5 percent stayed nearby. "It is unfair," he judged, to "burden the Indiana taxpayers" in this way.[94]

Among those applicants who declined the Purdue Rocket Lab for the 1967–1968 academic year, six went to Stanford and six to MIT. Among the others, two each chose Michigan and Wisconsin, and one each went to Caltech, Arizona, New Mexico, the University of Washington, Minnesota, Louisiana State University, Kentucky, Case Institute of Technology (Cleveland), Cornell, and the Carnegie Institute of Technology (Pittsburgh). One Purdue senior rejected Purdue's offer, saying he got "tired of Lafayette after close to four years here." The Rocket Lab faculty seemed to understand, but then someone added, sarcastically, "So he went to Ohio State."[95]

This kind of competition was a healthy sign of growth in university programs in aerospace engineering. Princeton's Jet Propulsion Center was Purdue's mightiest rival, having graduated seventy-six Guggenheim fellows since 1948, with twenty-nine graduate students and four undergraduates at work in its five major labs: for combustion research, solid-propellant rockets, shock tube research, and electrical and nuclear propulsion. It took pride in its role as a national leader, sending half of its PhD graduates to academic departments around the country. Princeton produced university professors. On a 1966 visit to Princeton, Bruce Reese listened to R. G. Jahn, director of its Electric Propulsion Plasma Dynamics Lab, lecture him on the elite "Princeton philosophy." Jahn was fearful that NASA's new national support for broad university aerospace education would lead to a "genuine debasement of the graduate degree." As a physicist, he represented a "typical 'scientific' point of view," judged Reese, emblematic of Princeton's theoretical approach. Jahn "emphasized the science part of engineering science and, as near as could be determined, would consider any implication that he is working on hardware an insult."[96] Hardware meant Purdue, of course.

## PROTESTS AND PROSPECTS

Zucrow spent the first years of his retirement watching the early Apollo missions on full display. The media excitement peaked with the Apollo 8 circumlunar flight (1968), the Apollo 11 Moon landing (1969), and the Apollo 13 near disaster (1970). In its centennial pocket calendar for the 1969–1970 academic year, Purdue even named itself the "Mother of Astronauts," thanks to all its astronaut alumni and to the "over 30 Purdue graduates" holding "key positions in the space program, and hundreds of Purdue men" contributing "to its successes."[97] Zucrow saved it in his records. But he was also wise to the impending crash of America's spaceflight ventures. He and his close colleagues sensed the forebodings as early as 1966, for example, at an AIAA "Zucrow" panel devoted to post-Apollo advanced propulsion systems for robotic and human interplanetary flight. The hopes were high, but NASA administrators failed to answer Zucrow's entreaties for future support.[98] Hopes faded as President Johnson scaled back the surge of NASA funding by 1967 and President Nixon canceled the later missions in 1970, along with plans for space stations and Moon colonies.

In this culminating age of Apollo, Bruce Lundin's 1957 predictions about the curse of the spaceflight "project" now proved prophetic. This was Apollo's greatest strength, and greatest weakness, its Achilles' heel. Defining Apollo as the ultimate American project had made it achievable but also terminal, the ultimate "end-point article" in Lundin's terms. Apollo was also the future President Eisenhower had warned about, the technocratic mobilization of vast resources of science and engineering for an ambitious, if limited, goal. America's scientists and engineers, writers and astronauts, joined the chorus of complaint. The NSF warned against the ongoing "Era of Big Science," all those "great projects" that flooded the university with inordinate millions of dollars and "giant installations," shifting higher education "out of balance" with its traditional freedoms. The American Association for the Advancement of Science critiqued Apollo's fascination with the "spectacular feats of modern technology," all at the expense of the slow "discourse" of basic science.[99] Purdue grappled with these trends. Yes, it eagerly pursued federal funding, but it also feared becoming a mere "federal satellite." The PRF even enlisted the new School of Humanities and the old School of Engineering to organize a prestigious national conference and book on science policy, how to confront the high costs of technological change in dehumanizing factory labor, industrial pollution, and excessive defense spending.[100]

Protests came from the heart of the professional aerospace engineering community, the AIAA, when one of its members remarked that he was "fed up with the complications and pain of our Monster Project Machine and Welfare Age." Several leading aerospace academics lamented the state of the field. Engineers were "too busy with their projects" to appreciate design creativity and social consciousness. The project system rewarded obedience over innovation, leading to the "triumph of the squares," a profession filled with timeservers who were "dull, uncultured, socially insensitive, and dollar-oriented." At the height of Apollo's success, writer Norman Mailer put it best, echoing Lundin's caution: "Somewhere in the center of NASA was the American disease: Focus on one problem to the exclusion of every other."[101] NASA astronauts and engineers were by no means oblivious. They were perfectly aware that Apollo had purchased their commitments to development over exploration. Or as Neil Armstrong complained, even before he flew to the Moon, NASA was more invested in the "space business," less in the exciting and "new problems" of aerospace research.[102]

This Apollo pessimism did not simply replace Apollo optimism. The two were knotted into the fabric of the time and its political and economic fault lines. By the late 1960s, the country was already turned inward more than upward, experiencing the movements for civil rights and youth protests.[103] University students agitated against the Vietnam War. Some faculty joined. Their administrations were under siege, especially to answer for their complicity in weapons development and sponsored defense research. In terms of higher education, 1968 also represented the collapse of the "postwar consensus" in government funding of university R&D, as the rates of increase since 1953 now fell flat. The campus protests helped break the trust and cooperation.[104] It all meant an end to the postwar S curve boom.

Purdue avoided significant protests thanks to its conservative culture. During the very week when Zucrow retired, the campus saw a rally of seven hundred people, sponsored by the Young Republicans and Young Americans for Freedom, in support of the Vietnam War. President Hovde delivered a commencement address in the spring of 1967 on the virtues of order and freedom, meaning "self-regulation," "self-discipline," and "self-control." William F. Buckley had a popular column in the campus newspaper. Conservative Republican stalwart Barry Goldwater gave a public speech in February of 1968 before a capacity crowd of six thousand in the Elliott Hall of Music all about the values of civic responsibility and small government.[105] Purdue did have a short season of discontent during the academic year of 1968–1969 with fall rallies on the Mall and marches by the Peace Union against the Vietnam War and defense R&D.

After massive spring sit-ins at the administration building and student union, President Hovde brought in the local police to arrest just over two hundred students. But they were protesting against a student fee increase, not against the war.[106]

In their own moments of radical action, leading schools like MIT and Caltech, in response to both student and faculty demands, set out to mobilize the liberal arts to raise a more progressive social and political consciousness for science and engineering education.[107] At least at first, Purdue preferred tradition. Warren Howland opened a national debate in the journal *Technology and Culture* to civilize rather than politicize. He still believed in the power of "liberalizing, eye-opening values" to help raise the moral bearing of engineers. His colleagues around the country responded with ridicule. Purdue and Howland were behind the times, they argued, parading half idealistic, half utilitarian technocratic values.[108] Within a few years, the criticisms divided the campus itself, as for example at the Third Annual Grissom-Chaffee Memorial Seminar, dedicated to the two fallen alumni. In his dialogue between the now radicalized liberal arts and traditional engineering, Purdue's new president, Arthur Hansen, recognized the "general malaise" afflicting America's "frustrated society," our people no longer impressed with scientific/technological achievements like Apollo. It was the last in the seminar series. The rift between the two cultures was growing.[109]

The Rocket Lab should have made a perfect political target for protests. But in simple logistical terms, it was not an easy place to get to, at the distant edge of campus, accessed by a narrow and dangerous road at a railroad overpass. There were no marches. After Zucrow, it still survived on military and industrial contracts from the Office of Naval Research, the Naval Ordnance Test Station (China Lake), the Air Force Office of Scientific Research, the Army Ballistic Research Lab, and the Army Missile Support Command, as well as Aerojet and Thiokol. Zucrow handed over his various contracts to his most trusted student, now colleague, Bruce Reese. Over the 1966–1967 academic year, Reese led the faculty and graduate students, in small groups of two or three, on a series of trips around the country in search of new contracts. They went near: to Wright-Patterson Air Force Base and NASA Lewis. They went far: to the Aerospace Corporation in Los Angeles, the Marquardt Company in Van Nuys, and TRW in Redondo Beach. To the Air Force Rocket Propulsion Lab at Edwards Air Force Base. To Thiokol Chemical Corporation and the Martin Company. To General Electric and Aerojet. To the Applied Physics Lab. To NASA Headquarters, and to its Langley and Marshall Space Flight Centers. The Air Force soon awarded Hoffman and Thompson one of the first major grants, $360,000 "to develop mathematical analyses and computer programs to optimize nozzle designs" for the new scramjet hypersonic air-breathing engine, for speeds up to Mach 25.[110] They were making their national debut, these successors of Purdue's Maurice Zucrow.

In the anti-military climate after 1968, classified military work among graduate students at the Rocket Lab became the rarer exception. The PRF stopped publicizing its contract defense work in the *Engineering Bulletin*. Professors relegated their classified studies to their confidential professional résumés, or asked that their military work remain confidential. In 1972 Reese and his teams renamed the Jet Propulsion Center, too beholden to the age of guided missiles and defense work, as the Thermal Sciences and Propulsion Center. Reese also helped to heal an old wound, the Zucrow–Clauser rupture from twenty years before, when he became head of the new School of Aeronautics and Astronautics in 1973, which offered a fresh vitality and unity with the School of Mechanical Engineering as well as cooperative faculty research and teaching on design projects and propulsion.[111]

Amid all these realignments, Princeton University was not so lucky. One of the most dramatic initiatives against military sponsored research and the project system came out of its Special Committee on Sponsored Research (1970), chaired by Thomas Kuhn, a philosopher of science. He was moved by the escalating violence in Southeast Asia, especially the Cambodian crisis, and by economic factors like cutbacks in science and social welfare funding, these amid continuing preferences for defense and aerospace. Kuhn boldly set out to amass details from both Princeton and the country's largest research universities on all federal "classified" and "weapons-related" research, as well as on their "Special Laboratories." His aim was to identify all the "problems" and "conflicts of interest," with a mind to divorce and divest the universities from the military services.[112]

A main target at his own campus was Princeton's Department of Aerospace and Mechanical Sciences. With good reason. It received DOD funding of $1,250,000 for fiscal year 1971 out of a total $3,000,000 in defense sponsored research to Princeton. The exchanges of correspondence and public debates became quite heated, which the engineers often framed as "attacks" by Kuhn and his liberal arts allies against them. Princeton's aerospace engineers mounted a defense of "one of the country's strongest departments," supporting vanguard new discoveries and new graduates in such advanced technologies as "propulsion, radar, stability and control, computers, and so on." R. G. Jahn, soon to be the dean of the School of Engineering and Applied Science, assured its graduate students that sponsored defense and aerospace research mattered: "Research is the lifeblood of any vital university." Without such "creative activity," teaching becomes "perfunctory and sterile."[113] Martin Summerfield raised a spirited critique of the Kuhn Committee's methods, branding it a prosecutorial inquisition. He vigorously defended teaching his students how to become "creative innovators and leaders of industrial technology." Above all, he celebrated "scientific engineering" and "practical engineering" as an integral whole, married to the "principle of applicability."[114] Maurice Zucrow would have approved.

Kuhn also went national, sending out surveys to the country's thirty-eight major research universities. Most replied, their presidents sympathetic to Kuhn's political views, their students likewise embroiled in the anti-war movement.[115] Purdue University was tenuous. President Hovde had F. N. Andrews write back, as dean of the Graduate School and vice president for research, as well as vice president and general manager of the PRF. Andrews confirmed that Purdue received DOD research dollars comparable to Princeton: in 1968, just over $3 million. Purdue received a similar amount from the Public Health Service and NSF, reflecting the increasing funds for nondefense R&D. The Atomic Energy Commission gave nearly $2 million. NASA gave half a million or less (this amount decreased nearly 27 percent from 1968 to 1969). In all, by fiscal year 1969 Purdue was delivering 868 federal research projects totaling $13.7 million, amounting to nearly 85 percent of Purdue research expenditures. Andrews did not despair these figures. He applauded them. The two decades since 1945, he wrote, "have constituted an era of rapid advancement in federal support for research and development." Sponsored federal research monies had spurred Purdue's growth, and Andrews wanted more.

Andrews also celebrated the Purdue way of doing things. There was still no formal university committee to approve or deny sponsored research, though the University Research Policy Committee set guidelines. Decisions were in the hands of several countervailing bodies: the Council of Deans and the Division of Sponsored Programs, the Industrial Communication Research Council, and the PRF and its Office of Contract Administration. Purdue worked by the separation and balance of powers. Of all Purdue research and training programs, a few were still classified, but there were "no weapons research" or "biological warfare studies." In deference to Stanley Meikle's PRF mandate from the 1930s, the university allowed no sponsored projects "unless initiated by interested and competent faculty members" and unless "within the broad educational objectives of the University."[116]

For Purdue graduates in industry, the economic downturn of the early 1970s hit much harder than any of these campus trends. Theirs was a story of America's "fall from grace," as told in family terms by David Beers, whose father was an engineering classmate of Neil Armstrong at Purdue. When Project Apollo waned, jobs did too. The 400,000 employees in the NASA-supported aerospace industry at the peak of Apollo dropped to 100,000 in 1970. J. P. Sellers lost his Rocketdyne job at this time. In metropolitan Los Angeles alone, 57,600 aerospace employees were out of work. For some of Zucrow's students, these shifts meant irreparable loss. William Cowdin lost his job at Lockheed in 1972 after twenty years in the business. He never returned.[117] Many aerospace engineers adapted and retooled, or at least tried to. Some responded to the new social imperatives with one of the best methods in their arsenal: systems engineering. It had worked quite well in the building of missiles and spacecraft, the Atlas and Titan ICBMs, and in the Apollo program. Bernard Schriever established Urban Systems Inc.

in 1968–1969 to apply systems engineering to reform America's inner cities, amassing an impressive team of civil, electrical, computer, and aerospace engineers to help rebuild housing and transportation. The company folded. It received no contracts. In this, it followed a national trend. System interfaces, mathematical models, and feedback loops were too elegant a solution to some of the country's difficult social challenges.[118]

From the Congressional Panel on Science and Technology Zucrow had warned about making such utopian leaps. He applauded the effort to "bring systems analyses to studying how cities should develop," how they might solve poverty and pollution, noise, congestion, and crime. But he saw little hope in applying the NASA moonshot paradigm to any number of earthly social problems, especially those involving politics and law. "It is like smoking LSD," he said. He advised applying the "systems approach" with great care, with due deference to the humanities and social sciences, as well as to the charity of the human touch. "We are ignorant of the social life around us," he added. "We are more ignorant than wise."[119]

Zucrow participated in the Earth-centered turn, as a member of NASA's Research Advisory Committee on Air-Breathing Propulsion Systems (1962–1967). The committee was actually part of NASA's return to aviation interests, once the Apollo surge was over. The committee was charged to improve the technical fundamentals of commercial jet travel without imperiling safety and profitability and was especially focused on reducing jet aircraft noise and exhaust pollution. Zucrow advised NASA on temperature-resistant materials for jet engine "inlets, fans, compressors, and nozzles," as well as on new kinds of high-energy fuels.[120] It was also a convenient way to steer contracts Purdue's way. The committee included some rather futuristic discussions about such initiatives as the supersonic transport (SST), vertical and/or short take-off and landing craft (V/STOL), and even several projects in hypersonic research engines for Mach 3–10 speeds. From this forum Zucrow "implored" the US Congress to invest in a "vigorous R&D effort to create a hypersonic Mach 10 airplane within ten years. He encouraged NASA to pursue research in engine design and fuel injection and mixing for the SST, and in "high temperature structures" and materials for such a hypersonic plane. After all, "no matter how fast man has flown he wants to fly faster." Yet he also advised NASA to balance supersonics and hypersonics with low-speed research too, as for example in the older but reliable technology of the helicopter. He ended his career as it had begun, with a compelling plea that he shared at NASA and before Congress: "One should always remember that 'propulsion paces progress.'"[121]

ALL OF THESE TWISTS AND TURNS OF AEROSPACE FORTUNE DID NOT DISAPPOINT Zucrow. He understood "continuous change" as the essential rule to history. This did not necessarily mean progress or decline. It was just change.[122] This might sound strange for such a New Deal progressive in his political leanings. But in his role as an engineer and professor, he set out to educate, not indoctrinate. Time was actually on his side. Zucrow's professorial career spanned the two decades of the country's most remarkable economic growth and prosperity, replete with a rising middle class, and the aerospace revolutions in guided missiles, commercial jet travel, and Project Apollo to the Moon. He lived at a time of coinciding economic and scientific/technological "bubbles."[123] His own insights and teachings helped to drive this ascending curve of progress. As Clair Beighley once said on his mentor's retirement in 1966, "Conclusions which you reached 20 years ago in the classroom are just now becoming accepted as fact."[124] Zucrow's own contributions to advanced propulsion for interplanetary flight, encompassed in his Sigma Xi lectures and his testimony before the US Congress, may have been a bubble all their own. They represented an early peak in "space propulsion science." They occupied a high point along the S curve of aerospace progress into the 1960s. Yet they did not last, lingering in relative stasis and atrophy over the next fifty years.[125] The trouble with bubbles is that they burst. Apollo's bubble certainly did. Which brings us back to the egg metaphor. At least one critic, a professor of English at Yale University, had an answer to the new phenomenon of "moon-shooting" in politics and culture, one he gnarled in a bizarre set of mixed metaphors: "The moon race is nothing less than the dying technological gasp of patriarchy: an ordinary man in a man-made uterus, perched on the tip of a huge phallus with tremendous thrust, ready to be born without woman on the dry surface of the moon or else to crack, like Humpty Dumpty."[126] So much for eggs. So much for evolution.

Some of Zucrow's colleagues from the defense and propulsion fields were less reluctant to give up hope. Herman Kahn, who had once calculated the odds of human survival in the nuclear age, throwing millions of people about like playthings, now turned more optimistically to our survival in the long term. If humanity could beat the odds of nuclear annihilation, it certainly stood a good chance of applying the fruits of science and technology to solve urgent crises in pollution, overpopulation, and poverty. On an upward, slowly rising S curve. Social scientists eventually joined the chorus, admitting that the old military–academic–industrial complex was actually a force for good in the world, compelling an upward "helix" of progress. One of Zucrow's colleagues from Aerojet, Rudi Beichel (a former member of the Nazi Peenemünde team), applied this enthusiasm to space propulsion, what he and a colleague called the rising parabolic line combining "gradual improvement" with technological "breakthroughs," like their own more specific vision for "various high-pressure advanced engines" and "dual fuel and dual nozzle thrust chambers." Zucrow did not have much confidence in the design, comparing the dual rotations to spinning wheels, and for which he coined the term "Beichel-cycles."[127] He also knew that parabolic leaps in good fortune could not last, that the S curve was bound to plateau. Sometimes change was just change.

# EPILOGUE

## The Enduring Life of a Consultant

*"To Maurice J. Zucrow, research engineer, teacher, and respected protagonist, for his exploitation of basic research in the field of jet propulsion; for his genius in stimulating and disciplining the creative endeavors of young research engineers; for the vigor of his intellect, the enthusiasm of his spirit and the prodigious and ever-renewed accomplishment of his mature years; for founding and directing the Purdue University Jet Propulsion Center; for participation in affairs that are vital to the competitive strength of our nation; for his notable contributions to the literature of jet propulsion and missile design, we present this fifth Vincent Bendix Award."*

*JOURNAL OF ENGINEERING EDUCATION*, OCTOBER 1960

MAURICE ZUCROW RETIRED FROM PURDUE IN 1966 AFTER TWENTY YEARS. IT was time, judging by his failing health: high blood pressure, kidney ailments, and a heart condition. It was also time judging by the students around him. A generation had come and gone since he began teaching. His first graduate students were now colleagues in the prime of their careers. And now Purdue was about to enroll their children, the very toddlers once featured in the *Purdue Alumnus* as the coming class of 1970. He had spent nearly twenty years in industry, somewhat by accident. He had spent twenty in teaching, by choice. He may have hoped for twenty in retirement. He only made it part of the way.

Over the next nine years, from his new home in Santa Barbara, California, Zucrow rested and worked. He remained something of the absent-minded professor, as when he failed his California driver's license test in the very first weeks of retirement. He neglected to exercise caution when changing lanes and missed parallel parking several times.[1] Though he eventually passed the exam, Lillian usually drove them places, around town, or down south to visit relatives in Los Angeles. Santa Barbara was his last horizon, looking westward from the Pacific shores. It was a fitting place to match the three romantic frontiers that had framed his life: Frederick Jackson Turner's frontier thesis, published in his first years at Harvard (1920); Vannevar Bush's, *Science: The Endless Frontier* (1945), praising civilian science at the end of the Second World War; and JFK's "New Frontier" and Project Apollo (1961), opening a new era of spaceflight. As to Apollo, Zucrow may have felt some gratification when he saw the Purdue course catalog for 1971, one that dramatically displayed an aerial view of campus and the smokestack, side by side with Buzz Aldrin on the Moon, standing in attention to the American flag. Smokestack and flag, two emblems of Zucrow's life. Further along in the pages, Purdue advertised its degree programs of Zucrow's own making: in "such types of propulsion systems as chemical rocket engines, air-breathing engines, electrical rocket engines, nuclear rocket engines, and space power plants."[2]

Retirement meant a new appreciation for small things. Rest came in the way of fishing. In 1967, he bought a Fenwick Feralite FF 80 fly rod at the tony Kerr's department store in Beverly Hills. He used it for fishing trips near his home and up north at Gold Beach in Oregon. He even kept the instructions, neatly folded in his wallet, on how to properly tie a Palomar fishing knot, promising "100 per cent line strength" for "even a fumble-fingered fisherman." Not all his joy was in little things. One was big. In a special deep sea excursion at Mazatlan, Mexico, in February of 1973, he caught a Marlin. It was 8 feet 9 inches long, weighing 135 pounds. A color Polaroid snapshot captured the moment, one of the few personal treasures he saved.[3]

He also walked. A lot. On the average, four to six miles a day in the neighborhoods of Santa Barbara and along the beach. True to his profession, he calculated the pace at roughly four miles an hour.[4] Walking proved dangerous. He broke his leg in a fall in 1969 and was hit by a car at a downtown crosswalk in 1971. He also spent several weeks in the hospital from physical ailments during May and June of 1972. Relaxation meant reading and learning. He took Spanish classes, inspired by the beauty of the colonial missions nearby. He returned to his Jewish heritage with donations to Israeli causes after the Six-Day War and with the book *Life Is with People*, a study of the shtetl communities of Eastern Europe. He once sent in money for a transcript to a KTTV news show (14 June 1967), "What Is a Jew?" The answer seemed to touch at something deep inside him. The Jew was someone "tempered by travail and straightened and hardened by an unrelenting determination to be free." Zucrow underlined the word "free." He was impressed with the notion that anyone was able to "mentally and morally" become a Jew so long as they had a love for the "freedom of the individual spirit."[5]

In this spirit, Zucrow also paid silent tribute to his father, Solomon, circling back to the life lessons he had learned as a youth in the East End of London and south side of Boston. It came in a poignant moment during his testimony to the US Congress, when he affirmed his belief in democracy and popular sovereignty: "I mean, law and life must go together, because if the law is inadequate it will die. Life must go on." Any laws promoting science and spaceflight need serve "life as we can project it in our feeble minds as far as we can."[6] Zucrow fortified the sentiment in a speech a year later: "A wise man, my father" once said that "when a civil law has outlived its usefulness, it becomes a dead letter and is abrogated." This was a reference to his father's *Adjustment of Law to Life in Rabbinic Literature* (1928). "There is no status quo. Social customs, civil law, public moral conceptions, economic theories, educational methods, and the like," were all open to revision to meet "new social and international conditions." For "change is the only permanent state known to man."[7]

Retirement offered new opportunities to read and think. Zucrow returned to his New Deal progressive roots. He took stands against the Vietnam War and the Watergate scandal. These were only ever in private forums. But they were deliberate, taking shape in campaign contributions for Alan Cranston and George McGovern. He became more of an environmentalist after the oil spill in 1969 that polluted the waters off his beloved Santa Barbara. He advised the Picatinny Arsenal, formerly the Dover Naval Air Rocket Test Station, on its "energy management program" for toxic waste removal. Where he once advised on launching guided missiles, he now advised on cleaning up TNT and lead azide residues. He also read and appreciated *Crisis in Black and White* (1964) on the plight of poverty and racial discrimination in the cities.[8] In his last years, he also witnessed several African American minority achievements by his lab and students. In 1970, Renaldo Jensen (USAF) graduated from the Rocket Lab with the PhD in aerospace and mechanical engineering. Douglas E. Abbott served as director of the Lab (1973–1977). Zucrow acolyte and civil rights advocate, J. P. Sellers, also taught a generation of African American engineers at Tuskegee University (1970–1984). One of his first students was Lonnie Johnson, inventor of a unique apparatus for jet propulsion: the Super Soaker water gun.

Most of all, Zucrow still worked. He wrote a new two-volume study of engineering science, *Gas Dynamics*, though it became a posthumous work, coauthored with his student Joe Hoffman. Here they translated his earlier books and course teachings for the computer age.[9] He also consulted. Besides his work for the US Congress, he traveled around the country, often once a month. His new business letterhead named him simply as "Consultant." He peppered his appointment books between 1969 and 1971 with all of the acronyms of America's defense establishments. Their tidy abbreviations fit nicely into the small boxes of his daily schedules. He consulted at MICOM and SAFSCOM at Huntsville (the Army Missile Command and Safeguard System Command) for a total of eighteen times over three years. He visited the ATAC and TACOM in Detroit (Army Tank-Automotive Center and Army Tank-Automotive and Armaments Command); the BRL and APG in Maryland (Ballistics Research Laboratory and Aberdeen Proving Grounds); the PA in Dover, New Jersey (Picatinny Arsenal); and the AMC (Army Materiel Command), ASAP (Army Scientific Advisory Panel), and MUCOM (Army Munitions Command). He also made short trips to nearby JPL at Pasadena (Jet Propulsion Lab, as a member of its Visiting Committee) and to AGC's Azusa, Sacramento, El Monte, and Downey sites (Aerojet-General Corporation, as a member of its Technical Advisory Board). He consulted for NASA's Ames Research Center and for the Navy's missile test sites at China Lake.[10] Several of these stops were visits to old friends, like William Zisch and Bernard Schriever at Aerojet; or old acquaintances, like Gen. Bernard Luczak in Washington, DC (former director of several Army missile test sites); or old students, like Ernie Petrick in Detroit. These were usually occasions for a "social gathering and a lively exchange," moments of "warm camaraderie."[11]

These consultations amounted to a third career, especially Zucrow's work for the Army's Missile Command (MICOM). Its technical director for research and development, John L. McDaniel, has left us a rare documentary record about Zucrow's contributions, emblematic of those that he made throughout his career. McDaniel praised Zucrow as a "vital factor in the phenomenal success" of the Army's propulsion technologies through the 1960s, this thanks to his ability to see the "total picture" in both detail and diagram. Zucrow's advice on "technical and management" issues, on immediate goals and far-reaching policies, had, by "conservative estimates," saved the government "many millions of dollars." These contributions covered weapons for battlefield use. Zucrow advised MICOM on its new light antitank weapon (LAW), a "free flight" infantry rocket. He also solved problems with the Army's heavy antitank weapon, the TOW missile (tube-launched, optically tracked, wire-guided). Its early rocket motor design spewed jets of flaming fuel. Zucrow proposed using the more reliable "double-base" material from the LAW. Thus resolved, the TOW became one of the Army's most successful battlefield weapons. The Lance tactical missile system, said McDaniel, also "owes much of its success to the efforts and recommendations of Dr. Zucrow." In the early phases of R&D, Zucrow insisted that the contractor implement new tests on the motor or suffer the loss of the company's "image." His "ability to be so forthright with the highest-level contractor executives," said McDaniel, was crucial. Zucrow also recommended a solution to a "leak in the propulsion system" by an improved pressure seal. He advised on an improved gas generator and on hiring new subcontractors for the feed system, thereby saving time and money. Presented with a severe combustion instability problem called "engine buzz," he found a unique solution in "several theoretical and experimental efforts," reducing the "program delay" by "fifty percent." For his efforts, MICOM awarded Zucrow with a trophy of sorts, a piece of shrapnel from a failed Lance experimental flight, the TP-16 "failure" on 2 May 1968, encased in a clear plastic block. It is on display at the Zucrow Labs today. On it were the words "Success through Failure. Presented to Dr. M. J. Zucrow, Consultant."[12]

These contributions also rose to the highest levels of the Army's strategic and advanced programs. Zucrow offered the Sentinel Project Office for Ballistic Missile

Defense "invaluable" technical and management help with the Sprint propulsion system. As a member of the Selection Board for the Sprint missile contractors, he "made significant technical contributions to Sprint propulsion which involves unexplored areas where there is little or no theory to guide experimentation." He "assisted in the selection of additives to propellant formations which have increased burning rates three-fold." He "recommended action for the integral joining of two high-strength fiberglass structures which must withstand severe flight loading." He designed a new approach to "mechanically join nozzles to motor cases" in order to save on weight. As a member of the Investigative Committee on Sprint Flight Failures, he asked all the right "searching and probing questions," leading to a series of six test-flight successes. For the Army's futuristic supersonic combustion ramjet (SCRAM), Zucrow "recommended specific basic studies and parametric studies" to begin the pathbreaking project. As chair of the Laser Technology Committee of MICOM, he directed current work and recommended "future programs in the application of lasers to Army missile guidance," as well as directing the "technical review board for the MICOM managed Army High Energy Laser Program."[13]

Zucrow also advised the Air Force. As consultant and "special adviser" for the Divisional Advisory Group, Aeronautical Systems Divisions, Wright-Patterson Air Force Base (1964–1968), he faced the very challenging "compressor surge and stall issues" with the new F-111 "Tactical Fighter Experimental." He advised technical design modifications to the fighter's inlet that resolved the problems.[14]

These itineraries reveal just how little had changed for Maurice Zucrow in retirement. He was still content to make history rather than write it. In this, he was following the rule of a Purdue colleague, historian Louis Martin Sears, who advised engineers to "fashion" history and to let the historians "interpret" it. Most engineers, even the most significant among them, followed the advice, leaving few historical interpretations of their own work. Memoirs or history books made little sense to them. Donald Putt, for one, remarked that he would "never want to take on a project like that."[15] Not all engineers followed the Sears standard. Frank Malina and Martin Summerfield were adept at writing themselves into the history, this to establish their own posterity and settle old scores against their rivals, especially Clark Millikan and Robert Goddard.[16] Several other leading engineer-historians eventually scrubbed Zucrow from their bibliographies and historical surveys.[17] Some of this was accidental more than malicious. Zucrow was at a distinct disadvantage for being so often first among the pioneers. This gave him influence and authority, but it also meant that later writers surpassed him in the fast-paced world of rocket propulsion. Zucrow's contributions were overshadowed by the very profession that he had helped to create. His works, like some lost civilization, were obscured by the new layers built above it. Telling his story here has been an exercise in forensic archeology.

In the first years of his retirement, Zucrow also wrote another book. He has never been given credit for it. The title page does not list him as author. Librarians do not catalog it by his name. And yet it was, in a sense, his magnum opus, the most "Zucrow" of all his books: *Elements of Aircraft and Missile Propulsion*, an *Engineering Design Handbook* of the Army Materiel Command (1968). He wrote it for credentialed engineers and scientists in the defense establishment, young people with bachelor of science degrees in engineering, physics, or chemistry and ready to learn the fundamentals of jet propulsion. Like his 1948 and 1958 textbooks, it covered a wide a range of topics, this one with updates on scramjet engines and hypersonic flight. He filled it, like he did his classroom chalkboards, with even more intricate formulas and diagrams. He added the full range of his favorite authors, including the works of his students, and the classic engineering reports from the vast landscape of national military, industrial, and research centers that he had crisscrossed since 1946. Here was his life's work. Here were the works of many lives.[18]

Zucrow also finished some old business at this time, donating his Walter rocket engine, the one for the Messerschmitt 163 given him by Wright Field, the one he tested out at the airport in one of the first operations of the Rocket Lab. Chuck Ehresman negotiated the gift to the National Air and Space Museum beginning in 1966. He sent it with an oak framed plate honoring Purdue. The engine and plaque sit today in the Udvar-Hazy exhibits, right next to an actual Messerschmitt 163.[19] There is a little bit of Zucrow at our nation's capital.

We also ought to remember Zucrow's legacy in the making of the high-pressure engine for the Space Shuttle.[20] After his retirement, Zucrow's new NASA-funded lab began a long line of direct research contributions to the coming space shuttle main engine (SSME). We still tend to think of the shuttle as mostly a reusable space plane. But it was also propelled by "the only reusable large liquid rocket engine ever developed." These engines operated successfully for each of the 135 shuttle flights. The space shuttle was a reusable plane built around reusable engines that never failed. There is no direct line from the Rocket Lab to the shuttle, of course. The lab was but one small corner of the R&D networks that contributed to it. The shuttle's heritage hardware was also complex and varied. But one of its premier hallmarks was that "the engine was required to operate at a high chamber pressure to minimize engine volume and weight." Namely, a chamber pressure at 2,994 psi, or 109 percent RPL (rated power level). This was not possible without the Rocket Lab's vanguard work.[21] The shuttle was a breakthrough technology in Zucrow's high-pressure combustion chambers. Robert Briggs, one of its project engineers and historians, has placed it "high on the list of humankind's greatest engineering achievements."[22]

In a poignant study of the SSME on its retirement, two reviewers made some powerful *longue durée* comparisons. "Even though the SSME weighed one-seventh as much as a locomotive engine," they wrote, "its high-pressure fuel turbopump alone delivered as much horsepower as 28 locomotives, while its high-pressure oxidizer turbopump delivered the equivalent horsepower for 11 more." Moreover, "although not much larger than an automobile engine," so they continued, "the SSME high-pressure fuel turbopump generates 70,000 horsepower or 70 horsepower for each pound of its weight, while an automobile engine generated approximately one-half horsepower for each pound of its weight."[23] Here was a fitting summary to our story, one that began way back with W. F. M. Goss and the Locomotive Lab, and with Harry Huebotter's Internal Combustion Engine Lab, and with Maurice Zucrow's Rocket Lab. The shuttle recalled Purdue's long-time prime movers that crossed continents, highways, atmospheres, and outer space.

Not one to reminisce or hold forth as such, Zucrow represented his life and times as a simple résumé. No more and no less. On retirement, he submitted a short paragraph of his career achievements to the compendium *American Men of Science*. It was a relatively long "short" paragraph, as far as the other biographical items in the series went. Only a small fraction of a page, in small print, of his degrees and positions and major achievements. In a later professional note to Harvard University, he summarized his career of research, teaching, and consulting in a simple way: "Most of this effort has been rewarding."[24]

Some of his consulting work brought him back to Purdue. This was part of his retirement package. The School of Engineering was obliged to pay for three visits a year, for research reviews at the Rocket Lab and for guest lectures in mechanical engineering courses. H. Doyle Thompson remembered one such occasion, a Zucrow talk to a junior-year fluid mechanics course. He told a story about a Venturi meter, one that Zucrow had designed to measure the flow in a gas line at a Commonwealth Edison power plant. It was a "multi-million-dollar installation," Zucrow said, and the meter would determine charges worth "thousands and millions of dollars." When it failed to work at the start of operations, he reluctantly agreed to have laborers dig up the pipe. He was the responsible engineer, after all. What they found were a pair of overalls on the meter, carelessly left there on a hot summer day by one of the workers. Zucrow's advice to the class: "Trust your engineering," when you know you've done it well. "I repeat, have confidence in your engineering."[25]

What mattered most were his students, his people, his students now become colleagues. After 1948 Zucrow rarely published any new work as sole author. Instead, he mostly worked in pairs with his colleagues and students, drawing them into his research agendas and expertise, fortifying their work, raising their own career profiles. They did not fail him, celebrating his life and theirs with letters and notes, many written in their own hands, reminiscing and thanking him. They filled two scrapbooks, marking his retirement in 1966 and his seventy-fifth birthday in 1974. These were small books of lives, really. We've been reading from them all along, but it is proper to offer a few more reminiscences here.

The tribute letters addressed all the many personas of Maurice Zucrow: Moishe, Maury, Doc, Morris, Dr. Zucrow, Maurice. Some were serious. William Pickering, director of JPL, named him as one of the "Founding Fathers" of America's "military and space programs." Others were fun-loving. J. Preston Layton of Princeton remembered, "Among the things you taught—a little scientific and technical knowledge, a lot of jokes, how to be slightly offensive but always a gentleman." Others relayed a deep emotional bond. Bruce Reese spoke of a son's heartfelt "love" for a father.[26] In sum, Zucrow was admired by his colleagues and beloved by his students, remembered by both as a person of warmth and humanity. Maybe the kindest praise came from R. B. Stewart, not known for his sentimentality. It came in one of R. B.'s oral history interviews, where he remembered Zucrow as that curious "graduate student" of 1928 "who helped send a man to the Moon" in 1969, none other than Neil Armstrong, another Purdue alumnus. Purdue had cultivated them both: one as a professor of jet propulsion, the other as Earth's astronaut "emissary."[27]

These kinds of compliments buoyed Maurice Zucrow in his last years. He received many of them the year before died, from a heart attack on 5 June 1975. He was the second of that small coterie of Purdue officials to pass away. G. Stanley Meikle had died in 1960. Andrey Potter survived Zucrow by four years (1979); Frederick Hovde by eight (1983). R. B. Stewart survived them all, passing away in 1988. Potter gave Zucrow one last private tribute. In a conversation with President Edward Elliott, they both agreed on one name as first on their list "of the three most brilliant men whom they had known in their years at Purdue." It was Zucrow. Karl Lark-Horowitz from Physics was second and R. B. Stewart third.[28]

In the end, Zucrow's was a life multiplied, not by fame or fortune, nor by any scholarly memorial, but by the students who graduated under him. "I think my outstanding contributions were in education: the large number of excellent young men I was privileged to associate with as their major professor."[29] These were the men who, in the legend of the Rocket Lab, were "made of gold." The original appellation was a bit less elegant. Leon Shenfils of Aerojet coined the phrase. He actually said that Zucrow's students had "balls of gold." It was a reference, with a double entendre and some Yiddish humor, to the two balls of gold attached to the bulls of the medieval Popes. Zucrow's students, in other words, went into the world exalted, with his imprimatur. Reese and Zucrow later had some fun with all of this, in a gold gift tie pin that Dr. Zucrow wore for his official portrait, now in the Chaffee Hall conference room, where Cecil Warner has a portrait too.[30]

The Rocket Lab tradition, the heritage of Maurice Zucrow, has remained strong and adaptable. In 1998, director Stephen Heister, with the collaboration of Charles Ehresman, organized a fiftieth anniversary reunion, bringing together the makers of the Rocket Lab from the early days onward. Purdue had already renamed the Maurice J.

Zucrow Laboratories in his honor. One administrator warned that the Memorial Committee should not "ignore the potential linkages between the laboratory name and a major endowment." Instead of Zucrow's, a donor's name might even be worth $10 million. That was quite a backhanded compliment. Ehresman and his colleagues pushed forward anyhow, adorning the new outdoor sign with Dr. Zucrow's name, Purdue's colors, and two decorative spindles, topped with finial gold balls. Loyal to his mentor, Ehresman wrote, "We have to maintain the lore and in a way the tradition of the Lab."[31] Heister and colleagues also published a mighty tribute to their predecessor in their college textbook *Rocket Propulsion* (2019), a study of the state of the art of rocket science, recalling Zucrow's own contributions from seventy years before.[32]

Several generations of teachers and students have crossed the Rocket Lab's pathways. By 2018, under Robert Lucht's directorship, it became "the largest academic propulsion lab in the world," with a budget topping $13 million and an "all-time high" of 144 graduate students.[33] Besides its hundreds of graduates who have filled the American aerospace industries and offices, the lab has also so far counted three astronauts as alumni. Jerry Ross (BSME 1970, MSME 1972) was its first, and a record-breaking one at that, with seven space flights (totaling over 1,393 hours) and nine space walks (extravehicular activity). He helped build the International Space Station. Professor Cecil Warner was the crucial mentor in Ross's life, advising his senior design project, recommending him for a research assistantship, and offering spiritual counsel.[34] Ross represented a fitting legacy to the Zucrow tradition, schooled in its methods and principles, writing his master's thesis under Warner and Ehresman on the thermodynamics of the ramjet engine. He remembered how his mechanical engineering degree from Purdue, culminating at the Zucrow Labs, gave him a "coin" of the realm. At the Propulsion Laboratory of Wright-Patterson Air Force Base, his next assignment, where he followed two other Zucrow alumni, Ross's Rocket Lab experience christened him as a graduate of promise and high expectation.[35]

Among the other Zucrow Labs astronaut alumni are Scott Tingle (MSME 1988), who specialized in fluid mechanics and propulsion and flew on a Russian Soyuz rocket to the International Space Station. In May of 2018, aboard the station, he awarded another Purdue astronaut alumnus, Drew Feustel (BS 1989, MS 1991), an honorary doctorate in space. Behind them in the picture was a Purdue streamer proclaiming "Boiler Up Zucrow Labs!" Perhaps most appropriately, as of January 2023, one Zucrow alumnus is an astronaut yet to fly, Loral O'Hara (MSAAE 2009), named as one of NASA's 2017 class of astronauts. She was also a member of the first "all-female team" in rocket propulsion at the Zucrow Labs. Here was a most fitting homage to the legacy of the Rocket Lab. Or, as Dr. Zucrow himself was fond of saying, "Propulsion paces progress." About all of this, he would have been proud, but not surprised.

# APPENDIX 1

## Zucrow PhD Students (as of 1965), With Graduation Dates and Positions

R. L. Duncan, PhD 1950
Vice President of Engineering, United Aircraft Corporation, East Hartford, Connecticut

W. J. Hesse, PhD 1951
Director, Nucleonics Systems Operation, Chance Vought Aircraft, Dallas, Texas

C. H. Trent, PhD 1951
Technical Director, Nuclear Propulsion Division, Aerojet-General Corporation, California

C. M. Beighley, PhD 1953
Senior Department Manager, Research Department, Aerojet-General Corporation, California

R. W. Graham, PhD 1953
Head, Cryogenics Section, Lewis Research Center, NASA, Cleveland, Ohio

S. V. Gunn, PhD 1953
Project Engineer (Project Rover), Rocketdyne, Canoga Park, California

B. A. Reese, PhD 1953, MSME 1948
Professor, Mechanical Engineering, Purdue University

E. L. Katz, PhD 1954
Assistant Manager, Electro-Optical Systems, Pasadena, California

J. H. Fisher, PhD 1954
Vice President and Manager, Energy Research Division, Electro-Optical Systems, Pasadena, California

E. N. Petrick, PhD 1955
Senior Research Engineer, Kelsey-Hayes Company, Detroit, Michigan

D. E. Robison, PhD 1955
Head, Advanced Technology Group, Liquid Rocket Plant, Aerojet-General Corporation, California

J. M. Botje, PhD 1956
Consultant, Heat Transfer, Missile and Space Systems, General Electric Corporation, Pennsylvania

A. B. Greenberg, PhD 1956
Head, Advanced Propulsion Group, Aerospace Corporation, El Segundo, California

R. D. Smith, PhD 1956
Application Engineer, United Technology Corporation, Sunnyvale, California

J. A. Bottorff, PhD 1957
Group Leader, Corporate Research Department, Aerojet-General Corporation, Azusa, California

J. R. Osborn, PhD 1957, MSME 1953, BSME 1950
Professor, Mechanical Engineering, Purdue University

H. Wolf, PhD 1958
Professor, Mechanical Engineering, University of Arkansas

J. P. Sellers Jr., PhD 1958
Principal Development Engineer, Rocketdyne, Canoga Park, California

J. P. M. Diamond, PhD 1958
Section Chief, Aerospace Corporation, El Segundo, California

A. R. Graham, PhD 1958
Head, Advanced Research Group, General Electric, Malta Station, New York

D. A. Charvonia, PhD 1959
Head, Thermodynamics Section, Space General Corporation, Glendale, California

D. G. Elliott, PhD 1959
Research Specialist, Jet Propulsion Laboratory, Pasadena, California

H. L. Wood, PhD 1960
Professor, Mechanical Engineering, Virginia Polytechnic Institute

D. L. Emmons Jr., PhD 1962
Senior Engineer, Boeing Airplane Company, Seattle, Washington

J. D. Hoffman, PhD 1963
Assistant Professor, Mechanical Engineering, Purdue University

M. R. L'Ecuyer, PhD 1964, MSME 1960, BSME 1959
Assistant Professor, Mechanical Engineering, Purdue University

J. G. Skifstad, PhD 1964, MSME 1959
Assistant Professor, Mechanical Engineering, Purdue University

H. D. Thompson, PhD 1964, MS Engineering 1962
Assistant Professor, Mechanical Engineering, Purdue University

# APPENDIX 2

## Career Paths and Contributions of Select Zucrow Students

I HAVE DRAWN THESE BIOGRAPHICAL MATERIALS ABOUT SOME OF DR. ZUCROW'S students from the oral history interviews listed in the notes; from folders 3–5, box 9, Maurice Zucrow Papers; and from press releases and confirmed internet searches.

### US Air Force

Ralph W. Harned (BSME 1949), a veteran of the Second World War, and Korean and Vietnam Wars, became director of Wright-Patterson Air Force Base (WPAFB) Rocket Propulsion Laboratory by 1951. He went on to serve as chief of the Solid Rocket Division at Edwards Air Force Base (AFB) and chief of Minuteman propulsion for the Ballistic Systems Division at Norton AFB.

Langdon Ayres (MSAE 1947) served at the Air Research and Development Command (ARDC) at WPAFB, later becoming a colonel and chair of the Space Systems Division, Air Force Systems Command, in charge of the Minuteman III propulsion system. He took on a second career at Aerojet-General after 1963.

James Albert Povalski (MSAE 1947) got a job in Air Force technical intelligence with Zucrow's help, then served in advanced planning for the Space Command's Space and Missile Systems Center, and on to McDonnell Douglas Astronautics Corporation, Huntington Beach, California.

Spencer Simmons Hunn (MSME 1949), a B-24 pilot with the Eighth Air Force in the Second World War, veteran of the raids against the Ploesti oil fields in Romania, served at the ARDC and at the Programming Division of Air Force Headquarters. He became program director of the Cheyenne Mountain Complex, in charge of the North American Air Defense Command (NORAD) control center, retiring as a brigadier general.

Howard Rodean (MSAE 1949) also went to the WPAFB Rocket Propulsion Lab, then on to Chance Vought Aircraft to work on the Cutlass and Crusader jet fighters. He later partnered with Walter Hesse on the Supersonic Low Altitude Missile, meant to be propelled by the Pluto nuclear ramjet. He also worked at Lawrence Livermore Radiation Lab, there applying his Purdue-acquired expertise in compressible fluid flow, heat transfer, and thermodynamics to the civilian adaptation of nuclear power in the Plowshare Program.

Lt. Col. J. A. Saavedra (MSAE 1952) worked in the Weapons Systems Directorate, then as an air technical liaison, and then became deputy chief for technology, Liquid Rocket Division, at the Air Force Rocket Propulsion Laboratory. By 1966, he was working in Systems Command at Andrews AFB on the management of space launch vehicles.

Col. Robert C. Thompson (MSAE 1949), a C-47 pilot in the Pacific during the Second World War, first worked on propulsion issues at the 1st Guided Missiles Wing at Eglin AFB (on the Matador, Falcon, Snark, and Rascal missiles), then on to the Air Defense Command and the Western Development District, where he "managed the total program" for the Atlas ICBM (1959–1964), as well as testing of the Minuteman ICBM (for the Advanced Ballistic Reentry System, Vandenburg AFB).

R. D. Smith (PhD 1956) first went to the Rocket Propulsion Lab at WPAFB to manage advanced propulsion projects for jet engines; then to the Dynasoar Weapons Systems Program Office to manage its booster rockets. He later worked at the United Technology Center as an assistant program manager for the Titan III Solid Rocket Program. As Smith once wrote to Zucrow: "Your absolute insistence in learning how to write has paid off a million times."

Vincent Capasso (MSAE 1954) moved on to the Air Force Institute of Technology at WPAFB, then became a propulsion engineer for the X-15 aerospace plane at Edwards AFB, and eventually director of the Flight Systems Laboratory at the NASA Flight Research Center, "responsible for all airborne avionics systems."

Col. Joe E. Zollinger (MSME 1953) managed the Directorate of Propulsion and Engineering of the Aeronautical Systems Division at WPAFB, specializing in the propulsion and airframe interface of America's fighter aircraft.

Col. Walter Moe (MSME 1959), a B-17 pilot in the Second World War and later a Vietnam War veteran, headed the Air Force and Advanced Research Projects Agency (ARPA) technical assessment team, managing the transition of the F-1 engines to the NASA Saturn Program in 1960. Col. Moe also worked with Rocketdyne on the propulsion systems for the Atlas and Thor ICBMS, directed the Liquid Rocket Division at Edwards AFB, and was a program director for the Titan III satellite launcher. He represented the Air Force on the NASA selection board for the main space shuttle engines; commanded the Air Force Propulsion Laboratory at WPAFB; and was a manager of the Pratt & Whitney Aircraft Military Division, working on the stealth fighter and bomber.

## US Navy

Richard L. Duncan, aeronautical engineering's first PhD in 1950 (BSEE 1937, MSAE 1948), a Navy veteran combat and test pilot, went to work for the Office of Naval Research upon graduation, assigned to advanced propulsion systems for high-speed aircraft and missiles, and in time became assistant director, Nuclear Propulsion Division of the Bureau of Aeronautics, where "he directed work on the Navy's aircraft nuclear power plant and other advanced nuclear systems." Duncan later became manager of advanced planning at Pratt & Whitney Aircraft Division, United Aircraft Corporation.

James Preston "Pres" Layton (MSAE 1955) was a graduate of New York University's premier Aeronautical Engineering program (BSAE 1941) and flew JATOs for the Navy in the Second World War, testing the Aerojet solid variety as a pilot in the Pacific and collaborating with Robert Goddard on testing liquid-propelled units at Annapolis, as well as in the field as a liaison officer. He worked at the Naval Bureau of Aeronautics after the war, then as a chief propulsion engineer for the Glenn Martin Company, working on the Demon ramjet and the Viking rocket at White Sands Proving Ground. When Princeton's Guggenheim-funded Jet Propulsion Center needed to build a rocket lab in the 1950s, it sent Layton to Purdue to work under Zucrow. He completed research for most of his master's thesis at the James Forrestal Center at Princeton, also working with Luigi Crocco, with support and funding from the US Navy and Air Force ARDC. He also learned Zucrow's methods and adapted them to establish Princeton's Aerospace Systems Laboratory, where he was its chief engineer, appointed as a research associate to "handle the complicated propulsion installations." Princeton, to this extent, became something of a Purdue Rocket Lab satellite.

## US Army

Kenneth C. Van Auken (MS Engineering 1952) became a colonel in the Army Ordnance Corps, with tours at the Pentagon, White Sands Missile Range, and Redstone Arsenal, also serving as program manager for the Dragon anti-tank missile system and as commander of the 60th Ordnance Group.

Lt. Col. Raoul Quantz (MS Engineering 1952) began his career as project engineer and manager in Army R&D for the Corporal, Sergeant, Redstone, and Pershing guided missiles, *Explorer* satellite, and Nike Zeus/X missiles systems.

Ernie Petrick (MSME 1948, PhD 1954) became technical director and chief scientist of the Army Tank-Automotive Command, a specialist in gas turbine and rocket motor design. He patented an ion engine, and at Detroit partnered with the automotive industry in the development of the disc brake.

## NACA/NASA

Several of Zucrow's top students (like dozens of others from Purdue's engineering programs) went to work at the NACA Lewis Flight Propulsion Laboratory in Cleveland. Anthony Fortini (BSAE 1951, MSAE 1953) collaborated with V. N. Huff and S. Gordon on experimental rocket fuels at Lewis and later worked at Boeing. Fortini and Huff later won the Space Act Award (1987) for their 1959 design of a unique rocket altitude test cell in the Rocket Engine Test Facility.

Bill Goette (BSME 1956) worked in the NACA Rocket Lab beginning in 1955, adapting Zucrow's methods there, helping to build the hardware for Cell 22, where he analyzed data from rocket thrust chamber testing. This included testing different propellant combinations and injector designs.

Robert Graham (PhD 1953) interned at NACA Lewis (Walter Olson was also on his dissertation committee), then launched his career there, at Zucrow's advice, rather than move on to General Electric or Westinghouse. Graham worked on axial flow compressors and rotating stall in the Compressor and Turbine Division but transferred to rocket work in 1956 based on his heat-transfer dissertation at Purdue.

## Industry

Zucrow students joined the aviation industries. Donald William "Bill" Craft (BSME 1945, MSME 1949) worked at General Electric's Aircraft Accessory Turbine Department (Lynn, Massachusetts), serving as a principal contributor for the "turbine components for the accessory power units on the X-15," transferring in 1962 to the Direct Energy Conversion unit, where he worked on hydrogen–oxygen fuel cells and water electrolysis systems (the product water removal system) for NASA's Gemini spacecraft.

Foster L. Gray (MSME 1955) was at Ling-Temco-Vought (Dallas) 1957–1960, "working on almost every phase" of propulsion, rocket propulsion, auxiliary power, heat transfer, and environmental control. In 1960 he moved on to Texas Instruments, helping to establish its new program in energy conversion at its Corporate Research Laboratories, devoted to the spaceflight future for thermoelectric heating and cooling, thermionics, and fuel cells.

John Ira "Johnny" Nestel (BSAE and BSME 1948), a member of the Sigma Alpha Mu fraternity who had flown B-29s in the Pacific front, joined the new Israeli national airlines, El Al, eventually, after 1961, founding Consolidated Airborne Systems Inc., a manufacturer of avionics, including fuel gauges and engine instrumentation.

Warren Trent (MSME 1948) became chief engineer in propulsion and thermodynamics at the McDonnell Aircraft Company and director of engineering technology for the McDonnell Douglas Corporation.

Edward C. Govignon (MSME 1948) started out as a research engineer at the M.W. Kellogg Company and went on to work as a project engineer at Reaction Motors Inc., working on the XLR-99 Pioneer engine that powered the X-15. He was then a staff engineer for the Rocket Operations Center (Utah) of the Thiokol Chemical Corporation.

William W. Brant (BSAE 1955, MS Engineering 1956), began his career at WPAFB and the Thiokol Chemical Corporation, where he worked on the development of propulsion systems for the Bomarc, Peacekeeper, and Trident missiles, rising to vice president and general manager of its Strategic Operations Division. In the 1990s he assisted the Russian and US governments in dismantling strategic missiles and recycling their parts and fuels.

J. P. Sellers, whom Zucrow singled out for his "pioneering research in film cooling," had an Ohio State University BSME (1943) and an MSAE (1948) and PhD (1958) from Purdue. At Purdue, and while interning at Aerojet (1949–1950), he invented and disseminated a unique "swirl means of film-coolant injection." In 1955 he went to work at Rocketdyne, where he joined teams in the design and development of regenerative cooling for the Atlas, Thor, J-25, and the Saturn J-2, F-1, and H-1 engines, as well as film cooling for the X-8 and F-1 engines.

## Top Secret Programs

Several Zucrow students disappeared into the special access or "black" world of aerospace secrets, into the ultimate Q and Omega clearances dealing with atomic energy.

Elliott Katz (PhD 1954), with a degree in applied physics and rocket propulsion, worked at Convair on the Atlas missile, a technical member of the scientific advisory group assisting E. O. Lawrence, Edward Teller, Enrico Fermi, and Theodore von Kármán. At the Aerospace Corporation he worked closely with ARPA on countermeasures for ICBM defense (the Advanced Ballistic Reentry System).

Robert Strickler (BSAE 1960, MSAES 1962, PhD 1968) started his postdoctoral career at the Aerospace Corporation, working on reentry vehicle technology and the Advanced Ballistic Reentry System. He served as a Congressional Science Fellow for the US Senate, advising on the Strategic Arms Limitation Treaty. In the 1980s and 1990s, he worked at TRW in its Defense Systems Group and its Ballistic Missiles Division, working on ICBM technology and the Strategic Defense Initiative, and later in its Environmental Safety Systems, working on the Radioactive Waste Management Program.

Paul Petty (BSAE 1952), a student of Zucrow, Max Jacob, and George Hawkins, worked at General Electric and the Perkin-Elmer Corporation on satellite surveillance as manager of thermodynamics and thermal protection for the cameras, including those on the Corona/Discoverer, Hexagon, and Gambit satellite systems (the latter from 1971 to 1986). He also helped manage the Hubble Space Telescope system.

Arthur Greenberg (BSME and BSAE 1950, MSAE 1952, PhD 1955) rose to become a vice president and general manager of government support operations at the Aerospace Corporation, where he helped direct "rapidly advancing technological issues in areas such as solar and nuclear energy, civilian applications of satellites, energy conservation technology, electric and hybrid automobiles, offshore structural analysis techniques, energy resource development opportunities, and the strategic petroleum reserve," as cited in his Purdue Distinguished Engineering Alumnus Award.

Not all Zucrow stories ended happily.

Richard James Burick (PhD 1967), a specialist in solid-propellant rockets, died in January 2003 from a self-inflicted gunshot wound, just having retired as deputy director of operations at Los Alamos National Laboratory, the lab then under investigation for a spy scandal and allegations of corruption.

## US Government

Several Zucrow students went on to high positions in the US government.

David Charvonia (PhD 1959) was special assistant to the deputy director for advanced research and technology and became director of the Advanced Technology Office, both in the Office of the Director of Defense Research and Engineering and the Office of the Secretary of Defense. He was one of the early developers of the Advanced Research Projects Agency Network (ARPANET) and an architect of the Pentagon's Technology Base Strategy and Total Information Awareness programs. He later served as director of the Mclean Research Center, Unisys Defense Systems.

George Schneiter (BSME 1959, MSME 1962, PhD 1966) started his career at the Aerospace Corporation in San Bernardino, California, working on ballistic missile reentry systems. In 1973 he joined the Office of the Secretary of Defense, where he served as a technical adviser for the Strategic Arms Limitation Talks. After 1986, he served as assistant deputy undersecretary, Strategic Aeronautical and Theater Nuclear Systems, and as director, Strategic and Space Systems, "responsible for acquisition oversight of all strategic offensive systems, ballistic missile defense systems, meteorological satellites, and space launch systems;" and as director, Strategic and Tactical Systems, Office of the Under Secretary of Defense (for Acquisition, Technology, and Logistics).

## US Missile and Space Program

Other graduates participated in the US space program.

With just a BSAE (1952), but the best of Zucrow's and Joe Liston's mentorship, William Cowdin joined the nascent "rocket industry," as he called it, hired as a rocket test engineer with Bell Aircraft in Niagara Falls, New York, then three years later joined Aerojet-General, and later Lockheed. He was the program manager on a number of programs, including the Titan II ICBM, Gemini, Dyna-Soar, and Agena spy satellite. "All of this was the result of my Purdue education," he said.

Edward B. "Mister Ed" Dobbins (MSME 1954) married Rosemary at Purdue, and together they settled in Huntsville after 1954, working at the Army Ballistic Missile Agency. He became director of the Army Rocket Propulsion Technology and Management Center, Redstone Arsenal. For the NASA Marshall Space Flight Center, he worked on the Saturn V separation stages. Rosemary became an accomplished NASA artist, drawing technical illustrations (unsigned) for many of the early NASA publications for Mercury, Gemini, and Apollo.

Koos Botje (PhD 1956) went to work at the General Electric Missile and Space Division, located in the middle of downtown Philadelphia at 3198 Chestnut Street, later at the $14 million Space Research Center near Valley Forge, where he designed film-cooling methods for reentry vehicles.

Bob Bowlin (MSME 1956) first went to the Rocket Lab at Edwards AFB as project engineer for the early Minuteman ICBM launch configuration, then on to General Electric as a rocket systems engineer (1959) working on plug nozzles and high-energy propellants. By 1964, he was a project engineer at the Schenectady R&D Center, where did "dynamic analysis of the lunar excursion module propulsion system."

A. R. Graham (PhD 1958) also joined General Electric, where he invented the segmented plug nozzle rocket engine concept and wrote the *NASA Plug Nozzle Handbook*, what he called his principal contribution to the rocket industry.

Richard W. Foster, winner of two American Rocket Society (ARS) student awards (1952 and 1955), enjoyed a wide-ranging career that began with summer jobs at Bell Aircraft, Reaction

Motors Inc., and General Electric. He went on to work for the Project Vanguard launches for the Naval Research Laboratory, then for the Redstone and Jupiter programs for the Army Ballistic Missile Agency and the Saturn I program for Apollo. He also helped pioneer renewable solar/hydrogen energy systems and the rocket-based combined cycle system, joining a ramjet, scramjet, and rocket for reusable spaceflight launches, a perfect homage to a Zucrow education.

J. Michael Murphy (BSAE 1957, MSAE 1959, PhD 1964) had a unique and interesting set of experiences after graduating with his MS from Purdue. He first took a six-week workshop at the Space Launch Division of United Aircraft in Hartford, Connecticut, "studying large Nova-class launch vehicles." Following that, with his recent commission as a 2nd lieutenant in the Army Corps of Engineers, he went on a six-month tour of duty with the Corps Missile and Space Office, where he worked on Corps of Engineer items for the Moon such as habitats and latrines, about which he commented, "Needless to say, 52 years later the United States still has not built any such items on the Moon." In 1990 Murphy won the AIAA Wyld Propulsion Award for leading the 110-person contractor evaluation team for NASA's shuttle return-to-flight program after the *Challenger* disaster.

## Universities

A few Rocket Lab alumni made their careers in academia.

Eldon Knuth (MSME 1950) became professor and chair of the Energy and Kinetics Department, UCLA.

Helmut Wolf (MSME 1950, PhD 1958) became the principal scientist in heat transfer at Rocketdyne's Research Lab but eventually settled in as a professor of mechanical engineering at the University of Arkansas, Fayetteville (1961–1988).

J. P. Sellers (PhD 1958) became a professor of mechanical engineering at Tuskegee University (1970–1984).

John M. Bonnell (MSME 1960, PhD 1964) took the opposite track, beginning his career as an assistant professor of mechanical engineering at the University of California, Santa Barbara, but soon joined Pratt & Whitney Aircraft, starting out as a project engineer in the Combustion Group, later receiving a patent for jet engine afterburner technology.

James H. Fisher (MSME 1951, PhD 1953) became an engineering professor at the University of Washington for a few years, but he too felt "the call of industry" and went on to work in the Propulsion Structures Laboratory at Ramo-Woolridge, then on to Electro-Optical Systems, and then in 1962 he became president of Application Research Corporation.

James Beverly Jones (MSME 1947, PhD 1951) taught mechanical engineering at Purdue and Virginia Polytechnic Institute and authored several leading textbooks on thermodynamics.

Loren R. Davis (MSME 1961) went on for his PhD at the University of Illinois and then taught in the Department of Mechanical Engineering at San Jose State College.

# APPENDIX 3

## Purdue Rocket Laboratory Courses Between 1946 and 1971

I HAVE BASED THESE LISTS ON THE INFORMATION IN THE PURDUE COURSE CATALOGS.

### Courses Created in 1946–1947

Steam and Gas Turbines (ME 139)
Gas Turbines and Jet Engines (AE 115)
Laboratory in Gas Turbines and Jet Engines (AE 117)
Principles of Gas Turbines and Jet Propulsion I (ME and AE 281)
Principles of Gas Turbines and Jet Propulsion II (ME and AE 282)
Seminar in Gas Turbines and Jet Propulsion (ME and AE 283)
Special Problems in Gas Turbines and Jet Propulsion (ME and AE 298)
Research in Gas Turbines and Jet Propulsion (ME and AE 299)

### Courses by 1958–1959

Steam and Gas Turbines (ME 433)
Gas Turbines and Jet Engine Power Plants (ME 451)
Guided Missile Systems (ME 550)
Rocket Propulsion Systems (ME 552)
Principles of Gas Turbines and Jet Propulsion (ME 650)
Principles of Gas Turbines and Jet Propulsion (ME 651)
Fundamentals of Rocketry (ME 652)
Special Projects in Gas Turbines and Jet Propulsion (ME 687)
Gas Turbines and Jet Propulsion Seminar (ME 691)
Research (MS) in Gas Turbines and Jet Propulsion (ME 698)
Research (PhD) Gas Turbines and Jet Propulsion (ME 699)

## New Courses Created by 1966–1971

Turbomachinery (ME 433)
Air Breathing Propulsion Systems/Jet Propulsion Power Plants (ME 451)
Propulsion Systems for Spaceflight (ME 452)
Thermodynamics (ME 500)
Statistical Thermodynamics (ME 501)
Heat and Mass Transfer (ME 505)
Two Phase Flow and Heat Transfer (ME 506)
Fluid Mechanics/Gas Dynamics (ME 510)
Magneto-Fluid Dynamics (ME 511)
Experimental Gas Dynamics (ME 514)
Mass Transfer (ME 515)
Cryogenic Engineering (ME 518)
Fundamentals of Chemical and Gas Dynamics Lasers (ME 520)
Combustion (ME 525)
Introduction to Missile and Space Systems Design (ME 550)
Jet Propulsions Systems (ME 551)
Rocket Propulsion (ME 552)
Principles of Turbomachinery (ME 557)
Foundations of Laser Theory (ME 595)
Advanced Thermodynamics (ME 600 and 601)
Nonequilibrium Thermodynamics (ME 602)
Diffusion of Heat and Mass (ME 604)
Advanced Heat Transmission/Radiation Heat Transfer (ME 605 and 606)
Convection of Heat and Mass (ME 605)
Advanced Fluid Dynamics (ME 610 and 611)
Advanced Gas Dynamics (ME 614)
Mass Transfer (ME 615)
Combustion (ME 625)
Propulsion Gas Dynamics I (ME 650)
Gas Turbines and Jet Propulsion (ME 651)
Airbreathing Jet Propulsion (ME 651)
Rocket Propulsion Systems/Rocket Jet Propulsion (ME 652)
Propulsion Gas Dynamics II (ME 653)
Space Propulsion Systems (ME 654)
Dynamics of Real Gases (ME 655)
Jet Propulsion Seminar (ME 691)
Optimization of Aerodynamic Shapes (ME 697)
Research in MS and PhD (ME 698 and 699)

# NOTES

FOR THE FULL LIST OF ARCHIVAL COLLECTIONS CITED IN THE CHAPTER NOTES, AS well as their abbreviations and abbreviated titles, please see the list of archival sources that follows the notes.

### INTRODUCTION

1. There is no mention of Zucrow in *Men of Space*, ed. Shirley Thomas, 6 vols. (Philadelphia: Chilton, 1960). "Crowned heads of missiledom" (including Theodore von Kármán, Wernher von Braun, and Bernard Schriever) is from Matt Ek's "Visit of John L. Sloop to Rocketdyne" (24 April 1974), 5, NASA History Office (NASA).
2. For some biographies of Zucrow's contemporaries, see John Johnson Jr., *Zwicky: The Outcast Genius Who Unmasked the Universe* (Cambridge: Harvard University Press, 2019); and Abigail Foerstner, *James Van Allen: The First Eight Billion Miles* (Iowa City: University of Iowa Press, 2009).
3. "Retirement Statement for Dr. Maurice J. Zucrow" (1966), in the M. J. Zucrow file, Department of Mechanical Engineering, Purdue University (Zucrow file).
4. Robert J. Gordon, *The Rise and Fall of American Growth: The U.S. Standard of Living since the Civil War* (Princeton: Princeton University Press, 2017).
5. J. R. Van Pelt, "The Road Ahead in Engineering," *Mechanical Engineering* (November 1949): 918.
6. C. E. Kenneth Mees, *The Path of Science* (New York: Wiley, 1946), 14. Derek J. de Solla Price, *Science since Babylon* (New Haven: Yale University Press, 1960), 183–92. David Kaiser, "Booms, Busts, and the World of Ideas," *Osiris* 27, no. 1 (January 2012): 276–79.
7. News release of 14 December 1948, Aviation and Rocketry, box 2, Guggenheim Foundation Records, Library of Congress. Marvin McFarland, preface to *Men of Space*, vol. 4, ix.
8. Remarks on May 18, 1959, 86th Congress, *Congressional Record*, vol. 105, part 6, 8276–95.
9. Joseph A. Schumpeter, *Capitalism, Socialism, and Democracy* (New York: Harper, 1942).
10. *Missile Design and Development* (August 1958): 15.
11. M. J. Zucrow, "The Problems of the Turbojet Engine as a Propulsion Engine for Supersonic Flight," *Aeronautical Engineering Review* 15, no. 12 (December 1956): 44.
12. "Prof. M. J. Zucrow," *Purdue Engineer* (March 1960): 50–51; and Edward Kaplan, *To Kill Nations: American Strategy in the Air-Atomic Age and the Rise of Mutually Assured Destruction* (Ithaca: Cornell University Press, 2015), 77–107.
13. Clarence Danhof, *Government Contracting and Technological Change* (Washington, DC: Brookings Institution, 1968). David M. Hart, *Forged Consensus: Science, Technology, and*

*Economic Policy in the United States, 1921–1953* (Princeton: Princeton University Press, 1998). Aaron L. Friedberg, *In the Shadow of the Garrison State: America's Anti-Statism and Its Cold War Grand Strategy* (Princeton: Princeton University Press, 2000). For a focused study, see Peter Morris, *The Management of Projects* (London: Thomas Telford, 1994).

14. Danhof, *Government Contracting and Technological Change*, 2–11, 132–35. Bruce L. R. Smith, *American Science Policy since World War II* (Washington, DC: Brookings Institution, 1990), 47. The remainder went for graduate training, institutional support, and facilities.
15. *Panel on Science and Technology Fourth Meeting. Hearings before the Committee on Science and Astronautics, US House of Representatives*, 87th Congress (21 and 22 March 1962), 60.
16. John Sloop, "Review of Rocket Research at Lewis" (9 February 1954), 6, Rocket Engine Research, box 502, RG 255, National Advisory Committee for Aeronautics.
17. H. Doyle Thompson interview with Michael G. Smith (17 January 2017), 22, Purdue University Archives and Special Collections (ASC). George P. Sutton, *History of Liquid Propellant Rocket Engines* (Reston: AIAA, 2005), 25.
18. Maurice Zucrow interview with Robert Eckles (5 June 1970), 27, ASC.
19. Paul Forman, "The Primacy of Science in Modernity, of Technology in Postmodernity, and of Ideology in the History of Technology," *History and Technology* 23, no. 1–2 (March/June 2007): 2.
20. For grand studies of technology, see George Basalla, *The Evolution of Technology* (New York: Columbia University Press, 1988); G. F. C. Rogers, *The Nature of Engineering: A Philosophy of Technology* (London: Palgrave, 1983); and Robert Pool, *Beyond Engineering: How Society Shapes Technology* (New York: Oxford University Press, 1997).
21. Walter G. Vincenti, *What Engineers Know and How They Know It: Analytical Studies from Aeronautical History* (Baltimore: Johns Hopkins University Press, 1990). Also see Arnold Pacey, *The Culture of Technology* (Cambridge: MIT Press, 1983); and Eugene S. Ferguson, *Engineering and the Mind's Eye* (Cambridge: MIT Press, 1999).
22. *Purdue Alumnus* (February 1966): 3; and (March 1966): 16. For an early use of the term with Zucrow in mind, see Dan Kimball's comments in *Aero Digest* (February 1948): 58.
23. On "engineering science," see John Anderson, "The Evolution of Aerodynamics in the Twentieth Century," in *Atmospheric Flight in the Twentieth Century*, ed. Peter Galison and Alex Roland (Dordrecht: Kluwer, 2000), 241–56.
24. Abe Silverstein interview with John Mauer (10–13 March 1989), 8, Abe Silverstein Papers, NASA Glenn Research Center. James Beverly Jones to Zucrow (14 November 1974), folders 3–5, box 9, Maurice J. Zucrow Papers (Zucrow).
25. Elliott Katz interview with Michael G. Smith (29 September 2016), 6, ASC.
26. Bruce Reese interview with Tracy Grimm (18 November 2015), 6, ASC.
27. Allan A. Needell, *Science, Cold War and the American State* (Amsterdam: Harwood, 2000). David H. DeVorkin, *Science with a Vengeance: How the Military Created the US Space Sciences after World War II* (New York: Springer Verlag, 1992).

28. T. A. Heppenheimer, *The Space Shuttle Decision: NASA's Search for a Reusable Space Vehicle* (Washington, DC: NASA, 1999), 14.
29. *Purdue Students Handbook*, vol. 25 (1904), in the Gilbert A. Young Papers.
30. See Michael Aaron Dennis, "A Change of State: The Political Cultures of Technical Practice at the MIT Instrumentation Laboratory and the Johns Hopkins University Applied Physics Laboratory, 1930–1945" (PhD diss., Johns Hopkins University, 1991).
31. Joseph Haberer, *Politics and the Community of Science* (New York: Van Nostrand, 1969); Roger L. Geiger, *Research and Relevant Knowledge: American Research Universities since World War II* (New York: Oxford University Press, 1993); and Daniel Lee Kleinman, *Politics on the Endless Frontier: Postwar Research Policy in the United States* (Durham, NC: Duke University Press, 1995).
32. *Panel on Science and Technology Sixth Meeting. Proceedings of the Committee on Science and Astronautics, US House of Representatives*, 89th Congress (26 and 27 January 1965), 21.
33. D. V. Gallery, "Guided Missiles," *Aero Digest* (December 1948): 48.

## CHAPTER 1

1. Max Nordau, "Address to the Fifth Zionist Congress" (1901), in *Max Nordau to His People: A Summons and a Challenge* (New York: Scopus, 1941), 126–27. George L. Mosse, "Max Nordau, Liberalism and the New Jew," *Journal of Contemporary History* 27, no. 4 (October 1992): 565–81.
2. Israel Zangwill, "Luftmensch," in his *Ghetto Comedies*, vol. 2 (New York: Macmillan, 1907), 255–80.
3. Nicolas Berg, *Luftmensch: Zur Geschichte einer Metapher* (Göttingen: Vandenhoeck, 2016), 67–84. Todd Samuel Presner, *Muscular Judaism: The Jewish Body and the Politics of Regeneration* (New York: Routledge, 2007), 108, 204–16.
4. Israel Zangwill, *The Melting Pot* (New York: Macmillan, 1920). Edna Nahshon, ed., *From the Ghetto to the Melting Pot: Israel Zangwill's Jewish Plays* (Detroit: Wayne State University Press, 2006).
5. Benjamin Nathans, *Beyond the Pale: The Jewish Encounter with Late Imperial Russia* (Berkeley: University of California Press, 2002).
6. Valerie Khiterer, *Jewish City or Inferno of Russian Israel? A History of the Jews in Kiev before February 1917* (Boston: Academic Studies, 2016), 420. Richard Stites, *Russian Popular Culture: Entertainment and Society since 1900* (New York: Cambridge University Press, 1992), 9–36.
7. Robert Paul Magosci and Yohanon Petrovsky-Shtern, *Jews and Ukrainians: A Millennium of Co-Existence* (Toronto: University of Toronto Press, 2016). John D. Klier and Shlomo Lambroza, *Pogroms: Anti-Jewish Violence in Modern Russian History* (New York: Cambridge University Press, 1992).
8. Lillian Zucrow, "Father," *Jewish Advocate* (19 February 1932): 6.
9. Dova was also known as Doba or Dora. Jonathan Eisenberg's conversation with Joanne

Bickow (16 October 2000), folder 3, box 2 (addition), Zucrow. Aharon Weissman, ed., *Stavisht*, trans. Ida Cohen Selavan (New York: Stavisht Society, 1961), 2–30; and Martha Lear, "The Roots People," *New York Times* (31 July 1994): E32.

10. These family reminiscences come from Jonathan Eisenberg's genealogical findings in his communications with Joanne Bickow (16 October 2000), folder 3; and Louis Bornstein (8 March 1997), folder 6; both in box 2 (addition), Zucrow.
11. Natan M. Meir, *Kiev, Jewish Metropolis: A History, 1859–1914* (Bloomington: Indiana University Press, 2010), 168, 176–80. Khiterer, *Jewish City*, 157, 198. Gennady Estraikh, "From Yehupets Jargonists to Kiev Modernists," *East European Jewish Affairs* 30, no. 1 (2000): 17–38.
12. The family facts come from Jonathan Eisenberg's findings in the file Runia Feiman Kanief Letters, folder 7, box 2 (addition), Zucrow. Zucrow interview with Eckles, 1.
13. Lloyd P. Gartner, *The Jewish Immigrant in England* (London: Allen & Unwin, 1960), xi, 272. Lillian Zucrow, "Father," 6. Bernard Norwood to Michael G. Smith (22 March 2017), Bernard Norwood folder, box 2 (addition), Zucrow.
14. Rebecca Kobrin, *Jewish Bialystok and Its Diaspora* (Bloomington: Indiana University Press, 2010).
15. W. W. Hutchings, *London Town Past and Present*, vol. II (London: Cassell, 1909), 948–66.
16. Chaim Lewis, *A Soho Address* (London: Gollancz, 1965), 74.
17. Ralph L. Finn, *Time Remembered: The Tale of an East End Jewish Boyhood* (London: Macdod, 1963), 16.
18. I have drawn these contexts from several memoirs: Lewis, *A Soho Address*, 96–97; Harry Blacker, *Just Like It Was: Memoirs of the Mittel East* (London: Vallentine, 1974); and William Goldman, *East End My Cradle* (London: Rob, 1988). Also see Selma Cantor Berrol, *East Side/East End: Eastern European Jews in London and New York, 1870–1920* (Westport: Praeger, 1994).
19. Susan Tananbaum, *Jewish Immigrants in London, 1880–1939* (New York: Routledge, 2015), 98–99.
20. Lillian Zucrow, "Father," 6. Solomon's letters of reference, from his patrons in the London Jewish community (dated October of 1917), are in folder 9, box 2 (addition), Zucrow. Stephen Sharot, "Reform and Liberal Judaism in London, 1840–1900," *Jewish Social Studies* 41, no. 3–4 (Summer–Autumn 1974): 214–15.
21. Hutchings, *London Town*, 948–66.
22. Tananbaum, *Jewish Immigrants*, 72–73. Lewis, *A Soho Address*, 14–15, 130. Gartner, *The Jewish Immigrant*, 172.
23. Gartner, *The Jewish Immigrant*, 99–105.
24. I've drawn these items from the 1912–1915 issues of the Foundation School's student magazine, the *Central*, from its library in London, England.
25. Tananbaum, *Jewish Immigrants*, 115–23. Michael Berkowitz and Ruti Ungar, eds., *Fighting Back: Jewish and Black Boxers in Britain* (London: University College, 2007).

26. "Debating Society," *Central* 4, no. 2 (April 1914): 56.
27. *Central* 3, no. 7 (Christmas 1912): 216.
28. Louis Heron, *Growing Up Poor in London* (London: Hamish Hamilton, 1973), 11, 205–6.
29. Selig Brodetsky, *Memoirs: From Ghetto to Israel* (London: Weidenfeld, 1960). Jacob Bronowski, *The Ascent of Man* (Boston: Little, Brown, 1974).
30. Quoted from Zucrow's enrollment file, kindly shared by the front office at the school.
31. Lewis, *A Soho Address*, 68.
32. Ralph Finn, *Time Remembered* (London: Macdonald, 1963). *Central* 4, no. 7 (Christmas 1915): 193, 208–10.
33. Pauline Holmes, *A Tercentenary History of the Boston Public Latin School, 1635–1935* (New York: Johnson Reprint, 1969), 297.
34. Zucrow interview with Eckles, 1–2.
35. David O. Levine, *The American College and the Culture of Aspiration, 1915–1940* (Ithaca: Cornell University Press, 1986), 23–28. Carol S. Gruber, *Mars and Minerva: World War I and the Uses of the Higher Learning in America* (Baton Rouge: Louisiana State University Press, 1976). Charles Franklin Thwing, *The American Colleges and Universities in the Great War, 1914–1919: A History* (New York: Macmillan, 1920).
36. 9 January 1918 letter, in SATC General, Student Army Training Corps Records (SATC), Digital Collections and Archives, Tufts University. William Murray Hepburn and Louis Martin Sears, *Purdue University: Fifty Years of Progress* (Indianapolis: Hollenbeck, 1925), 142–57. Fred Goldsmith, ed., *Purdue University Alumni Record and Campus Encyclopedia* (Lafayette, IN: Purdue University, 1929), 626. Thwing, *The American Colleges and Universities*, 96–98.
37. *Students Army Training Corps Regulations* (Washington, DC: GPO, 1918), 5–12. Gruber, *Mars and Minerva*, 217, 239–43.
38. War Department and Tufts correspondence (January–March 1918), in SATC Engineers, SATC.
39. SATC Roster (1918), SATC. Advisory Board, *Committee on Education and Special Training* (Washington, DC: War Department, 1919), 135–38.
40. Gruber, *Mars and Minerva*, 232. Parke Rexford Kolbe, *The Colleges in War Time and After: A Contemporary Account of the Effect of the War* (New York: Appleton, 1919), 76.
41. *Students Army Training Corps Regulations*, 13–15. Correspondence with War Department and Boston colleges (December 1917–April 1918), in SATC Roll and Voucher, and SATC General, SATC. For the regulations governing contracts, see Kolbe, *The Colleges in War Time and After*, 292–95.
42. From the 12 October 1918 contract, and "Accounting Memorandum, War Department Committee on Education and Special Training" (18 June 1919), SATC General, SATC.
43. President Herman C. Bumpus of Tufts to the Polytechnic Preparatory Country Day School (28 October 1918), in SATC General, SATC.

44. "Experiences with the Vocational Sections of the SATC (1918)," folder 28, box 7, A. A. Potter Papers (Potter). A. A. Potter interview with Robert Eckles (16 July 1969), 36, ASC.
45. C. R. Mann, *The American Spirit in Education* (Washington, DC: Department of the Interior, 1919), 55–63. The quotes are from Thwing, *The American Colleges and Universities*, 5, 57, 251.
46. Solomon taught Talmud at Hebrew Teachers College from 1921 to 1932. "Organization of the Hebrew Teachers College," in *Hebrew Teachers College Register, 1954–1955* (Brookline, MA: Hebrew Teachers College, 1954), 7.
47. See Arthur A. Goren, "Ben Halpern: At Home in Exile," in *The "Other" New York Jewish Intellectuals*, ed. Carole S. Kessner (New York: New York University Press, 1994), 76–77.
48. *Hebrew College Bulletin* 5, no. 4 (June 1975): 20. Lillian Zucrow, "Father," 6.
49. Joseph Shalom Shubow's comments in the *Jewish Advocate* (19 February 1932): 6, folder 9, box 1 (addition), Zucrow.
50. Solomon Zucrow, "Specimens of Hyperbole, Dubious Testimony, and Yarns Found in the Talmud," part of a series from 5 November 1925 to 7 January 1926 in the *Jewish Advocate*. The copy is in folder 9, box 2 (addition), Zucrow. Gartner, *The Jewish Immigrant*, 168–72.
51. Solomon Zucrow, *Adjustment of Law to Life in Rabbinic Literature* (Boston: Stratford, 1928), ii–iii, 160–86.
52. Solomon Zucrow, *Women, Slaves and the Ignorant in Rabbinic Literature* (Boston: Stratford, 1932), 16–22, 194–95, 230.
53. Lillian Zucrow, "Father," 6.
54. The story was told by Leslie Bornstein Stacks (the daughter of Maurice's sister, Lillian), in Dr. Maurice Tuchman's letter to Jonathan Eisenberg (8 December 1997), folder 4, box 2 (addition), Zucrow.
55. Zucrow interview with Eckles, 2.
56. *Harvard University Catalogue: 1920–1921* (Cambridge: Harvard University, 1920), 171, 326–65; and H. J. Hughes, "The Harvard Engineering School," *Harvard Graduates' Magazine* (September 1920): 71–78.
57. Mann, *The American Spirit in Education*, 49–54. Senior year included courses in accounting and business administration.
58. Zucrow interview with Eckles, 2–4.
59. Robert H. Gardner, "The Harvard Liberal Club of Boston," *Harvard Graduates' Magazine* (December 1920): 230; and "The Spring Term" (June 1921): 594.
60. Richard Norton Smith, *The Harvard Century: The Making of a University to a Nation* (New York: Simon and Schuster, 1986), 87. Levine, *The American College and the Culture of Aspiration*, 137–51. Also see Marcia Graham Synnott, *The Half-Opened Door: Discrimination and Admissions at Harvard, Yale, and Princeton, 1900–1970* (Westport: Greenwood, 1979).
61. "Prof. M. J. Zucrow," *Purdue Engineer* (March 1960): 50–51.
62. Zucrow interview with Eckles, 10–11.

63. "Literary Notes," *Harvard Graduates' Magazine* (March 1921): 498. M. J. Zucrow, "The Engineering Graduate and his Career as a Professional Engineer" (1947), 7, Zucrow Talks, box 1 (addition), Zucrow.
64. Zucrow, "The Engineering Graduate," 2–5. George S. Morison, *The New Epoch: As Developed by the Manufacture of Power* (Boston: Houghton Mifflin, 1903). Edwin T. Layton Jr., *The Revolt of the Engineers: Social Responsibility and the American Engineering Profession* (Cleveland: Case Western Reserve University Press, 1971), viii–ix.
65. Transcript of Ben and Sam Smullin (28 July 1996), 14–15, folder 2, box 2 (addition), Zucrow.
66. For contexts, see David A. Hollinger, *Science, Jews, and Secular Culture: Studies in Mid-Twentieth-Century American Intellectual History* (Princeton: Princeton University Press, 1996).
67. The letter is in Zucrow Talks, box 1 (addition), Zucrow. "Nathan Marein," *Boston Globe* (19 March 1982): 1.

## CHAPTER 2

1. Bruce Mazlish, ed., *The Railroad and the Space Program: An Exploration in Historical Analogy* (Cambridge: MIT Press, 1965). Wolfgang Schivelbusch, *The Railway Journey: The Industrialization of Time and Space in the Nineteenth Century* (Oakland: University of California Press, 2014).
2. William Murray Hepburn and Louis Martin Sears, *Purdue University: Fifty Years of Progress* (Indianapolis: Hollenbeck, 1925), 3–7.
3. John F. Stover, *A History of American Railroads* (Boston: Rand McNally, 1967), 23–27.
4. Fred Goldsmith, ed., *Purdue University Alumni Record and Campus Encyclopedia* (Lafayette, IN: Purdue University, 1929), 573. Gil Stein, "The True Boilermakers," *Purdue Engineer* (October 1958): 22–24.
5. Barton C. Hacker, "Technology and Research," and "Engineering and Science," in *Encyclopedia of the American Military: Studies of the History, Traditions, Policies, Institutions, and Roles of the Armed Forces in War and Peace*, ed. John E. Jessup and Louise B. Ketz (New York: Scribner's, 1994), 1373–1444.
6. Potter interview with Eckles, 32.
7. See W. F. M. Goss, "An Experimental Locomotive," *Transactions of the American Society of Mechanical Engineers (ASME)* 13 (1892): 427–37; his "Locomotive Testing Plants," *Transactions ASME* 25 (1904): 827–67; and his *Locomotive Performance: The Result of a Series of Researches Conducted by the Engineering Laboratory of Purdue University* (New York: Wiley, 1907), 1–5.
8. Tom Morrison, *The American Steam Locomotive in the Twentieth Century* (Jefferson, NC: McFarland, 2018), 67–71.
9. James H. Smart and W. F. M. Goss, "The Burned Testing Laboratory at Purdue University," *Railway Review* (3 February 1894): 76–77.
10. William L. Withuhn, *American Steam Locomotives: Design and Development, 1880–1960*

(Bloomington: Indiana University Press, 2019), 119–20, 124–25. W. F. M. Goss, "Steam Pipes within Locomotive Smoke-Boxes as a Means of Superheating," *Railway Review* (28 July 1894): 427; and his "Form and Character of the Exhaust-Steam Jet," *Proceedings of the American Railway Master Mechanics' Association* 29 (1896): 59–143.

11. W. F. M. Goss, "Atmospheric Resistance to the Motion of Railway Trains," *Proceedings of the Western Railway Club* 10, no. 8 (25 April 1898): 347–77.
12. "The Proposed Locomotive Tests at Purdue," *Railway Review* (1 December 1894): 676. Smart's comments are in *Proceedings of the American Railway Master Mechanics' Association* 29 (1896): 293.
13. *New York Times* (23 June 1912): 6. Thomas P. Hughes, *American Genesis: A Century of Invention and Technological Enthusiasm* (New York: Viking, 1989).
14. David B. Danbaum, *The Resisted Revolution: Urban America and the Industrialization of Agriculture* (Ames: Iowa State University Press, 1979).
15. Quoted in H. C. Knoblauch, et al., *State Agricultural Experiment Stations* (Washington, DC: Department of Agriculture, 1962), 159–63.
16. E. W. Allen, "Administration of Experiment Station Work by Project," *Proceedings of the Association of American Agricultural Colleges and Experiment Stations* 28 (1915): 234–39.
17. *Proceedings of the Association of American Agricultural Colleges and Experiment Stations* 28 (1915): 119. *Classified List of Projects in Agricultural Economics and Rural Sociology* (Washington, DC: Office of Experiment Stations, 1932), 6.
18. E. W. Allen, "Work and Expenditures of the Agricultural Experiment Stations," in *Report on Agricultural Experiment Stations*, US Department of Agriculture (Washington, DC: GPO, 1916), 19, 29.
19. Fred Whitford, *For the Good of the Farmer: A Biography of John Harrison Skinner* (West Lafayette, IN: Purdue University Press, 2013), 329–35. Roger James Wood, "Science, Education, and the Political Economy in Indiana" (PhD diss., Purdue University, 1993), 162, 215–16, 221.
20. *Proceedings of the Association of American Agricultural Colleges and Experiment Stations* 29 (1915): 72–82, 256–95.
21. Bruce Seely, "Research, Engineering, and Science in American Engineering Colleges," *Technology and Culture* 34, no. 2 (April 1993): 344–54.
22. Elihu Root, "The Need for Organization in Scientific Research," *Bulletin of the National Research Council* 1 (October 1919): 10.
23. Vernon Kellogg, "The University and Research," *Science* 54, no. 1384 (8 July 1921): 19. Ira Remsen, "The Development of the University in America," *New York Times* (2 September 1911): 10.
24. Zucrow interview with Eckles, 5–6. Harvard College, Class of 1922, *Fiftieth Anniversary Report* (Cambridge: Harvard University, 1972), 584–85.
25. Robert Babcock and Robert B. Stewart, eds., *Purdue University, 1922–1932* (Lafayette, IN: Purdue University, 1933).

26. W. S. von Bernuth, "Estimate of Personnel Situation" (1926), folder 14, box 6, Horton B. Knoll Papers (Knoll).
27. Robert H. Knapp and Hubert B. Goodrich, *Origins of American Scientists* (Chicago: University of Chicago Press, 1952), 270–93. Robert Merton, *Social Theory and Social Structure* (New York: Free Press, 1949), 329.
28. David O. Levine, *The American College and the Culture of Aspiration, 1915–1940* (Ithaca: Cornell University Press, 1986), 39, 127. Also see James Axtell, *Wisdom's Workshop: The Rise of the Modern University* (Princeton: Princeton University Press, 2016); and Roger L. Geiger, *To Advance Knowledge: The Growth of American Research Universities, 1900–1940* (New York: Oxford University Press, 1986).
29. Babcock and Stewart, *Purdue University, 1922–1932*, viii–ix. Frank K. Burrin, *Edward Charles Elliott, Educator* (Lafayette, IN: Purdue University Studies, 1970).
30. Edward C. Elliott, M. M. Chambers, and William Ashbrook, *The Government of Higher Education* (New York: American Book, 1935). Edwin B. Stevens and Edward C. Elliott, *Unit Costs of Higher Education* (New York: Macmillan, 1925).
31. Robert B. Stewart and Roy Leon, *Debt Financing of Plant Additions for State Colleges and Universities* (West Lafayette, IN: Purdue Research Foundation, 1948). Ruth W. Freehafer, *R. B. Stewart and Purdue University* (West Lafayette, IN: Purdue University Press, 1983).
32. Potter interview with Eckles, 33–36. Robert B. Eckles, *The Dean: A Biography of A. A. Potter* (West Lafayette, IN: Purdue University, 1974).
33. "Engineering Experiment Station Organization" (1917), folder 9, box 6, Knoll.
34. By 1922, Purdue was one of twenty-two land-grant institutions with an EES. Bruce Seely, "Reinventing the Wheel," in *Engineering in a Land-Grant Context: The Past, Present, and Future of an Idea*, ed. Alan Marcus (West Lafayette, IN: Purdue University Press, 2005), 171. Society for the Promotion of Engineering Education, *A Study of the Supplementary Activities of the Engineering Colleges* (Lancaster, PA: Lancaster Press, 1926), 12.
35. A. A. Potter, "Decennial Review of the Accomplishments of the Engineering Schools and Departments, 1922–1932" (1933), folder 1, box 6; and his "Industrial Research Facilities of Purdue University" (1933), folder 18, box 2; both in Knoll.
36. H. L. Solberg interview with Robert Eckles (5 June 1970), 25, ASC. Potter interview with Eckles, 35–37.
37. A. A. Potter, "Purdue's Contribution to Engineering Knowledge" (1927), folder 10, box 1, Knoll.
38. Goldsmith, *Purdue University Alumni Record*, 594–95. *Purdue Exponent* (22 April 1947): 1. "Purdue University Builds New Plant," *Power Plant Engineering* (1 September 1925): 876–83.
39. *Power Plant Engineering* (1 October 1925): 983; (1 November 1926): 1137; and (1 July 1926): 728.
40. The imagery and quotes are in *Purdue Alumnus* (January 1962): 24–25; (March 1962): 1; (May 1964): front cover; and (May 1967): 1. James Mullins, ed., *A Purdue Icon: Creation, Life, and Legacy* (West Lafayette, IN: Purdue University Press, 2017). Scholer Corporation,

*The Building of a Red Brick Campus: The Growth of Purdue as Recalled by Walter Scholer* (Lafayette, IN: Tippecanoe County Historical Association, 1983).

41. Robert Glick interview with Michael G. Smith (20 January 2017), 18, ASC.
42. These articles appeared in the leading journal *Power* (May to December of 1928).
43. *Engineering News* [Purdue University] (January 1963): 3–4. Solberg interview with Eckles, 6–8.
44. H. L. Solberg, G. A. Hawkins, and A. A. Potter, "Characteristics of a High-Pressure Series Steam Generator," *Transactions ASME* 54 (1932): 9–27.
45. George E. Lien, ed., *The Behaviour of Superheater Alloys in High Temperature, High Pressure Steam* (New York: American Society of Mechanical Engineers, 1968).
46. M. J. Zucrow and A. A. Potter, "Smoothing Out the Load with a Ruth's Steam Accumulator," *Power* (18 December 1926): 554–56. M. J. Zucrow, "High Pressure Makes Progress," *Power Plant Engineering* (15 March 1926): 359–61.
47. *Purdue Engineer* (June 1946): 11. Zucrow interview with Eckles, 7. In the 1940s, Potter also chose Zucrow as one of his editors at the leading journal *Industry and Power*.
48. F. W. Greve and W. E. Stanley, "Coefficients of Discharge of Sprinkler Nozzles," *Bulletin of the Purdue University Engineering Experiment Station (EES),* no. 3 (November 1919). M. J. Zucrow, "A Method for Comparing Sewage Sprinkler Nozzles," *Engineering News Record* (18 December 1924); and M. J. Zucrow, with F. W. Greve and R. B. Wiley, "Characteristics of Sewage Sprinkle Nozzles," *Bulletin of the Purdue University EES,* no. 20 (June 1925). F. W. Greve and M. J. Zucrow, "Measurement of Pipe Flow by the Coordinate Method," *Journal of the American Waterworks Association* 13, no. 5 (March 1925): 306–11. M. P. O'Brien and M. J. Zucrow, "European High Specific Speed Hydraulic Turbines," *Power Plant Engineering* (1 November 1926): 1166–70.
49. O. C. Berry and C. S. Kegerreis, "The Carburation of Gasoline," *Bulletin of the Purdue University EES,* no. 4 (April 1920). H. A. Huebotter, "Hints on Automobile Maintenance," *Circular of the Purdue University Engineering Extension Department,* no. 3 (1924). Both Kegerreis and Huebotter also published widely in the journal *Motive Power* (1930–1931).
50. *Purdue Engineering Review* (November 1927): 14. Zucrow interview with Eckles, 6–7. M. J. Zucrow, C. S. Kegerreis, and O. Chenoweth, "Commercial Carburetor Characteristics," *Bulletin of the Purdue University EES,* no. 21 (August 1925); M. J. Zucrow, C. S. Kegerreis, and H. A. Huebotter, "Carburetion of Kerosene," *Bulletin of the Purdue University EES,* no. 27 (March 1927). M. J. Zucrow and H. A. Huebotter, "Torsional Strength of Splined Shafts," *Society of Automotive Engineers (SAE) Transactions* 22 (September of 1927): 99–100.
51. C. S. Kegerreis, "Research Work in the Carburetion Department" (18 October 1923), folder 10, box 1, Knoll. Memorandum of A. A. Potter, "Inventory of the Contributions of the Purdue EES, 1922–1929" (26 September 1929), folder 13, box 1, Knoll.
52. Kegerreis, "Research Work in the Carburetion Department."
53. Solberg interview with Eckles, 12.

54. H. A. Huebotter, *Mechanics of the Gasoline Engine* (New York: McGraw-Hill, 1923), v–vi.
55. Priscilla Decker, "Pioneer in Jet Propulsion," *Campus Copy* [Purdue University] (March 1965): 8.
56. Solberg interview with Eckles, 2–4.
57. See R. G. Dukes, chair, "The Committee on Graduate Study, 1921–1928" (15 July 1928), 1–4, folder 8, box 28, Charles M. Ehresman Papers (Ehresman).
58. Decker, "Pioneer in Jet Propulsion," 8. Zucrow interview with Eckles, 11–15. The book was either Alfred North Whitehead, *Science and the Modern World* (1925); or his *An Enquiry Concerning the Principles of Natural Knowledge* (1925).
59. Stephen Hilgartner, *Science on Stage: Expert Advice as Public Drama* (Stanford: Stanford University Press, 2000), 5–9.
60. Harvard College, Class of 1922, *Twenty-Fifth Anniversary Report* (Cambridge: Harvard University, 1947), 1140–41.
61. Goldsmith, *Purdue University Alumni Record*, 370. *Purdue Alumnus* (March 1960): 15; (February 1966): 3; and (March 1966): 16.
62. M. J. Zucrow, "Discharge Characteristics of Submerged Jets" (PhD diss., Purdue University, 1928); also published in *Bulletin of the Purdue University EES* 12, no. 4 (June 1928). "Properties of Submerged Carburetor Jets Studied at Purdue," *Automotive Industries* (6 October 1928): 480–83. M. J. Zucrow, "Flow Characteristics of Submerged Jets," *Transactions ASME* 51(1929): 213–18; and M. J. Zucrow, "Fuel-Mixture Distribution," *SAE Transactions* 24 (1929): 162–64.
63. For earlier uses of the term "jet propulsion" in aviation and maritime technology, see Edgar Buckingham, *Jet Propulsion for Airplanes* (Washington, DC: GPO, 1923); along with *Aerial Age Weekly* (July 1923): 317; and *Aviation* (16 May 1921): 624.
64. Edward W. Constant II, *The Origins of the Turbojet Revolution* (Baltimore: Johns Hopkins University Press, 1980), 99–105.
65. Opie Chenoweth, "Supercharged Engine Performance, Calculated and Actual," *SAE Transactions* 22 (1927): 256–64; and his work in *SAE Transactions* 25–26 (1930): 592–607; and 33 (1938): 472–84. Constant, *The Origins of the Turbojet Revolution*, 63, 99, 142.
66. For use of the term "internal aerodynamics," see the *Bulletin of Purdue University* (West Lafayette, IN: Purdue University, February 1947), 84; and Rudolf Hermann, *Supersonic Inlet Diffusers and Introduction to Internal Aerodynamics* (Minneapolis: Honeywell Regulator Co., 1956).
67. Zucrow, "Discharge Characteristics of Submerged Jets," 1, 68–71. Two of his sources included W. L. Cowley and H. Levy, *Aeronautics in Theory and Experiments* (London: Edward Arnold, 1920), 84–97; and Edwin Bidwell Wilson, *Aeronautics* (New York: Wiley, 1920), 183–98. On dimensional analysis, see Vincenti, *What Engineers Know*, 140–50.
68. Opie Chenoweth interview with John Sloop (7 June 1974), 6, NASA.
69. H. M. Jacklin, "Improving Engine Performance," *SAE Journal* 22, no. 5 (May 1928): 554; and

23, no. 2 (August 1928): 200. G. A. Young, "Research Facilities of the School of Mechanical Engineering" (6 January 1938), folder 9, box 6, Knoll.

70. *SAE Journal* 24, no. 5 (May 1929): 486; and *Transactions ASME* 58 (1936): 56–57; 61 (May 1939): 359–61; and 62 (May 1940): 304.
71. Harvard College, *Twenty-Fifth Anniversary Report*, 1140–41.
72. Deborah Douglas, "The End of 'Try and Fly': The Origins and Evolution of American Aeronautical Engineering Education through World War II," in Marcus, *Engineering in a Land-Grant Context*, 78–85.
73. "Purdue's Lindy," *Purdue Exponent* (11 September 1928): 3.
74. Geiger, *To Advance Knowledge*, 123–25, 162; Rebecca Lowen, *Creating the Cold War University: The Transformation of Stanford* (Berkeley: University of California Press, 1997), 23. Edward Embree, "In Order of their Eminence: An Appraisal of American Universities," *Atlantic Monthly* 155, no. 6 (June 1935): 652–54.
75. Judith R. Goodstein, *Millikan's School: A History of the California Institute of Technology* (New York: Norton, 1991); and Lowen, *Creating the Cold War University*, 45–48.
76. William K. LeBold, Edward C. Thoma, John W. Gillis, and George A. Hawkins, *A Study of the Purdue University Engineering Graduate* (West Lafayette, IN: Purdue University, 1960), 33–35. Richard Hallion, *Legacy of Flight: The Guggenheim Contribution to American Aviation* (Seattle: University of Washington Press, 1977), 45–46.
77. Clarence Danhof, *Government Contracting and Technological Change* (Washington, DC: Brookings Institution, 1968), 24–29. Nancy Petrovic, "Design for Decline: Executive Management and the Decline of NASA" (PhD diss., University of Maryland, 1982), 6, 95, 108. Lowen, *Creating the Cold War University*, 45–48.
78. Ethel M. De Haven, *History of Separation of Research and Development from the Air Materiel Command* (Dayton, Ohio: Wright-Patterson Air Force Base, 1954), 3–17, 162.
79. A. S. Reynolds, "Aviation at Purdue," *Purdue Engineering Review* (January 1928): 13. Goldsmith, *Purdue University Alumni Record*, 574. Potter, "Industrial Research Facilities of Purdue University."
80. Weldon Worth, "Lubrication and Cooling Problems of Aircraft Engines," *SAE Transactions* 40–41 (1937): 315–24.
81. *Purdue Engineering Review* (January 1928): 21; and (January 1932): 99.
82. G. Stanley Meikle, "Research" (1956), chap. 4, pp. 9–10, folder 4, box 2, G. Stanley Meikle Papers (Meikle). "Ross Memorandum" (1926), box 9, Potter. Fred C. Kelly, *David Ross: Modern Pioneer* (New York: Knopf, 1946).
83. David Ross, "Industrial Research from the Manufacturer's Viewpoint," 10–11; and Andrey Potter, "Purdue University and Indiana Industry," 20–24; both in "Proceedings of Industrial Conference, Held at Purdue University, June 1, 1926," *Bulletin of the Purdue University Engineering Extension Department*, no. 15 (June 1926).
84. Meikle, "Research," chap. 12, p. 7. Irving Langmuir, E. Q. Adams, and G. S. Meikle, "Flow of Heat through Furnace Walls: The Shape Factor," *Transactions of the American Electro-*

*chemical Society* 24 (1913): 53–76. G. Stanley Meikle, "The Hot Cathode Argon Gas-Filled Rectifier," *General Electric Review* 19, no. 4 (April 1916): 297–304.

85. Meikle, "Research," chap. 4, pp. 10–16, 22–24; chap. 11, p. 9. "Twenty-Five Years of Progress," *PRF Horizons* 2, no. 9 (May 1956): 2–3.
86. Meikle, "Research," chap. 4, pp. 1–2; chap. 12, p. 8.
87. Meikle, "Research," draft chapter marked "xxxx," pp. 4, 8, 14–15; chap. 12, p. 9. R. B. Stewart interview with Ruth Freehafer (28 March 1977), 178, ASC.
88. Archie Palmer, *University Research and Patent Policies, Practices and Procedures* (Washington, DC: NAS, 1962), 34–35, 43–46, 130–31.
89. Stanley DeGraff, "Institutionalizing Entrepreneurship: A History of Sponsored Research at the University of Michigan," *American Educational History Journal* 33, no. 1 (2006): 137–39. Seely, "Reinventing the Wheel," 172–73.
90. Palmer, *University Research and Patent Policies*, 30–31, 102–3, 209. Robert Stephenson, "A History of the Ohio State University Research Foundation, 1936–1969," OSU Research Foundation (September 1969), 7–10.
91. Stewart interview with Freehafer, 92.
92. The University of Akron, under the leadership of George Zook, received the prize. Hallion, *Legacy of Flight*, 67.
93. On Moore and Lark-Horovitz, see *Purdue Engineering Review* (May 1927): 22; and (May 1928): 14.
94. "Aeronautical Education and Research Development of Facilities," 120, in Research Director's Calendar, folder 11, box 2, PRF Records.
95. I thank Purdue student Megan McGue for this information. George T. Mitchell, *Dr. George: An Account of the Life of a Country Doctor* (Carbondale: Southern Illinois University Press, 1994), 140.
96. K. D. Wood, Joseph Liston, and P. C. Emmons, "A Proposed Aeronautical Laboratory for Purdue University" (15 June 1938), in "A History of Aeronautical Education and Research at Purdue University, for Period 1937–1950" by Elmer F. Bruhn, vol. 2, ASC.
97. Michael G. Smith, *Rockets and Revolution: A Cultural History of Early Spaceflight* (Lincoln: University of Nebraska Press, 2014), 253–92.
98. "Rocket Plane Visualized Flying at 1,200 Miles an Hour," *Los Angeles Times* (28 March 1935). Michael H. Gorn, *The Universal Man: Theodore von Kármán's Life in Aeronautics* (Washington, DC: Smithsonian Institution, 1992).
99. Shirley Thomas, "Theodore von Kármán's Caltech Students," in *History of Rocketry and Astronautics*, AAS History Series (San Diego: AAS, 1997), 3–39.
100. Alex Roland, *Model Research: The National Advisory Committee for Aeronautics, 1915–1958*, vol. 1 (Washington, DC: GPO, 1985), 166. John Holmfeld, "The Site Selection for the NACA Engine Research Laboratory: A Meeting of Science and Politics" (master's thesis, Case Institute of Technology, 1967), 13–16, 32–40.
101. "NACA Proposed Site," research director's calendar, folder 11, box 2, PRF Records.

102. Franklin Cooper, "Location and Extent of Industrial Research Activity in the United States," in *Industrial Research* (Washington, DC: National Resources Planning Board, 1941), 173–87. Knapp and Goodrich, *Origins of American Scientists*, 15.
103. Reese interview with Grimm, 6.
104. M. J. Zucrow, "Utilizing Low-Grade Fuels in Otto-Cycle Engines," *Motive Power* 3, no. 1 (January 1932): 9–12. M. J. Zucrow, "Spark-Ignition Fuel Oil Engines Discussed from Several Angles," *SAE Journal* 34, no. 4 (March 1934): 28.
105. Zucrow interview with Eckles, 15–16; Harvard College, *Twenty-Fifth Anniversary Report*, 1140–41; and from "Nomination of M. J. Zucrow for Honorary Doctorate in Engineering" (1969), Zucrow file. M. J. Zucrow, "Dynamics of Reciprocating Machinery," undated, folder 2, box 5, Zucrow.
106. Jacob invented a steam meter, a profitable gas-burning heater, and an electric meter to measure the flow of fluids in pipes. He was a winner of the prestigious Longstreth Award from the Franklin Institute (1921).
107. M. J. Zucrow: "Instrumentation in Industrial Power Plants," *Industrial Power* (July 1936): 33–34; and (September 1936): 31–33; "Automatic-Control Regulators," *Transactions ASME* 59 (1937): 132–35; "The Measurement of the Flow of Fluids in Pipes," *Proceedings of the Midwest Power Conference* 1 (1938): 88–96; "Influence of Steam Flow Metering on Piping Design," *Mechanical Engineering* (July 1939): 550–51; and Albert Spitzglass and Maurice Zucrow, "Application of Automatic Control in the Oil Industry," *Transactions ASME* 61 (1939): 350–51. He was also a member of the Special Research Advisory Committee on Fluid Meters (ASME, 1938–1946).
108. "Experts Tell Plans of Huge Power Project," clipping from the *Davenport Democrat and Leader* (13 January 1942), black scrapbook, box 1 (addition), Zucrow. M. J. Zucrow, "The Old Municipal Steam Electric Plant," *American City* (August 1939): 57. The legal dispute was between the First Iowa Hydroelectric Cooperative, the Iowa State Executive Council, and the Federal Power Commission. The goal was to build a massive earthen dam on the Cedar River at Moscow, Iowa, with an 8-mile open canal, several smaller dams, and two artificial lakes, one that was 11 miles long and 2.5 miles wide.
109. S. J. G. Taylor, "The 'Ring Balance' Flow Meter," *Metallurgia* 39, no. 234 (April 1949): 305–8. "Nomination of M. J. Zucrow for Honorary Doctorate in Engineering" (1969), Zucrow file.
110. "Thermometric Lag Determination for Cylindrical Primary Elements," *Instruments* 15, no. 10 (October 1942): 398–402.
111. Cecil Warner, "Thermal Sciences and Propulsion Center," folder 2, box 29, 2, Ehresman.
112. M. J. Zucrow and A. C. Bates, "Bibliography of the Vibration of Shafts, Vibration Measurements, and the Design of Crankshafts," *Bulletin of the Purdue University EES*, no. 39 (November 1931). Zucrow was also a member of ASME's Power Test Code Committee on Steam Jet Compressors and its Research Committee on Fluid Mechanics.

113. E. H., "Writings of Interest," *University Review* (February 1927): 4. Potter, "Decennial Review of the Accomplishments of the Engineering Schools and Departments."
114. Larry W. Owens, "Straight Thinking: Vannevar Bush and the Culture of American Engineering" (PhD diss., Princeton University, 1987), 2–24. For the long historical context, also see Matthew Wisnioski, "Liberal Education has Failed," *Technology and Culture* 50, no. 4 (October 2009): 753–82.
115. Potter interview with Eckles, 36, 39–40. Potter, "Decennial Review of the Accomplishments of the Engineering Schools and Departments"; and J. C. L. Fish, professor of civil engineering at Stanford, to A. A. Potter (12 November 1925), folder 13, both in box 6, Knoll. Walton C. John, *A Study of Engineering Curricula* (Lancaster, PA: Lancaster Press, 1927), 55. I thank Kate Herrin for sharing her research and insights about Potter's teaching strategies.
116. *Purdue Engineer* (December 1946): 34.
117. Janette Morris, "Thoughts Take Wings," *Scrivener* (September 1940), 53–56. Joe Schwager, "Whirling Propellers," *Scrivener* (September 1941), 49–54.
118. Warren E. Howland, "Cultural Opportunities at Purdue" (1934), 1–21, and associated documents, in folder 5, box 8, Knoll; and his "Coordination of Non-Technical Studies in General Education" (2 May 1935), Warren E. Howland Papers.
119. R. B. Stewart, "The Heritage of Free Education" (7 June 1943), box 8, R. B. Stewart Papers (Stewart). Editorial, "A Purdue Tradition," *Scrivener* (September 1940), 5.
120. *Purdue Engineer* (April 1947): 12. A. A. Potter, foreword to *A Manual of the William Freeman Myrick Goss Library of the History of Engineering and Associated Collections*, curated by William Murray Hepburn (Lafayette, IN: Purdue University, 1947), 7. Potter report to Hovde (2 January 1946), Hovde, Frederick L., box 31, Frederick L. Hovde Papers (Hovde).
121. Clifford C. Furnas, *The Next Hundred Years: The Unfinished Business of Science* (New York: Reynal, 1936), 340, 400.
122. *Campus Copy* (November 1947): 6. Wendy Greenhouse, Gregg Hertzlieb, and Michael Wright, *The Art of George Ames Aldrich* (Bloomington: Indiana University Press, 2013), 5. Someone stole the painting from the Purdue Memorial Union in 1984.
123. Potter, "Decennial Review of the Accomplishments of the Engineering Schools and Departments." The quotes are from A. A. Potter, "Purdue University and the Utilities of Indiana" (1927), folder 10, box 1, Knoll.
124. W. S. von Bernuth, "Estimate of Personnel Situation" (1926), folder 14, box 6, Knoll. *Purdue Alumnus* (November 1950): 2–3, 43. *Purdue Exponent* (4 March 1950): 3. Potter interview with Eckles, 37–38.
125. Young, "Research Facilities of the School of Mechanical Engineering."
126. *Rock Island News Digest* (May 1944): 9. The company named its first "rocket" train in 1937; thirteen rocket trains were in operation across the Midwest and West by 1947.
127. *Purdue Engineer* (December 1930): 90; (May 1931): 212; (November 1936): 24; and (January 1940): 53.

128. Israel Selkowitz, "Shall Purdue Have a Robot Mascot?," *Purdue Exponent* (26 September 1939): 2. The Boilermaker Special, a small model of a railroad engine cab, became the official mascot of the university in 1942.

CHAPTER 3

1. *Aerial Age Weekly* (16 June 1919): 678. *New York Times* (13 July 1922): 32.
2. Benoît Godin, "Research and Development," *Science and Public Policy* 33, no. 1 (February 2006): 59–76.
3. Frank Wattendorf interview with the Goddard Space Flight Center (26 June 1973), 27, NASA.
4. *Mechanical Engineering* (November 1947): 965. *Mechanical Engineering* (November 1945): 775.
5. James Hershberg, *James B. Conant: Harvard to Hiroshima and the Making of the Nuclear Age* (New York: Knopf, 1993), 127. National Academy of Sciences, *Federal Support of Basic Research in Institutions of Higher Learning* (Washington, DC: GPO, 1964), 34.
6. Don K. Price, *Government and Science* (New York: New York University Press, 1954), 43–47. G. Pascal Zachary, *Endless Frontier: Vannevar Bush, Engineer of the American Century* (New York: Free Press, 2018).
7. *New York Times* (3 January 1943): A32.
8. William A. Hanley correspondence with Vannevar Bush (letters of 29 May, 2 June, and 5 June 1945), Purdue University, box 94, Vannevar Bush Papers (Bush). Cameron Day, "Purdue's Rocket Bomb Prexy," *Pic* (January 1948): 26–28; *Purdue Alumnus* (January 1948): 14–15; and Robert Topping, *The Hovde Years* (West Lafayette, IN: Purdue Research Foundation, 1980).
9. Frederick Hovde, "Report on the NDRC Mission, London, England" (1 March to 1 August 1941), Liaison Great Britain, box 6, entry 7, RG 277, Office of Scientific Research and Development.
10. Topping, *The Hovde Years*, 164. Hershberg, *James B. Conant*, 147–50, 178–87. "Purdue's Rocket Man," *Time* 45, no. 10 (3 September 1945). Septimus H. Paul, *Nuclear Rivals: Anglo-American Atomic Relations, 1941–1952* (Columbus: Ohio State University Press, 2000), 26; and Kevin Ruane, *Churchill and the Bomb in War and Cold War* (London: Bloomsbury, 2016).
11. Speeches, box O, Hovde.
12. James Bryant Conant, "The National Defense Research Committee," *Harvard Alumni Bulletin* (18 October 1941), 2–3. A. Hunter Dupree, "The Great Instauration of 1940," in *The Twentieth Century Sciences: Studies in the Biography of Ideas*, ed. Gerald Holton (New York: Norton, 1972), 443–67; and Nathan Reingold, "Vannevar Bush's New Deal for Research: Or the Triumph of the Old Order," *Historical Studies in the Physical and Biological Sciences* 17, no. 2 (1987): 309–10.
13. Vannevar Bush, "Research and the War Effort" (26 January 1943), NDRC, box L, Hovde.

Irvin Stewart, *Organizing Scientific Research for War: The Administrative History of the Office of Scientific Research and Development* (Boston: Little, Brown, 1948), 5–52.

14. Bush to Hovde (25 February 1942); and Conant to the NDRC (18 June 1942); both in Hovde World War II Correspondence, box L, Hovde.
15. Hershberg, *James B. Conant*, 146, 156–71, 211.
16. "Summary Technical Report" (January 1946), 9–11, OSRD folder 2, box 37, Hovde.
17. Frederick Hovde, "The Coordination of War Research" (9 May 1943), Speeches, box O, Hovde.
18. John E. Burchard, *Rockets, Guns and Targets* (Boston: Little, Brown, 1948), 23–29. Frederick Hovde, "American Science in War," *Monthly Review* (March 1942): 59–60.
19. Hanley–Bush correspondence (letters of 29 May, 2 June, and 5 June 1945). For praise of Hovde's work on rockets, see the correspondence of Charles Lauritsen (28 December 1945 and 11 January 1946), box 1, Charles Lauritsen Papers; and Ralph Gibson's letter (15 March 1948), Research and Development, box 16, Hovde.
20. Edward Moreland (NDRC executive director) to Hovde (7 September 1943), OSRD folder 1, box 37, Hovde.
21. Transcript of the Conference of War Department Ordnance officers with directors of NDRC Division 3 (30 September 1943), Conference on Rockets, box K2320, RG 156, Records of the Office of the Chief of Ordnance.
22. James P. Baxter, *Scientists against Time* (Boston: Little, Brown, 1946), 202–20. Maj. H. G. Jones (Office of the Chief of Ordnance), "Development of Rocket Ammunition," *Mechanical Engineering* (April 1946): 317–20. G. M. Barnes, *Weapons of World War II* (New York: Van Nostrand, 1947), 177–95.
23. Vannevar Bush, *Pieces of the Action* (New York: Morrow, 1970), 43. Irvin Stewart *Organizing Scientific Research*, 69.
24. Burchard, *Rockets, Guns and Targets*, 51–58, 230–38.
25. Transcript of the Conference of War Department Ordnance Officers (30 September 1943).
26. Burchard, *Rockets, Guns and Targets*, 48–49. Walter Boyne, foreword to *Captured Eagles*, by Frederick Johnsen (London: Bloomsbury, 2014), 9, 51.
27. Minutes and reports, Special OSRD Committee on Jet Propulsion Systems (February–March 1944), OSRD, box 10, RG 38, Records of the Office of the Chief of Naval Operations. John McKinney Tucker, "Technologies of Intelligence and Their Relation to National Security Policy" (PhD diss., Virginia Tech, 2013), 172, 211, 221.
28. Bush to Hovde (25 February 1942 and 25 November 1942), Hovde World War II, box L, Hovde.
29. Bush to Hovde (28 November 1945), and Hugh Spencer to Hovde (19 November 1945), OSRD folder 1, box 37, Hovde.
30. Correspondence of Hovde and Lauritsen (28 December 1945 and 11 January 1946), box 1, Charles Lauritsen Papers. "Summary Technical Report" (January 1946), 9–11, OSRD folder 2, box 37, Hovde.
31. Hovde to R. B. Richmond of the General Radio Company (10 August 1946), Science and

Technology, box 10, Hovde. "Pres. F. L. Hovde Known for Rocket Work," *Purdue Alumnus* (April 1958): 7.

32. Irvin Stewart, *Organizing Scientific Research for War*, 18–19, 191.
33. Don K. Price, "The Scientific Establishment," in *Scientists and National Policy-Making*, ed. Robert Gilpin and Christopher Wright (New York: Columbia University Press, 1964), 33–47. See also Dupree, "The Great Instauration of 1940," 458; and Reingold, "Vannevar Bush's New Deal," 314. Bush, *Pieces of the Action*, 39, 63–65, 139, 282–93.
34. Carroll Pursell, "Science Agencies in World War II," in *The Sciences in the American Context: New Perspectives*, ed. Nathan Reingold (Washington, DC: Smithsonian Institution, 1979), 361–63. Clarence Danhof, *Government Contracting and Technological Change* (Washington, DC: Brookings Institution, 1968), 305.
35. Larry Owens, "The Counterproductive Management of Science in the Second World War," *Business History Review* 68, no. 4 (Winter 1994): 515–76.
36. Sanford Lakoff, "Accountability and the Research Universities," in *The State of Academic Science*, ed. Bruce L. R. Smith and Joseph J. Karlesky, vol. 2 (New York: Change Magazine Press, 1978), 173. National Academy of Sciences, *Federal Support of Basic Research*, 17, 36.
37. R. B. Stewart interview with Robert Eckles (29 September 1969), 28–29, ASC. Rebecca Lowen, *Creating the Cold War University: The Transformation of Stanford* (Berkeley: University of California Press, 1997), 58–64.
38. R. B. was the main author of the "Proposed Plan of Allowance for Indirect Costs and Contingencies (Now Called Overhead)." Irvin Stewart referred to R. B. and his team rather coldly as "those gentlemen." Irvin Stewart, *Organizing Scientific Research for War*, 208–21. Over half of the government contracts received excessive overhead, thanks to multiple contracts operating over several years.
39. Bureau of Navy Personnel to Stewart (18 August 1942), unmarked folder, box 23, Stewart.
40. Stewart interview with Eckles, 28–31.
41. Stewart interview with Eckles, 31–34. R. B. Stewart to undersecretary of war Robert Patterson and undersecretary of the Navy James Forrestal (6 August 1943), unmarked folder, box 23, Stewart.
42. "Report of the Joint Army and Navy Board for Training Unit Contracts" (30 November 1945), 20–21, box 8, Stewart. See also R. B. Stewart, "Government Research Contracts," in *Scientific Research: Its Administration and Organization*, ed. George P. Bush and Lowell Hattery (Washington, DC: American University Press, 1950), 29–34.
43. R. B. Stewart, as quoted in Gary Totten, "If You Would See His Monument, Look Around: R. B. Stewart, the Man Who Built Purdue," unmarked manila folder, box 24, Stewart.
44. V. Ray Cardozier, *Colleges and Universities in World War II* (Westport: Praeger, 1993), 1, 13–20, 29–57, 90–121. Frank K. Burrin, *Edward Charles Elliott, Educator* (West Lafayette, IN: Purdue University Studies, 1970).
45. Potter interview with Eckles, 28.

46. Henry H. Armsby, *Engineering, Science, and Management War Training* (Washington, DC: US Office of Education, 1946), ix, 3–14, 28, 78, 94–95, 148.
47. E. A. Loew letter, dean, College of Engineering, University of Washington (21 August 1945), binder 3, box 12, Potter.
48. Cardozier, *Colleges and Universities*, 168–82. Armsby, *Engineering, Science, and Management War Training*, 18, 32, 46–47.
49. Solberg interview with Eckles, 18. "Army Ordnance Project Citation," *Purdue Patrol* (September 1944): 1.
50. "Meeting of the Board of Directors" of the PRF (8 May 1946), Analysis and Revisions, box 2, Meikle. Further details are in the PRF files (1942–1945), box 198, RG 227, Office of Scientific Research and Development. *Purdue Exponent* (14 May 1948): 1.
51. Frank Malina, "Memorandum on the Future of Jet Propulsion Research at the California Institute of Technology" (November 1945), Miscellaneous, box 74, Theodore von Kármán Papers (von Kármán). *Research and Development at the Jet Propulsion Laboratory, GALCIT* (Pasadena: Caltech, June 1946), 4–8.
52. Remarks of Maj. Haley in the "Outline of Training Program," 2–3, Aerojet Training Program black binder, (1943), box 14, Ehresman. Aerojet became the Aerojet-General Corporation in 1945, after General Tire purchased a controlling share of its stock.
53. Frances M. Christeson, "Aerojet Daily Journal" (beginning 1 April 1944), 14. This is the second black Record book, box 5, Andrew Haley Papers (Haley).
54. Fraser MacDonald, *Escape from Earth: A Secret History of the Space Rocket* (New York: Public Affairs, 2019), 85–95. George Pendle, *Strange Angel: The Otherworldly Life of Rocket Scientist John Whiteside Parsons* (Orlando: Harcourt, 2005).
55. Delphine Bechtel, ed., *The Holocaust in Ukraine: New Sources and Perspectives* (Washington, DC: Holocaust Memorial Museum, 2013).
56. *Power Plant Engineering* (September 1945): 130–36. *Institution of Mechanical Engineers* 158, no. 1 (June 1948): 103–15.
57. Zucrow interview with Eckles, 16–18. Harvard College, Class of 1922, *Twenty-Fifth Anniversary Report* (Cambridge: Harvard University, 1947), 1140–41. Zucrow began work at Aerojet in January of 1943.
58. *Report from Aerojet: The Power of the Future* (Azusa, CA: Aerojet Corporation, 1945), 1–2. Bernie L. Dorman, et al., *Aerojet: The Creative Company* (San Francisco: Cooper, 1997), i, XI-8 to XI-22.
59. Frances M. Christeson, "Aerojet Daily Journal" (beginning 1 January 1944), 20–22, 27. This is the first black Record book, box 5, Haley.
60. F. J. Malina, "The US Army Air Corps Jet Propulsion Research Project, GALCIT Project No. 1," in *History of Rocketry and Astronautics*, AAS History Series (San Diego: AAS, 1986), 194–95. Frank Malina interview with Mary Terrall (14 December 1978), 12–13, Caltech. Martin Summerfield interview with J. D. Hunley (7 September 1994), 32–33, Martin

Summerfield file, NASA. M. J. Zucrow, "Contributions of Aerojet to the War Effort and the Jet Propulsion Field" (15 March 1946), box 1 (addition), Zucrow.

61. "Nomination of M. J. Zucrow for Honorary Doctorate in Engineering" (1969), Zucrow file. On the origins of the project engineer in the aircraft industry, see Benjamin Pinney, "Projects, Management, and Protean Times: Engineering Enterprise in the United States, 1870–1960" (PhD diss., MIT, 2001), 216.
62. "Notes of Conferences Held with Lt. Commander Warfel," Report No. 22 (24–27 May 1943), 28, box 3, Haley.
63. "Report of Research and Development Investigations" (1 March 1944), 5–11; and Report 27-A, prepared by J. M. Carter and F. Zwicky, director of research; both in box 3, Haley.
64. "Biographies of Personnel," box 5, Haley. Zucrow met a fellow Purdue graduate at Aerojet, Herman Coplen (BSME 1937), also an expert in flow control devices and instrumentation for public utility plants. Dorman built a career at Aerojet, later directing its R&D for the Titan and Polaris missiles and becoming vice president for long-range planning. He received an honorary doctorate from Purdue in 1960.
65. Christeson, "Aerojet Daily Journal" (beginning 1 January 1944), 1–70.
66. Christeson, "Aerojet Daily Journal" (beginning 1 April 1944), 8, 24–28.
67. Christeson, "Aerojet Daily Journal" (beginning 1 January 1944), 32–35. "Report #7 on Process Control at the Propellant Plant" (15 March 1945), Aerojet History file, box 5, Haley.
68. Minutes of "Change Order Meeting in Major Haley's Office," 15 January 1944, Aerojet Daily Journal (January–March 1944), box 5, Haley. "Methods and Procedures," memorandum from A. H. Rude to A. G. Haley and D. Kimball (12 July 1945), Aerojet History file, box 5, Haley.
69. For most facts and contexts about Aerojet's technical achievements, I have relied on Frank H. Winter and George S. James, "Highlights of 50 Years of Aerojet, a Pioneering American Rocket Company, 1942–1992," *Acta Astronautica* 35, no. 9–11 (1995): 677–81; and C. M. Ehresman, "The First Operational Liquid Rocket for Aircraft Developed and Produced in Quantity in the United States," paper presented at the 36th AIAA Joint Propulsion Conference (Huntsville, AL, 17–19 July 2000).
70. Zucrow's related inventions at Aerojet included the following: "Solid Propellant Cartridge Construction, a solid propellant cartridge which provides controlled burning under high temperature conditions. Solid Fuel for Jet Propulsion Motors, a method of manufacturing a propellant charge where a finely divided inorganic oxidizer is dry milled into a synthetic elastomer. Combination Solid and Liquid Propellant Unit, a jet motor that can use both liquid and solid propellants at the same time." These and the Zucrow inventions listed in later notes are from "Invention Disclosures (Unpatented)," Aerojet-General Corporation (24 September 1964), Zucrow file. Under Zucrow's tutelage, Aerojet also designed the solid-fuel booster motor for the Project Bumblebee surface-to-air missile at the Applied Physics Laboratory in fall of 1945, which achieved a series of successful test flights by early fall 1946.
71. "Notes of Conferences Held with Lt. Commander Warfel" (24–27 May 1943), 7, 21–22,

26–27. Zucrow helped engineer the "improved design changes to the 25 ALD-1000 droppable JATO (in a second run of 100 after the first) in a contract between October 1943 and August 1944. See the Aerojet Engineering Corporation, "Rocket Power Plants" (1948), folder 2, box 15, Ehresman. Among Zucrow's related inventions at Aerojet were the following: "Film Cooled Motor, means for cooling the interior wall of a jet combustion chamber by surface evaporation of a coolant. Jet Motor Construction, an insulation lined jet motor combustion chamber permitting operation at high temperatures with no auxiliary cooling. Liquid Cooled Jet Nozzle and Motor, cooling apparatus for the use with a jet thrust motor that directs a vaporizable liquid unto the exhaust nozzle of the motor. Film Cooled Motor, regenerative cooling of a jet motor combustion chamber and exhaust nozzle combined with transpiration of cooling fluid into the combustion chamber."

72. Zucrow interview with Eckles, 16–18. Clark Millikan, "The Application of Jet Propulsion to Assisted-Takeoff and Super-Performance of Military Aircraft," Report No. 21 (28 May 1943), 1–2, box 3, Haley.
73. "Notes of Conferences Held with Lt. Commander Warfel" (24–27 May 1943), 31. C. M. Ehresman, "Liquid Rocket Propulsion Applied to Manned Aircraft in Historical Perspective," *Journal of the British Interplanetary Society* 46 (1993): 255–68.
74. "Conferences on Project X (Secret)," Report No. 30 (7 June 1943), 1–2, 8–9, box 3, Haley. Zucrow's team (including A. Hollander and E. F. Mayer) was under Ernest Vogt's and Eric L. Pridonoff's supervision. Zucrow's related inventions included "Plastic Compounds for Resisting Inorganic Oxidizing Acids, an acid resistant synthetic plastic material composed of polyvinyl chloride and hexachlorobutadiene. Plastic Compounds for Resisting Inorganic Oxidizing Acids, an acid resistant synthetic plastic composed of polyvinyl chloride, pentachloro-ethane and chloropropane wax. Elastomeric Compounds for Resisting Inorganic Oxidizing Acids, an improved elastomeric material for use in corrosive environments consisting of a compound of polyvinyl chloride, tetrachloroethane and chloropropane wax. Elastomeric Compounds for Resisting Inorganic Acids, an acid resistant synthetic plastic having good mechanical properties composed of polyvinyl chloride and hexachlorobutadiene."
75. Ernest Vogt, "Photographs of Aerotojet Mock-Up," Report No. 25 (15 May 1943), 2, box 3, Haley.
76. Chandler Ross to Zucrow (15 November 1974), folders 3–5, box 9, Zucrow. Martin Summerfield's unsuccessful Aerotojet was coded as the XCALR-2000A-1, meant to provide 2,000 pounds of thrust.
77. The XCAL-200 was "pressure fed and using RFNA and mono-ethyl-aniline with cast aluminum combustion chamber and copper nozzle," providing 200 pounds of thrust. Winter and James, "Highlights of 50 Years of Aerojet," 683. Christeson, "Aerojet Daily Journal" (beginning 1 April 1944), 73. Bill Norton, *American Aircraft Development of the Second World War: Research, Experimentation and Modification 1939–1945* (Stroud, England: Fonthill

Media, 2019), 159–61. Tony Chong, *Flying Wings & Radical Things: Northrop's Secret Aerospace Projects & Concepts 1939–1994* (Forest Lake, MN: Specialty Press, 2016), 28–29.

78. The 38-ALDW-1500 used regenerative cooling for 1,500 pounds of thrust. "Summary of Liquid Development Work Accomplished under Navy Bureau of Aeronautics Contract No. NOa(s) 1330," Report No. 134 (18 October 1944), box 4, Haley.
79. M. J. Zucrow, "Final Report: Investigation of High Tensile Stainless Steels as Construction Material for Liquid Jet Propulsion Tanks," Report No. 117 (31 August 1944), box 4, Haley. This report was part of the US Navy contract for the 38-ALDW-1500 JATO engine. Christeson, "Aerojet Daily Journal" (beginning 1 April 1944), 22–23, 28–29, 60. While Zucrow and his teams were redesigning it, the No. 38 was known as the 45-ALDW-1500. See "Summary of Liquid Development Work," Report No. 134.
80. R. C. Truax, "Liquid Propellant Rocket Development by the US Navy during World War II," in *History of Rocketry and Astronautics*, AAS History Series (San Diego: Univelt, 1991), 61–67. Winter and James, "Highlights of 50 Years of Aerojet," 681. *Aero Digest* (1 August 1945): 42.
81. "Summary of Liquid Development Work," Report No. 134.
82. Rolf Sabersky to Zucrow (1974), folders 3–5, box 9, Zucrow. M. Summerfield and R. Sabersky, "Some Considerations on the Heat Transfer in a Liquid Cooled Jet Motor," Aerojet Engineering Corporation, Report No. 58 (7 March 1944), in *Jet Propulsion: A Reference Text*, ed. Hsue-shen Tsien (Pasadena: Jet Propulsion Laboratory, 1946), 376.
83. "Development of 1,500 Pound Fuel Cooled Jet Motor," Report No. 93 (7 July 1944), box 4, Haley.
84. For the sharing between Aerojet and GALCIT, see "Final Report: Development and Application of Plastics for Use in Contact with Monoethylaniline and Mixed Acids," Report No. 111 (15 July 1944), 2–5; and "Final Report: Investigation of the Corrosion Resistance of Metals in Contact with Liquid Propellants," Report No. 121 (14 August 1944); both in box 4, Haley.
85. "Case History of Assisted Take-Off Devices," Intelligence T-2, Wright Field (February 1946), 2, binder item no. 3, box 14, Ehresman. "Daily Record of the Ordnance Department Liaison Office" (8–9 December 1943), CIT Reports, box 2319, RG 156. Minutes and reports, Special OSRD Committee on Jet Propulsion Systems (February–March 1944), OSRD, box 10, RG 38.
86. Zucrow, "Contributions of Aerojet to the War Effort and the Jet Propulsion Field." At Aerojet, Zucrow invented a "Pump Drive for Jet Propulsion Motor, a jet propulsion system including a turbine driven pumping means for feeding propellants into the combustion chamber."
87. Martin Summerfield to Frank Winter (22 July 1993), in the Martin Summerfield folder, NASA. On the cooperation of GALCIT-JPL and Aerojet with Bell Aircraft on the X-1 Project MX-524, see Andrew Haley's letter (17 January 1945), Aerojet History file, box 5,

Haley. D. S. Fahrney, *The History of Pilotless Aircraft and Guided Missiles* (Washington, DC: US Navy, 1958), 1248.

88. US Air Force, *SAC Missile Chronology, 1939–1988* (Offutt AFB, NE: Strategic Air Command, 1990), 3. Aerojet Engineering Corporation, "Rocket Power Plants" (1948), folder 2, box 15, Ehresman. Also see Ray C. Stiff, "Storable Liquid Rockets," Paper No. 67-977, presented at the AIAA 4th Annual Meeting (October 1967), 4–6.
89. Zucrow, "Contributions of Aerojet to the War Effort and the Jet Propulsion Field."
90. Winter and James, "Highlights of 50 Years of Aerojet," 692. Frank Winter, "The East Parking Lot Rocket Experiments of North American Aviation, Inc.," in *History of Rocketry and Astronautics*, AAS History Series (San Diego: AAS, 2007), 191–218. Thomas A. Heppenheimer, "The Navaho Program and the Main Line of American Liquid Rocketry," *Air Power History* 44, no. 2 (Summer 1997): 6–9.
91. Andrew Hamilton and John B. Jackson, *UCLA on the Move: During Fifty Golden Years 1919–1969* (Los Angeles: Ward Ritchie, 1969), 102. United States Office of Education, *Training for War Industries under EDT-ESMDT-ESMWT* (Berkeley: University of California, 1945).
92. "Principles of Jet Propulsion" (announcement for the ESMWT program), box 1, record series 587, UCLA University Archives. Zucrow to Solberg (28 May 1946), Zucrow file. Zucrow interview with Eckles, 16–18.
93. There are two available copies. M. J. Zucrow, *Principles of Jet Propulsion* (for ESMWT course No. 949A2), UCLA Engineering Library (Call No. TL782.Z8); and the bound, typed lecture notes, Principles of Jet Propulsion course at the University of California for the War Training program, 1943–1944, folders 1–2, box 8, Zucrow.
94. Dorman, et al., *Aerojet*, XI-1; and Clayton R. Koppes, *JPL and the American Space Program: A History of the Jet Propulsion Laboratory* (New Haven: Yale University Press, 1982), 18–20. O. K. Walsh and R. J. Sippel, *Publications of the Jet Propulsion Laboratory, July 1960 through June 1961* (Pasadena: NASA, 1961), 36, 51, 76.
95. Malina interview with Terrall, 12–13.
96. "Nomination of M. J. Zucrow for Honorary Doctorate in Engineering" (1969); Zucrow to H. L. Stolberg (28 May and 12 June 1946); both in Zucrow file. Minutes of Project Squid Technical Committee (11 December 1946), Minutes, box 6, RG 72, Records of the Bureau of Aeronautics (USN).
97. Quotes from "Pres. F. L. Hovde Known for Rocket Work," 7; and Frances M. Christeson, "Aerojet Daily Journal" (beginning 1 January 1944), 18. Levin H. Campbell, *The Industry-Ordnance Team* (New York: Whittlesey House, 1946), 145.
98. James W. Bragg, *Development of the Corporal: The Embryo of the Army Missile Program*, vol. 1, *Narrative* (Huntsville, AL: Army Ballistic Missile Agency, 1961), 60–63. D. S. Fahrney, *The History of Pilotless Aircraft and Guided Missiles* (Washington, DC: US Navy, 1958), 1318–23. Stiff, "Storable Liquid Rockets," 4–6.

99. Minutes of the meeting of the Ordnance Department Advisory Committee on Guided Missiles, Conference on Rockets (24 June 1946), box 2320, RG 156. The first quote is from Winter and James, "Highlights of 50 Years of Aerojet," 691. Mary T. Cagle, *Development, Production, and Deployment of the Nike Ajax Guided Missile System, 1945–1959* (Redstone Arsenal, AL: Army Rocket and Guided Missile Agency, 1959), 41–48. The second quote is from Stiff, "Storable Liquid Rockets," 4–6.
100. US Navy BuAer and ONR reports (12 and 15 July 1948) and "Ad Hoc Subcommittee on Long Range Rockets, Minutes of 1st Meeting" (20 July 1948), folder 13, box 203, RG 330, Records of the Office of the Secretary of Defense. Farhney, *The History of Pilotless Aircraft and Guided Missiles*, 1133. John L. Sloop, *Liquid Hydrogen as a Propulsion Fuel, 1945–1959* (Washington, DC: GPO, 1978), 13–23, 38–56.
101. Zucrow to Stolberg (28 May and 12 June 1946), Zucrow file. Minutes of Project Squid Technical Committee (11 December 1946), Minutes, box 6, RG 72. George H. Osborn, Robert Gordon, and Herman Coplen, "Liquid Hydrogen Rocket Engine Development at Aerojet, 1944–1950," in *Essays on the History of Rocketry and Astronautics: Proceedings of the Third through the Sixth History Symposia of the International Academy of Astronautics* (Washington, DC: NASA, 1977), 279–324; and Sloop, *Liquid Hydrogen as a Propulsion Fuel*, 289. Stephenson, "A History of the Ohio State University Research Foundation," 25–33.
102. Mary R. Self, *History of the Development of Guided Missiles, 1946–1950* (Dayton, OH: Wright-Patterson Air Force Base, 1951), 116–18.
103. Topping, *The Hovde Years*, 162–63.
104. Biographical materials, Hovde Purdue Appointment, box L, Hovde.
105. Hanley–Bush correspondence (29 May, 2 June, and 5 June 1945).
106. Day, "Purdue's Rocket Bomb Prexy," 26–28; and *Purdue Alumnus* (January 1948): 14–15.
107. Fritz Zwicky, "Report on Certain Phases of War Research in Germany," Summary Report, Intelligence T-2, Air Materiel Command, Wright Field, Dayton (19 December 1946), 4, 173–78.
108. Leslie E. Simon, *German Research in World War II: An Analysis of the Conduct of Research* (New York: Wiley, 1948), 41–53, 72–80, 103–9, 196–211. Zucrow favorably reviewed it in *Mechanical Engineering* (December 1947): 1058–59.
109. W. von Braun, "Survey of Development of Liquid Rockets in Germany and Their Future Prospects," first reported publicly in W. G. A. Perring, "A Review of German Long-Range Rocket Development," *Journal of the Royal Aeronautical Society* 50, no. 427 (July 1946).
110. *Aero Digest* (1 August 1945): 78; and (15 January 1946): 99. Wolfgang E. Samuel, *American Raiders* (Jackson: University of Mississippi Press, 2004).
111. "Research Is Tomorrow's Insurance," *Aero Digest* (1 October 1945): 35.
112. *Mechanical Engineering* (August 1945): 546.
113. On his paper, the first in a series over four evening sessions, see *Mechanical Engineering* (August 1945): 546–49; and *Aero Digest* (1 August 1945): 42. *New York Times* (2 September 1945): E9.

114. M. J. Zucrow's "Jet Propulsion and Rocket Motors Point to New Horizons," *Industry and Power* (December 1945): 64–66; "Jet Propulsion and Rockets for Assisted Take-Off," *Transactions ASME* (April 1946): 177–88; and his "Technical Aspects of Rockets and Assisted Take-Off," *Aero Digest* (April 1946): 88–101. Also see Peter Bielkowicz, "Evolution of Energy in Jet and Rocket Propulsion," *Aircraft Engineering* (March 1946): 90–92; and H. S. Seifert, M. M. Mills, and M. Summerfield, "The Physics of Rockets," *American Journal of Physics* 15, no. 1 (January 1947): 1–21.
115. M. J. Zucrow, "The Rocket Powerplant," *SAE Transactions* 54, no. 7 (July 1946): 375–88, based on the last chapter of Zucrow's *Principles of Jet Propulsion*. The paper was presented at a meeting of the Southern California Section of the Society of Automotive Engineers, Los Angeles (11 January 1946).
116. Minutes (24 May 1943 and 28 November 1944), Board of Directors files, American Rocket Society, at the American Institute of Aeronautics and Astronautics (AIAA). G. Edward Pendray to Theodore von Kármán (8 December 1944), ARS General Correspondence, box 38, von Kármán.
117. Armsby, *Engineering, Science, and Management War Training*, 63–33.

## CHAPTER 4

1. "Research and Development and Classification of Materiel," War Department circular (14 May 1947), in Collation of Air Technical Intelligence Information, IRIS no. 142231, AFHRA. *Mechanical Engineering* (July 1946): 679; and (September 1946): 785.
2. Don K. Price, *Government and Science* (New York: New York University Press, 1954), 1. Oppenheimer quoted in a memorandum (21 December 1950) in *Impacts of the Early Cold War on the Formulation of US Science Policy: Selected Memoranda of William T. Golden, October 1950–April 1951*, ed. William A. Blanpied (Washington, DC: American Association for the Advancement of Science, 1995), 38.
3. Vannevar Bush, ed., *Science, the Endless Frontier* (Washington, DC: GPO, 1945), 18–19. For contexts, see G. Pascal Zachary, *Endless Frontier: Vannevar Bush, Engineer of the American Century* (New York: Free Press, 2018).
4. Bush, *Science, the Endless Frontier*, 72.
5. *Mechanical Engineering* (September 1945): 567. See the actual report with Potter's note (5 October 1945), box 13, George A. Hawkins Papers.
6. Frank Wattendorf interview with the Goddard Space Flight Center (26 June 1973), 50–51, NASA; and Francis Clauser interview with Peter Westwick (13 April 2011), 4–11, Huntington Library Archives (Huntington). Tucker, "Technologies of Intelligence," 291–92.
7. Michael Aaron Dennis, "Reconstructing Sociotechnical Order: Vannevar Bush and US Science Policy," in *States of Knowledge: The Co-production of Science and the Social Order*, ed. Sheila Jasanoff (New York: Routledge, 2004), 225–53.
8. Minutes of the Committee on Post-War Research (June to December 1944), Post-War Research, box 10, RG 38.

9. Rebecca Lowen, *Creating the Cold War University: The Transformation of Stanford* (Berkeley: University of California Press, 1997), 77, 99, 109–20. Weldon Gibson, *SRI: The Founding Years* (Los Altos: California Publishing Services, 1980).
10. Archie Palmer, *University Research and Patent Policies, Practices and Procedures* (Washington, DC: NAS, 1962), 31–32.
11. T. Drury, *War History of the Naval Research Laboratory* (Washington, DC: GPO, 1948).
12. Booz, Allen, and Hamilton, *Review of Navy R&D Management* (Washington, DC: Department of the Navy, 1976), 13–14, 341–44.
13. Harvey M. Sapolsky, *Science and the Navy: The History of the Office of Naval Research* (Princeton: Princeton University Press, 1990), 19–29, 42–43, 119.
14. Office of Naval Research, *Annual Report* (1949), 1.
15. John E. Burchard, *Rockets, Guns and Targets* (Boston: Little, Brown, 1948), 211–16. F. L. Hovde to Alan Valentine, president of the University of Rochester (30 July 1945) in Hovde folder, box 31, Hovde. Bruce Old, "The Evolution of the Office of Naval Research," *Physics Today* (August 1961): 30–35.
16. Hovde to G. S. Meikle (23 October 1945); Potter to Hovde (17 December 1945); and E. F. Bruhn to Potter (13 December 1945); all in Hovde folder, box 31, Hovde.
17. ONR reports (May 1946 and January 1947), ONR, box 7; and the memos, directives, and calls for contracts (18 March 1946), Project Squid Correspondence, vol. 1, box 8; all in RG 72.
18. "Initial Program Report" of Project Squid (1946), Initial Program Reports, box 6, RG 72.
19. "Trip Report" (5–7 February 1946), and the "Proposed Propulsive Devices Laboratory" (1 March 1946), Project Squid Correspondence, vol. 1, box 8, RG 72.
20. Technical Committee minutes (11 December 1946), Minutes, box 6, RG 72.
21. Minutes of meeting to discuss BuAer-ORI Propulsion Program (29 March 1946), folder 3, box 95, von Kármán.
22. Memorandum on Section T contracts (3 April 1946), Project Squid Correspondence, vol. 1, box 8, RG 72.
23. Memos and directives (18 March 1946), Project Squid, box 8; and Technical Committee minutes (3 June 1946), Minutes, box 6; both in RG 72.
24. Princeton University, *Introductory Report on Project Squid*, Report No. 1 (Princeton: Princeton University, 31 December 1946), 17.
25. Lawrence Hafstad, "Problems in the Coordination of Federal Research," in *Scientific Research: Its Administration and Organization*, ed. George P. Bush and Lowell Hattery (Washington, DC: American University Press, 1950), 17–19.
26. Hovde letter (20 March 1946) and Meikle's letters (23 March and 9 April 1946), in Project Squid Correspondence, vol. 1, box 8, RG 72. Also note Frederick Hovde and Clarence N. Hickman, "Rocket Fundamentals," an official history of Division 3 (OSRD), in Burchard, *Rockets, Guns and Targets*, 9–13.

27. Hovde to Meikle (23 October 1945), in Hovde, Frederick L., box 31; and his note to M. J. Zucrow (July 1948), Research and Development, box 16; both in Hovde.
28. Harvard College, Class of 1922, *Twenty-Fifth Anniversary Report* (Cambridge: Harvard University, 1947), 1140–41.
29. Solberg to Zucrow (9 and 26 April, and 4 and 20 May 1946); Potter to Zucrow (15 May 1946); and Solberg to Zucrow (26 April 1946); all in Zucrow file.
30. Zucrow interview with Eckles, 11, 19–20.
31. Bernie L. Dorman, et al., *Aerojet: The Creative Company* (San Francisco: Cooper, 1997), XI-2. Chandler Ross, "Life at Aerojet-General University: A Memoir" (1981), folder 1, box 21, Ehresman.
32. Homer J. Stewart interview with John Greenberg (October–November 1982), 69–70, Caltech. Fraser MacDonald, *Escape from Earth: A Secret History of the Space Rocket* (New York: Public Affairs, 2019), 50–55, 122, 173–74, 190, 268. Clauser interview with Westwick, 61.
33. R. B. Stewart's comments before the Indiana Ways and Means Committee (1948), 7–10, box 8, Stewart.
34. *Purdue Alumnus* (January 1948): 2; and (June–July 1948): 5. *Purdue Exponent* (1 October 1946): 1; and (12 October 1946): 1.
35. Stewart interview with Eckles, 35–36. Ruth W. Freehafer, *R. B. Stewart and Purdue University* (West Lafayette, IN: Purdue University Press, 1983). For contexts, see Jennifer Mittelstadt, *The Rise of the Military Welfare State* (Cambridge: Harvard University Press, 2015).
36. *Purdue Exponent* (October 1946 and March 1948), which regularly reported "the Ratio."
37. "Nomination of M. J. Zucrow for Honorary Doctorate in Engineering" (1969); and Solberg and Zucrow exchange (20 and 28 May 1946), Zucrow file. Zucrow and his courses first appear in *Bulletin of Purdue University* (West Lafayette, IN: Purdue University, February 1947). See appendix 3 in this volume for the full array of courses.
38. E. F. Bruhn to Zucrow (29 July 1946), Zucrow file. E. F Bruhn, "Six Months Progress Report of the School of Engineering" (January 1946), 1–2, unnumbered technical report, ASC.
39. E. F. Bruhn, "A History of Aeronautical Education and Research at Purdue University, for Period 1937–1950," vol. 2, ASC.
40. Frank Malina, "Memorandum on the Future of Jet Propulsion Research at the California Institute of Technology" (November 1945), 12–16, CIT JPL, box 74, von Kármán. *Aero Digest* (September 1947): 72. L. R. Hafstad, director of research, Applied Physics Lab, "GM Familiarization Course" (25 October 1946), folder 3, box 1, Richard Porter Papers (Porter).
41. Potter to Solberg (15 May 1946); Solberg to Zucrow (9 April 1946); both in Zucrow file.
42. M. J. Zucrow promotion materials (Nomination for Honorary Doctor of Engineering Degree), in the H. L. Solberg file, Department of Mechanical Engineering, Purdue University.
43. "New Leadership Beginning 1946," in "Report of President and Research Director, Meeting of Board of Directors" (14 May 1952), box 4; G. S. Meikle, confidential report (12 May 1948), Research Director's Calendars, bound folder, box 2; and "Report of President and

Research Director" (18 May 1950), Research Director's Reports, 1950–52, box 3; all in the PRF Records.

44. D. B. Prentice, "College Antecedents to Successful Engineers," *Mechanical Engineering* (May 1949): 397–98. Robert H. Knapp and Joseph J. Greenbaum, *The Younger American Scholar: His Collegiate Origins* (Chicago: University of Chicago Press, 1953), 93. Hovde to Vannevar Bush (3 December 1946), OSRD folder 2, box 37, Hovde.
45. Zucrow exchange with Solberg (15 May and 11 and 19 June 1946), Zucrow file.
46. Zucrow's comments in *Journal of Applied Mechanics* 17, no. 2 (June 1950): 225–26. Maurice J. Zucrow, "Jet Propulsion," in *Mechanical Engineers' Handbook*, 6th ed., ed. Theodore Baumeister (New York: McGraw-Hill, 1958), 11-116 to 11-129; and his "Thermal-Jet and Rocket-Jet Propulsion," *Handbook of Fluid Dynamics*, ed. Victor Streeter (New York: McGraw-Hill, 1961), 23-1 to 23-41. John Wiley folder, box 1 (addition), Zucrow.
47. M. J. Zucrow, *Principles of Jet Propulsion and Gas Turbines* (New York: Wiley, 1948), 464–65.
48. Lionel S. Marks, "Jet Propulsion Theories," *Scientific Monthly* 61, no. 3 (September 1945): 240–41. He had in mind such works as Thomas N. Dalton, *Jet Propulsion* (self-pub., 1945); and Albert L. Murphy, *Rockets, Dynamotors, Jet Motors* (Los Angeles: Wetzel, 1944).
49. Hsue-shen Tsien, ed., *Jet Propulsion: A Reference Text* (Pasadena: Jet Propulsion Laboratory, 1946). The US Army Air Forces kept the 1944 and 1945 editions classified.
50. Iris Chang, *Thread of the Silkworm* (New York: Basic Books, 1995), 102, 117, 147; and Wattendorf interview with Goddard, 38.
51. "Release of Jet Propulsion Textbook, Caltech, 1945–1946," box 2895, RG 342, Records of US Air Force Commands, Activities, and Organizations.
52. George Sutton, "Rocket Propulsion Elements" (UCLA Engineering Extension, XL-198, 1947), call no. TL.782,S86r, pp. IV–5; and his *Rocket Propulsion Elements: An Introduction to the Engineering of Rockets* (New York: Wiley, 1949). Jeffrey Kluger "Rocket Man," *Time* 179, no. 24 (18 June 2012): 44–48.
53. Solberg and Zucrow exchanges (9 April, 20 and 28 May, and 11 and 12 June 1946), in Zucrow file.
54. H. L. Solberg nominations of M. J. Zucrow for the Sigma Xi Award (6 January 1955), for the Bendix Award (22 February 1957), and for the Distinguished Professor Award, in Zucrow file. Warner to Hovde (4 December 1946), Cecil Warner Scrapbook, 9, ASC.
55. Headquarters Air Materiel Command, *Air Technical Intelligence General Concepts and Activities* (Dayton, OH: Wright Field, June–July 1947), 3–4.
56. Richard F. Gompertz to Zucrow (26 November 1974), folders 3–5, box 9, Zucrow. Richard F. Gompertz, "Rocket-Engine Flight Testing," *Journal of the American Rocket Society* (December 1950): 169–76. Lloyd Mallan, *Men, Rockets and Space Rats* (New York: Messner, 1961).
57. John I. Nestel to Zucrow (11 November 1974), folders 3–5, box 9, Zucrow.
58. Robert Glick interview with Michael G. Smith (20 January 2017), 13–14, ASC.
59. See M. J. Zucrow's series of works: "Potentialities and Division Problems of the Turbo-Prop Engine," *Journal of the American Rocket Society* (September 1949): 116–28; "Theory and

Characteristics of Jet Engines," in *Proceedings of the 1951 Summer School for Thermodynamics Teachers*, ed. Ralph Rotty and Donald Renwick (East Lansing: Michigan State College, 1951), 224–61; "Jet Propulsion Instruction and the Undergraduate Curriculum in Aero Engineering," *Journal of Engineering Education* 38, no. 7 (March 1948): 482–84; "Fundamental Characteristics of Jet Propulsion Engines," *Virginia Journal of Science* 4 (April 1953): 41–56; and "The Problem of the Turbojet Engine as a Propulsion Engine for Supersonic Flight, *Aeronautical Engineering Review* 15, no. 12 (December 1956): 44–52.

60. Zucrow, *Principles*, 509–10.
61. Leslie Stuart, *The Cold War and American Science: The Military–Industrial–Academic Complex at MIT and Stanford* (New York: Columbia University Press, 1993), 85. Booz, Allen, and Hamilton, *Review of Navy R&D Management*, 337. Mary R. Self, *History of the Development of Guided Missiles, 1946–1950* (Dayton, OH: Wright-Patterson Air Force Base, 1951), 49, 74, 77–79.
62. "Purdue Gun-Launched Guided Missiles Program" (21 December 1946), in Bruhn, "A History of Aeronautical Education and Research."
63. Technical Committee minutes (26 March 1947) and Policy Committee minutes (14 May 1947), Minutes, box 6, RG 72.
64. Andrew G. Haley, *Rocketry and Space Exploration: The International Story* (New York: Van Nostrand, 1958), 36–37, 43. Lee Correy [Harry Stine], *Contraband Rocket* (New York: Ace Books, 1956), 17. I thank Roger Strater for this reference.
65. Zucrow interview with Eckles, 20–21. M. J. Zucrow, "Report on Progress and Planning for Gas Turbines and Jet Propulsion Program at Purdue University" (August 1951), 4, folder 2, box 2, Jerry Ross Papers (Ross). Unless otherwise cited, I have drawn details about the Rocket Lab from the monthly progress memorandums of the PRF and Purdue University (between 1947 and 1953), box 6, RG 72.
66. Notice from the chief, Power Plant Division, BuAer, USN (23 October 1946), folder 3, box 95, von Kármán.
67. See the exchange of Hovde, Meikle, and Frank Parker (summer 1949), folder 4, box 95, von Kármán.
68. I have cited material from the Chaffee Hall copy of a revised version of his thesis: Walter Hesse, "The Design and Instrumentation of the Rocket Test Facility at Purdue University" (June 1948), 1–12. His original thesis was "Design and Instrumentation of a Rocket Test Pit" (MSME, January 1948); both in the Chaffee Hall Office files, Maurice J. Zucrow Laboratories, Chaffee.
69. Hesse, "The Design and Instrumentation of the Rocket Test Facility," 1–12.
70. Edward C. Govignon, "A Study of Some Possible Methods for Improving the Performance of a Rocket Propelled Missile" (MSME, June 1948), 1–6.
71. From the monthly progress memorandums of the PRF and Purdue University (between 1947 and 1953), box 6, RG 72.
72. Cecil F. Warner, *Thermodynamic Fundamentals for Engineers* (Ames, IA: Littlefield, Adams,

1953) and *Mechanical Engineering Laboratory* with C. W. Messersmith (New York: Wiley, 1950). Irwin Weisenberg interview with John L. Sloop (1 May 1974), 5–6; and "History of Rocket Research Laboratory at the Ohio State University" (1961–1962), John L. Sloop Papers (Sloop).

73. Technical Survey Group, "Comments on Research Programs" (31 May 1948), supplement to "Project Squid Status Report" (May 1946–May 1948), 12, Chaffee.
74. Frank Parker, "Project Squid Proposed Research Program for Fiscal 1949" (1 February 1948), 1–19; Atlantic Research Corporation, "Review of the Squid Program and Recommendations for the Year Beginning September 30, 1948" (30 April 1949), 1–7; both in Chaffee.
75. Details of 1946 work in Project Squid phases, folders 9 and 10; agenda for Minutes Policy Committee meeting (12 May 1949), folder 12; "Project Squid Conference Report" (25 January 1950), folder 5; Robert Courant to dean Hugh Taylor at Princeton (16 May 1950), folder 5; all in box 95, von Kármán.
76. Project Squid monthly progress memorandum (November 1948), folder 10, box 95, von Kármán.
77. Eldon Knuth, "Squid Project: Inside Story of Purdue Jet Research," *Purdue Engineer* (November 1948): 8–11, 24.
78. Price, *Government and Science*, 71–72. Peter Westwick, *The National Labs: Science in an American System, 1947–1974* (Cambridge: Harvard University Press, 2002). The quote is from *Reviews of Data on Research and Development* 23 (October 1960): 2–4.
79. *Aero Digest* (March 1947): 59.
80. Princeton University news release (14 December 1948), Aviation and Rocketry, box 2, Guggenheim Foundation Records.
81. Rolf Sabersky interview with Shelley Irwin (3 and 12 April 1990), 17, Caltech. *Missiles and Rockets* (December 1957): 71; and (June 1957): 23. Martin Summerfield, "Chemical Propulsion," in *Space Flight Report to the Nation*, ed. Jerry Grey and Vivian Grey (New York: Basic Books, 1962), 73.
82. Zucrow, "Report on Progress and Planning" (August 1951), 7, 15.
83. Summerfield to Zucrow (29 December 1953), Correspondence, box 2, Porter. J. Michael Murphy interview with Tracy Grimm (14 October 2015), 13, ASC. The pieces in question are in *ARS Journal* (March 1959); and *Journal of Jet Propulsion* 28, no. 10 (October 1958).
84. James B. Conant, "The University and the State," *Journal of Higher Education* 18, no. 6 (June 1947): 284; and James Hershberg, *James B. Conant: Harvard to Hiroshima and the Making of the Nuclear Age* (New York: Knopf, 1993), 397. Harold A. Bolz interview with Robert B. Sutton (13 April 1983), Ohio State University Archives. For a similar criticism of MIT arrogance, see Silverstein interview with Mauer, 99.
85. *Titanic*, directed by Jean Negulesco (Twentieth Century Fox, 1953). Louis Friedman in-

terview with Amir Alexander and Peter Westwick (28 January 2015), 5, Huntington. Wattendorf interview with Goddard, 37. Tom Wolfe, *The Right Stuff* (New York: Farrar, Straus and Giroux, 1979), 132, 230–33.

86. *Panel on Science and Technology Fourth Meeting. Hearings before the Committee on Science and Astronautics, US House of Representatives*, 87th Congress (21 and 22 March 1962), 99. Minutes (27 September 1960), AIAA.
87. *New York Times* (2 September 1945): 72. Thomas Haigh, et al., *ENIAC in Action: Making and Remaking the Modern Computer* (Cambridge: MIT Press, 2016).
88. Princeton University, *Introductory Report on Project Squid*, 3. Arnold Goldburg, "Sponsored Research: Project Squid," *Princeton Alumni Weekly* (13 May 1960): 10–12.
89. Policy Committee minutes (6 February, 26 March, 4 May, and 30 June 1947), Minutes, box 6, RG 72. Robert B. Stewart, et al., *Advisory Committee on Contractual and Administrative Procedures for Research and Development for the Department of the Army*, Final Report (Washington, DC: Department of the Army, 15 October 1948), 11–35.
90. Technical Survey Group, "Comments on Research Programs" (31 May 1948), 5–8.
91. Atlantic Research Corporation, "Review of the Squid Program," 38, 53–64.
92. These quotes are from William Golden, "Memo of 21 December 1950," in *Impacts of the Early Cold War*, 39; *The First Five Years of the ARDC* (Baltimore: Air Research and Development Command, 1955), 19; and *Research Reviews (ONR)* (9 March 1950): 16.
93. Frank Gibney, "The Missile Mess," *Harper's Magazine* (1 January 1960): 31.
94. Technical Committee minutes (11 December 1946), Minutes, box 6, RG 72.
95. *Record of the Conference on Problems of Heat Transfer in Rocket Motors (Princeton University, 2–3 December 1949)*, Project Squid Technical Report No. 19 (1 July 1950), 1–5, 12.
96. Minutes of meeting to discuss BuAer-ORI Propulsion Program (29 March 1946). Also see several memorandums and directives of 18 March 1946, Project Squid Correspondence, vol. 1, box 8, RG 72.
97. "Technical Library Abstracts" (1947), box 5, RG 72.
98. "Report of Publications Sub-Committee, Project Squid (1946–1947)," box 6, RG 72. Theodore von Kármán to Luigi Crocco (4 March 1949), folder 4, box 95, von Kármán.
99. Neil Armstrong as quoted by Henry Yang, "Teaching and Research," *Purdue Alumnus* (May 1992): 48. James R. Hansen, *First Man: The Life of Neil A. Armstrong* (New York: Simon and Schuster, 2005), 57–63.
100. Walter Hesse, "Analysis of Turbojet Thrust in Flight" (PhD diss., Purdue University, August 1951).
101. R. W. Graham interview with Sandra Johnson (20 September 2005), 2–3, NASA.
102. Project Squid Correspondence, vol. 1. Editorial, *Missiles Away* 3, no. 2 (Summer 1955): 4–5.
103. "Project Squid Conference Report" (25 January 1950), folder 5; Policy Committee minutes (12 May 1949), folder 12; both in box 95, von Kármán.

104. "Security Problem" report (1948), and the Policy Committee minutes (12 May 1949), folder 12, box 95, von Kármán. Executive Order 10501, Safeguarding Official Information in the Interests of the Defense of the United States (1953).
105. Paul Petty interview with Tracy Grimm (27 August 2013), 17–18, 24–34, ASC. Elliott Katz interview with Peter Westwick (14 June 2010), 10, Huntington. Robert Strickler interview with Tracy Grimm (9 October 2015), 24, ASC.
106. "Handling of 'Secret' Classified Documents or Material" (1 June 1960), Security, box 298; Walter T. Olson (chief, Fuels and Combustion Research Division, NACA Lewis), letter of 13 February 1953, R&D Board, box 503; both in RG 255. Security violations (1951–1958), box 1, Bruce Lundin Papers, NASA Glenn Research Center.
107. PRF to the School of Mechanical Engineering (11 March 1965), Zucrow file; and correspondence and employment information, box 9, Zucrow. On the library's system, see the Small Arms Ordinance Research Project, unprocessed box 1, no. 20080107, ASC.
108. Potter to Detlev Bronke at Johns Hopkins University (19 February 1951), National Science Foundation, box 30, Potter.
109. S. J. Woolf, "Dr. Bush Sees a Boundless Future for Science," *New York Times* (2 September 1945): 72.
110. Hovde speech to the Fourth International Conference on the Mechanics and Chemistry of Solid Propellants (23 June 1965), Navy Department, box 37, Hovde.
111. W. H. Auden, "Under Which Lyre: A Reactionary Tract for the Times," *Harper's Magazine* (June 1947): 508–9, originally recited at Harvard University as a Phi Beta Kappa poem in June of 1946. Russell Kirk, "A Conscript on Education," *South Atlantic Quarterly* 44 (January 1945): 97–98.
112. C. C. Furnas, ed., *Research in Industry: Its Organization and Management* (Princeton: Van Nostrand, 1948), 2. Benoît Godin, "Research and Development," *Science and Public Policy* 33, no. 1 (February 2006): 64.

## CHAPTER 5

1. Daniel Kevles, "Korea, Science, and the State," in *Big Science: The Growth of Large-Scale Research*, ed. Peter Galison and Bruce William Hevly (Stanford: Stanford University Press, 1992), 320–21, 327.
2. Robert H. Knapp and Hubert B. Goodrich, *Origins of American Scientists* (New York: Russell, 1952), 1–2.
3. Aaron L. Friedberg, *In the Shadow of the Garrison State: America's Anti-Statism and Its Cold War Grand Strategy* (Princeton: Princeton University Press, 2000). Carroll Pursell, *Technology in Postwar America: A History* (New York: Columbia University Press, 2007), 3.
4. Harold Lasswell, "The Garrison State," *American Journal of Sociology* 46, no. 4 (January 1941): 455–68. Daniel Yergin, *Shattered Peace: The Origins of the Cold War* (New York: Houghton Mifflin, 1978).

5. Booz, Allen, and Hamilton, *Review of Navy R&D Management* (Washington, DC: Department of the Navy, 1976), 191–95. Clarence Danhof, *Government Contracting and Technological Change* (Washington, DC: Brookings Institution, 1968), 302.
6. Director of Central Intelligence, "National Security Council Status of Projects" (25 May 1953), classified as top secret. E. M. Purcell, "The Case for the Needles Experiment," *New Scientist* 13, no. 272 (1 February 1962): 245–47.
7. Minutes of the Committee on Guided Missiles (24 June 1946), 7–10, Conference on Rockets, box 2320, RG 156.
8. Vice Adm. John Sides, "Ten Years of Missile Progress," *Astronautics and Aeronautics* (September 1957): 24.
9. Vannevar Bush (JRDB) to B. B. Hickenlooper, chair of the US Senate Joint Committee on Atomic Energy (14 May 1947), 1–4, in Collation of Air Technical Intelligence Information, IRIS no. 142231, AFHRA.
10. Bush to Hickenlooper, 1–4. Vannevar Bush, *Pieces of the Action* (New York: Morrow, 1970), 52. *Research and Development Board: History and Functions* (Washington, DC: GPO, 1948), 1–5.
11. John Ponturo, *Analytical Support for the Joint Chiefs of Staff* (Washington, DC: Institute for Defense Analyses, 1979), vii–xii, 23–24.
12. Harvey M. Sapolsky, *Science and the Navy: The History of the Office of Naval Research* (Princeton: Princeton University Press, 1990), 34–35.
13. Michael J. Hogan, *A Cross of Iron: Harry S. Truman and the Origins of the National Security State, 1945–1954* (New York: Cambridge University Press, 1998), 223–31, 263. Kelly Moore, *Disrupting Science: Social Movements, American Scientists, and the Politics of the Military, 1945–1975* (Princeton: Princeton University Press, 2008), 22–37. Elliott V. Converse III, *Rearming for the Cold War, 1945–1960* (Washington, DC: Department of Defense, 2012), 22–41.
14. Martin J. Collins, *Cold War Laboratory: RAND, the Air Force, and the American State, 1945–1950* (Washington, DC: Smithsonian Institution, 2002), 192.
15. Don K. Price, *Government and Science* (New York: New York University Press, 1954), 143–51; Kevles, "Korea, Science, and the State," 317–19; Booz, Allen, and Hamilton, *Review of Navy R&D Management*, 20–21; Converse, *Rearming for the Cold War*, 22–24; and Danhof, *Government Contracting and Technological Change*, 142.
16. Christopher Wright, "Science and the Establishment of Science Affairs," in *Scientists and National Policy-Making*, ed. Robert Gilpin and Christopher Wright (New York: Columbia University Press, 1964), 293–96.
17. Bush to Hickenlooper, 5–7. For a model project card form, see John Steelman, *The Federal Research Program*, vol. 2 of *Science and Public Policy* (Washington, DC: GPO, 1947), 303.
18. Converse, *Rearming for the Cold War*, 36. On the RDB "project" system of cards and procedures, see folders 3–5, box 573; folders 1–5, box 574; and folders 2–6, box 575, all in RG 330.
19. George Lupton, "Engineering on Government Contracts," *Machine Design* (9 August 1956): 76–80. Converse, *Rearming for the Cold War*, 43–45.

20. War Department and Navy Department, *Explanation of Principles for Determination of Costs under Government Research and Development Contracts with Educational Institutions* (Washington, DC: Office for Emergency Management, 1947). These were the standards also codified in the Army Procurement Regulations (1 November 1947), from Public Law 413, 80th Congress, Armed Services Procurement Act (1947), and in War Department Circular No. 126, "Research and Development and Classification of Materiel." Each military service was responsible for its own operation and enforcement.
21. Robert B. Stewart, et al., *Advisory Committee on Contractual and Administrative Procedures for Research and Development for the Department of the Army*, Final Report (Washington, DC: Department of the Army, 15 October 1948), 10–35. Danhof, *Government Contracting and Technological Change*, 321. Price, *Government and Science*, 143, 203. Lawrence Hafstad, "Problems in the Coordination of Federal Research," in *Scientific Research: Its Administration and Organization*, ed. George P. Bush and Lowell Hattery (Washington, DC: American University Press, 1950), 18.
22. R. B. Stewart, "Government Research Contracts," in *Scientific Research*, Bush and Hattery, 29–34.
23. *Reviews of Data on Research and Development* 2 (March 1957): 1–4; and 7 (September 1957): 4.
24. William A. Hanley exchange with Vannevar Bush (29 May, 2 June, and 5 June 1945), Purdue University, box 94, Bush.
25. The Guided Missiles Committee (GMC) was reconstituted in 1946 within the JRDB and held seven meetings before it became officially part of RDB on 30 September 1947. Karl Kellerman, "Memorandum for Executive Secretary RDB" (17 June 1948), folder 6, box 203, RG 330.
26. "Purdue's Rocket Man," *Time* 45, no. 10 (3 September 1945).
27. Biographical materials, Hovde Purdue Appointment, box L, Hovde; and *Purdue Alumnus* (January 1948): 14–15. Day, "Purdue's Rocket Bomb Prexy," 26–28.
28. *Rivet* 4, no. 2 (31 November 1949): 11.
29. Vannevar Bush to Hovde (5 February 1947), Research and Development, box 16, Hovde. Bush and Hovde maintained close contacts in these years (1946–1948). See the letters in Hovde, Frederick L., box 52, Bush.
30. Hovde to Robert Wilson of Standard Oil (28 December 1949), Research and Development, box 16, Hovde.
31. Bush, *Pieces of the Action*, 83, 109, 307. Frank Wattendorf interview with the Goddard Space Flight Center (26 June 1973), 52, NASA. Stewart interview with Greenberg, 21–22. Grayson Merrill, *The Reminiscences of Captain Grayson Merrill* (Annapolis: US Naval Institute, 1997), 151–53.
32. Mary R. Self, *History of the Development of Guided Missiles, 1946–1950* (Dayton, OH: Wright-Patterson Air Force Base, 1951), 40. Vannevar Bush, *Modern Arms and Free Men:*

*A Discussion of the Role of Science in Preserving Democracy* (New York: Simon and Schuster, 1949).

33. Edmund Beard, *Developing the ICBM: A Study in Bureaucratic Politics* (New York: Columbia University Press, 1976), 30–31. Robert L. Perry, "The Atlas, Thor, and Titan," *Technology and Culture* 4, no. 4 (Autumn 1963): 466–77. Max Rosenberg, *The Air Force and the National Guided Missile Program, 1944–1950* (Washington, DC: USAF, 1965), 34–35, 75–84. Self, *History of the Development of Guided Missiles*, 45, 109–12.
34. Fred Darwin, "Memorandum for the Executive Secretary RDB" (14 October 1949), folder 2, box 407, RG 330.
35. "Interim Statement of Policy to RDB by GMC" (10 February 1949), folder 2, box 487, RG 330. Minutes of the Ad Hoc Subcommittee on Long Range Rockets (20 July 1948), folder 13, box 203, RG 330.
36. Rosenberg, *The Air Force and the National Guided Missile Program*, 100–10. Christopher Gainor, *The Bomb and America's Missile Age* (Baltimore: Johns Hopkins University Press, 2018), 74–76, 84–85. Enid Curtis Bok Schoettle, "The Establishment of NASA," in *Knowledge and Power*, ed. Sanford Lakoff (New York: Free Press, 1966), 164–66.
37. Hovde to Vannevar Bush (29 September 1947); and Rear Adm. D. V. Gallery, chief of naval operations, to Hovde (20 December 1948); both in Research and Development, box 16, Hovde.
38. Hovde correspondence with Robert Wilson of Standard Oil (28 December 1949); and Clark Millikan (16 April 1948); both in Research and Development, box 16, Hovde. D. S. Fahrney, *The History of Pilotless Aircraft and Guided Missiles* (Washington, DC: US Navy, 1958), 1277–97.
39. Planning Consultants, "Sample Project Report on AAF Project MX-774-B" (9 April 1947), folder 4, box 541, RG 330.
40. Kellerman, "Memorandum for Executive Secretary RDB" (17 June 1948).
41. "Report of Technical Evaluation Group" (9 June 1948), folder 13, box 202, RG 330. Self, *History of the Development of Guided Missiles*, 21–38. Converse, *Rearming for the Cold War*, 211–44.
42. "Interim Statement of Policy to RDB by GMC" (10 February 1949). Technical Evaluation Group, "Report on Nike" (10 December 1948), TEG, box 210, RG 330; as well as "Technical Evaluation Group" (17 March 1949), TEG, box 209, RG 330. RDB staff report for the GMC, "Program Guidance in the Field of Guided Missiles Research and Development" (28 October 1952), Agenda, box 191, RG 330.
43. See Stephen B. Johnson, *The United States Air Force and the Culture of Innovation, 1945–1965* (Washington, DC: USAF, 2002), 2–4, 17–19, 60, 103–7. Full-fledged systems engineering, as perfected by the Ramo-Woolridge Corporation and the Aerospace Corporation, introduced the quantitative methods of cost schedules, input–output scales, system equations, and feedback loops.

44. "Outline for Report on National Program" (28 May 1948), TEG, box 209, RG 330.
45. "Report of Technical Evaluation Group" (9 June 1948).
46. Karl Kellerman, "Coordination Accomplishments of Committee on Guided Missiles" (24 August 1948), folder 6, box 203, RG 330.
47. *Research and Development Board*, 1–5.
48. F. L. Hovde and the GMC, "The National Guided Missiles Program" (15 December 1948), 1–10, folder 8, box 200, RG 330.
49. "Interim Statement of Policy to RDB by Committee on Guided Missiles" (10 February 1949).
50. Rear Adm. J. A. Furer's comments (5 December 1944), Post-War Research, box 10 (OSRD Jet Propulsion Committee), RG 38. Duplication was accepted practice, according to Converse, *Rearming for the Cold War*, 340–43.
51. GMC memorandum (15 October 1947), folder 9, box 202, RG 330. "Outline for Report on National Program" (28 May 1948), TEG, box 209, RG 330. "Report of Technical Evaluation Group" (9 June 1948). Minutes of the Ad Hoc Subcommittee on Long Range Rockets (20 July 1948).
52. Maj. Gen. L. C. Craigie, memorandum for the Air Force Secretary RDB (28 June 1948), in folder 1, box 204, RG 330.
53. Gainor, *The Bomb and America's Missile Age*, 93.
54. Hovde to Robert Wilson of Standard Oil (28 December 1949).
55. Hovde, GMC, "The National Guided Missiles Program," 3–4, 24. "Interim Statement of Policy to RDB by GMC" (10 February 1949).
56. Robert Gilpin, *American Scientists and Nuclear Weapons Policy* (Princeton: Princeton University Press, 1962).
57. *American Education and International Tensions* (Washington, DC: National Education Association, 1949), 15, 37–38, 53. *Higher Learning and the World Crisis* (Washington, DC: National Education Association, 1949).
58. *Subversive Influence in the Educational Process: Hearings Before the Subcommittee to Investigate the Administration of the Internal Security Act and Other Internal Security Laws to the Committee on the Judiciary*, United States Senate, 82nd Congress (Washington, DC: GPO, 1952).
59. *Purdue Exponent* (1 July 1949): 1. *Purdue Alumnus* (November 1946): 12; and (June–July 1949): 4–5. Ronnie Dugger, *Our Invaded Universities: Form, Reform and New Starts* (New York: Norton, 1974).
60. *Purdue Alumnus* (September–October 1949): 6–8.
61. Stewart interview with Eckles, 40–41.
62. M. J. Zucrow, "The Engineer and Democracy," 4–5, 7–8, talk to the Purdue chapter of Pi Tau Sigma fraternity (8 November 1950), box 1 (addition), Zucrow.
63. *Rivet* 2, no. 3 (December 1947): 3.
64. "Report of the Panel on Propulsion and Fuels MPF" (June 1948), folder 10, box 203, RG 330.

65. Minutes of the 1 December 1949 meeting, MPF, box 218, RG 330. Converse, *Rearming for the Cold War*, 22–41.
66. Walter T. Olson interview with John L. Sloop (11 July 1974), 44, NASA.
67. Zucrow's MPF materials "reporting procedure" and the minutes, memorandums, mailing lists and reports of the panel from August 1948 to September 1950 are in MPF, box 218, RG 330. M. J. Zucrow, "Program Guidance Report and Technical Estimates" (21 October 1952), GMC, box 192, RG 330. M. J. Zucrow memorandum "Unitary Plan on Liquid Rocket Test Stands" (13 May 1948), folder 6, box 218, RG 330.
68. Minutes of the 4 November 1949 meeting, MPF, box 218, RG 330.
69. Summarized from the minutes of the second through seventh meetings of the panel, 15 April 1948 through 17 December 1948, MPF, box 218, RG 330. These included MX-606 GAPA (ground-to-air pilotless aircraft), the Nike SAM-A-7, Project Wizard MX-794, the Firebird MX-799, the Matador MX-771, the Snark MX-775, and the Navaho MX-770.
70. M. J. Zucrow, "Program Guidance Report and Technical Estimates" (8 December 1950), folder 8; M. J. Zucrow, minutes of the eighth meeting, MPF (24–25 February 1949), folder 3; both in box 218, RG 330.
71. Minutes of the 1 November 1950 meeting, MPF, box 218, RG 330.
72. Minutes of the 24–25 March 1949, 21–22 April 1949, and 4 November 1949 meetings, MPF, box 218, RG 330.
73. Denis H. O'Brien, "A Survey of Supersonic Diffusers Applicable to Ram-Jets" (MS, June 1948), with confidential research supported by the US Air Force. Donald William Craft, "The Ramjet Engine" (MS, February 1949).
74. Memorandum of 8 April 1949 and minutes of the 1 December 1949 meeting, MPF, box 218, RG 330.
75. M. J. Zucrow, "Report on Survey of National Ramjet Program" (17 March 1950), GMC Agenda, box 187, RG 330. Memorandums and reports (June to October 1952), in Minutes MPF, box 218, RG 330. "Program Guidance Report," MPF (15 December 1949), folder 8, box 218, RG 330.
76. M. J. Zucrow (chair of the MPF) to Capt. D. S. Fahrney (28 February 1949), with a study and recommendation on "the desirability of large liquid rocket propulsion systems" (24–25 February 1949) and memorandums and reports of the panel between August 1948 and September of 1950, MPF, box 218, RG 330.
77. Minutes of the 1 November 1950 meeting, MPF, box 218, RG 330.
78. M. J. Zucrow, "Program Guidance Report and Technical Estimates" (8 December 1950). *Panel on Science and Technology Third Meeting. Report of the Committee on Science and Astronautics, US House of Representatives*, 87th Congress (17 April 1961), 61.
79. "National Effort on Liquid Propellant Rocket Engines," report to the GMC for presentation (8 June 1951), Agenda, box 218, RG 330. The survey was complete as a report by 31 May 1951. "The Unitary Plan for Liquid Rocket Test Stands" was ready earlier, on 17 August 1950. Both were prepared by the MPF and edited and signed by M. J. Zucrow.

80. Clark Millikan (chair of the GMC), "Summary of Guided Missile Program Accomplishments" (27 June 1950), folder 6, box 203, RG 330. Dr. J. E. Lipp, "Presentation of USAF Long Range Rocket Program," report to the GMC of the RDB, with discussion (26 January 1951), folder 1, box 396, RG 330. See Gainor, *The Bomb and America's Missile Age*, 116, for a different perspective.
81. For a defense of R&D duplication in these terms, see M. J. Zucrow, "Program Guidance Report and Technical Estimates" (8 December 1950); and the summary evaluation in M. J. Zucrow, "Program Guidance Report and Technical Estimates" (21 October 1952), GMC, box 192, RG 330.
82. See "Report of MPF" (June 1948), folder 10, box 203, RG 330. M. J. Zucrow, "Research and Engineering Problems Pertinent to Rocket Engine Development" (15 March 1949), folder 6, box 218, RG 330; "Program Guidance Report," MPF (15 December 1949), folder 8, box 218, RG 330.
83. *Replies from Executive Departments and Federal Agencies to Inquiry Regarding Use of Advisory Committees. Committee on Government Operations, US House of Representatives*, 84th Congress (1 November 1956), 2297–99.
84. Price, *Government and Science*, 131–51. Stephen Hilgartner, *Science on Stage: Expert Advice as Public Drama* (Stanford: Stanford University Press, 2000), 51.
85. R. L. Gilpatric's memorandum to secretary of the Air Force Thomas Finletter (26 November 1952); Gen. Nathan Twining's memorandum to Finletter (24 December 1952); both in Reading file December 1952, box 58, Nathan Twining Papers (Twining).
86. James C. Fletcher (11 November 1974) and W. H. Pickering (15 December 1974) to Zucrow, folders 3–5, box 9, Zucrow.
87. *Panel on Science and Technology Fifth Meeting. Proceedings before the Committee on Science and Astronautics, US House of Representatives*, 88th Congress (22 and 23 January 1963), 36. For corroboration, see Edwin Speakman, "Research and Development for National Defense," *Proceedings of the IRE* (July 1952): 772–75; and Lee DuBridge, "Policy and the Scientists," *Foreign Affairs* 41, no. 3 (April 1963): 571–88.
88. GMC memorandum (15 October 1947), folder 9, box 202, RG 330. Kellerman, "Coordination Accomplishments of Committee on Guided Missiles" (24 August 1948).
89. Technical Evaluation Group, "Report on Nike" (10 December 1948), TEG folder (31), box 210, RG 330.
90. Bush to Hickenlooper, 5–7. Warner Schilling, "Scientists, Foreign Policy and Politics," in *Scientists and National Policy-Making*, Gilpin and Wright, 156.
91. Robert A. Lovett memorandum on the formation of the Ad Hoc Study Group (4 December 1952); and R. L. Gilpatric memorandum to Finletter (26 November 1952); both in Reading file December 1952, box 58, Twining.
92. James Isemann, "To Detect, To Deter, To Defend" (PhD diss., Kansas State University, 2009), 134–37. Hovde was one of seven on the committee.
93. Thomas A. Heppenheimer, "The Navaho Program and the Main Line of American Liquid

Rocketry," *Air Power History* 44, no. 2 (Summer 1997): 4–17. Dale D. Myers, "The Navaho Cruise Missile," *Acta Astronautica* 26, no. 8–10 (1992): 741–48.

94. DOD directive of 2 July 1951 and USAF memorandum of 30 July 1951, for release 6 August 1951, RDB-GM, box 348, RG 330.
95. "Presentation of Atlas Long-Range Surface-to-Surface Missile Project" (21–22 May 1952), GMC meeting agenda, Agenda, box 191, RG 330. This document was based on Zucrow's MPF, "Program Guidance Report and Technical Estimates," approved by the GMC on 8 December 1950, "for information and guidance of the Military Departments."
96. Col. R. L. Johnston (USAF), "Initial Presentation of Atlas Project" (21–22 May 1952) for the GMC, folder 14, box 396, RG 330. Jacob Neufeld, *The Development of Ballistic Missiles in the United States Air Force, 1945–1960* (Washington, DC: Office of Air Force History, 1990), 71–79.
97. M. J. Zucrow, "Program Guidance Report and Technical Estimates" (21 October 1952), GMC, box 192, RG 330. George Lemmer, *The Air Force and Strategic Deterrence* (Washington, DC: USAF, 1967), 5, confirms my readings of these documents; Gainor, *The Bomb and America's Missile Age*, 128, offers a different perspective.
98. RDB staff report for the GMC, "Program Guidance in the Field of Guided Missiles Rearch and Development" (28 October 1952), Agenda, box 191, RG 330.
99. For the full story, see Gainor, *The Bomb and America's Missile Age*, 112–29; and John Clayton Lonnquest, "The Face of Atlas: General Bernard Schriever and the Development of the Atlas Intercontinental Ballistic Missile" (PhD diss., Duke University, 1996).
100. "Report of the USAF Scientific Advisory Board's Ad Hoc Committee on Project Atlas" (30 December 1952), Millikan Report, folder 14, box 396, RG 330.
101. Neufeld, *The Development of Ballistic Missiles*, iii–iv, 44–50; Gen. Bernard A. Schriever interview with Carol Butler (15 April 1999), 2–31, NASA.
102. Convair advertisement, *Astronautics* (August 1957), 11.
103. ARDC memorandum "Rocket Test Stand Requirements" (April 1954), Propulsion 1954–1958, IRIS no. 00919673, AFHRA.
104. Stewart interview with Greenberg, 94–95. *Panel on Science and Technology Fourth Meeting. Hearings before the Committee on Science and Astronautics, US House of Representatives*, 87th Congress (21 and 22 March 1962), 18, 61–62.
105. For historians who appreciate the scientists and engineers, see Robert L. Perry, *The Ballistic Missile Decisions* (Santa Monica, CA: RAND Corp., 1967), 7–10, 28; Lonnquest, "The Face of Atlas," 26–41; and Gainor, *The Bomb and America's Missile Age*, 141.
106. Col. Edward N. Hall interview with Jack Neufeld (11 July 1989), 3–8, IRIS no. 01105443, AFHRA. Edward N. Hall, *The Art of Destructive Management: What Hath Man Wrought* (New York: Vantage Press, 1984), 21–30.
107. E. N. Hall, "Philosophy of Weapon Systems Development" (30 November 1954); and the ARDC memorandum "Rocket Test Stand Requirements" (April 1954); both in Propulsion 1954–1958, IRIS no. 00919673, AFHRA.

108. Robert Lusser, "Reliability of Guided Missiles" (29 July 1953), General Correspondence 1953, box 96, Wernher von Braun Papers (von Braun).
109. Ralph Gibson, "Supersonic Guided-Missile Progress," *Aero Digest* (July 1949): 40–44; and (August 1949): 48–53. He presented "The Philosophy of Guided Missiles" to the spring banquet of the ARS-IAS at Purdue on 22 May 1952, as reported in detail by Ed Hardesty, "Weapon with a Future," *Purdue Engineer* (December 1952): 12–13.
110. Donald Ritchie, "Rockets That See and Think," *Rocketscience* 6, no. 1 (March 1952): 14–19. Lee Correy [Harry Stine], *Contraband Rocket* (New York: Ace Books, 1956), 13, 99, 113, 131.
111. Herbert Hoover, *The Hoover Commission Report on Organization of the Executive Branch of the Government* (New York: McGraw-Hill, 1949), 187–95. Booz, Allen, and Hamilton, *Review of Navy R&D Management* (Washington, DC: Department of the Navy, 1976), 23–24. John C. Reis, *The Management of Defense: Organization and Control of the U.S. Armed Services* (Baltimore: Johns Hopkins University Press, 1964).
112. *Report of the Rockefeller Committee on Department of Defense Organization, Senate Armed Services Committee*, 83rd Congress (Washington, DC: GPO, 1953). Booz, Allen, and Hamilton, *Review of Navy R&D Management*, 35–52, 199. For the full context, see Nicholas M. Sambaluk, *The Other Space Race: Eisenhower and the Quest for Aerospace Security* (Annapolis: Naval Institute Press, 2015).
113. Vannevar Bush, "Report of the President," in *Yearbook 52, 1952–1953*, Carnegie Institution of Washington (Washington, DC: Carnegie Institution, 1953), 7.
114. *Sponsored Research Policy of Colleges and Universities* (Washington, DC: American Council on Education, 1954). The quotes are from Eric Hodgins, "The Strange State of American Research," *Fortune* (April 1955): 113–14; and *The Mighty Force of Research* (New York: McGraw-Hill, 1956), 1–3, 12–16.
115. *Panel on Science and Technology Seventh Meeting. Proceedings of the Committee on Science and Astronautics, US House of Representatives*, 89th Congress (25, 26, and 27 January 1966), 35–46, 115.
116. *Panel on Science and Technology Third Meeting. Report of the Committee on Science and Astronautics, US House of Representatives*, 87th Congress (17 April 1961), 53. Stewart interview with Greenberg, 14–15, 94–95.
117. William A. Blanpied, ed., *Impacts of the Early Cold War on the Formulation of US Science Policy: Selected Memoranda of William T. Golden, October 1950–April 1951* (Washington, DC: American Association for the Advancement of Science, 1995), 37. *Purdue Engineer* (October 1958): 22.
118. "Bomarc Evaluation Panel," box 26, Hovde.

## CHAPTER 6

1. T. A. Heppenheimer, *The Space Shuttle Decision: NASA's Search for a Reusable Space Vehicle* (Washington, DC: NASA, 1999), 14.

2. Hathaway Instrument Co. advertisement in *Mechanical Engineering* (December 1947), 53. G. Edward Pendray, *The Coming Age of Rocket Power* (New York: Harper and Brothers, 1945).
3. Lee Correy [Harry Stine], *Contraband Rocket* (New York: Ace Books, 1956), 102.
4. E. N. Heinemann, "New Developments in High-Speed Aircraft," *Mechanical Engineering* (October 1947): 806–9. Bell Aircraft Corporation, "New Horizons in Aviation," *Purdue Engineer* (April 1951): 2.
5. Zucrow delivered "Liquid-Propellant Power Plants" to the ASME convention (18–19 June 1947), published in *Transactions ASME* 69 (1947): 847; and *Journal of the American Rocket Society* (December 1947): 26. *Indianapolis News* (19 June 1947): 4; *Purdue Alumnus* (November 1947): 14–15. M. J. Zucrow, "Fundamentals of Jet Propulsion," *Purdue Alumnus* (February 1949): 5–8.
6. *Panel on Science and Technology Third Meeting. Report of the Committee on Science and Astronautics, US House of Representatives*, 87th Congress (17 April 1961), 61. Frank Malina, "Is the Sky the Limit?," *Army Ordnance* 31, no. 157 (July–August 1946): 45–48; and F. J. Malina and Martin Summerfield, "The Problem of Escape from the Earth by Rocket," *Journal of the Aeronautical Sciences* 14 (1947): 471–80.
7. John P. Sellers, "The Problem of Escape from the Earth and the Stepped-Rocket" (MSAE, February 1948). The copy at Chaffee Hall bears the edits of Dr. Zucrow. J. P. Sellers, "Influence of Earth's Gravitation on the Requirements of the Vertical Trajectory Rocket with Special Reference to Escape," *Journal of the American Rocket Society* (September 1948): 126–43. These studies were part of the wider Earth Satellite Vehicle Program.
8. *Journal and Courier* (18 June 1947): 13. *Indianapolis News* (28 June 1947): 1.
9. *Purdue Exponent* (1 February 1947): 1; and (11 March 1947): 1.
10. F. H. Clauser et al., *Preliminary Design of an Experimental World-Circling Spaceship* (Santa Monica, CA: RAND Corp., 1946), ii, 1–16, 221–36; and F. H. Clauser, "Flight Beyond the Earth's Atmosphere," *SAE Quarterly* 2–4 (October 1948): 563–70.
11. Robert L. Perry, *Origins of the USAF Space Program, 1945–1956* (Los Angeles: Air Force Systems Division, 1961), ix–x, 6–37. R. Cargill Hall, "Early US Satellite Proposals," *Technology and Culture* 4, no. 4 (1963): 410–34. Allan A. Needell, *Science, Cold War and the American State* (Amsterdam: Harwood, 2000), 97–127.
12. Sydney Chapman, *IGY: Year of Discovery; The Story of the International Geophysical Year* (Ann Arbor: University of Michigan Press, 1959). Walter Sullivan, *Assault on the Unknown: The International Geophysical Year* (New York: McGraw-Hill, 1961).
13. "Report of the President" (May 1952), AIAA.
14. Von Braun to Grosse (21 June 1952), A. V. Grosse, box 42, von Braun. Michael Neufeld, *Von Braun: Dreamer of Space, Engineer of War* (New York: Knopf, 2007), 261–68; and Sean N. Kalic, *US Presidents and the Militarization of Space, 1946–1967* (College Station: Texas A&M University Press, 2012), 29–30.
15. "General Correspondence 1952 A–Z," box 1, von Braun.

16. *Journal of the American Rocket Society* (January–February 1953): 41, 45.
17. James Harford to Maurice Zucrow (13 November 1974), folders 3–5, box 9, Zucrow.
18. Werner von Kirchner to Maurice Zucrow (15 December 1974), folders 3–5, box 9, Zucrow.
19. M. J. Zucrow, "Les moteurs de fusées à propergols liquids," in *Fusées at Recherche Aeronautique* 1, no. 1 (June 1956): 11–12, 35–43; and in *Matériaux et techniques pour les engins—fusées et l'aviation: Recueil des conférences prononcées lors du Premier Congrès International des Matériaux pour L'Aviation et les Projectiles-Fusées* (Paris: AERA, 1956).
20. *Purdue Alumnus* (Summer 1956): 18.
21. I have based these paragraphs on "Review of ARS News" in *Journal of the American Rocket Society* and in *Journal of Jet Propulsion* (between March 1950 and December 1956). Princeton formed a student section in November of 1952 as part of the New York Section.
22. Charles Trent, for example, presented his joint work with Zucrow (on hydrocarbons and WFNA as propellants) at the St. Louis Section of the ARS-ASME in June 1950.
23. The two papers are in *Fusées et engins guidés: Recueil des rapports du Premier Congrès International des Fusées et Engins Guidés, 1956–1957*, vol. IV (Paris: AERA, 1959), 97–103, 279–93; and in *Fusées at Recherche Aeronautique* 1, no. 3 (December 1956): 279–94; and 3, no. 2 (July 1958): 97–105. Osborn and A. C. Pinchak coauthored another paper with Zucrow (on combustion oscillations) for a second Paris Congress in June of 1959 in *Technique des fusées et des satellites: Recueil des rapports et conférences du Deuxième Congrès International des fusées et des Satellites* (Paris: AERA, 1959).
24. Philip M. Diamond's correspondence with Andrew Haley (23 November 1955), Geographic Sections, box 18, Haley.
25. *Journal of Jet Propulsion* (February 1956): 121; and (May 1956): 5.
26. Crabtree to Charles Ehresman (25 September 1996), folder 1, box 28, Ehresman.
27. Communication between Michael G. Smith and J. P. Sellers's son, John P. Sellers III (18 January 2021).
28. D. V. Gallery, "Guided Missiles," *Aero Digest* (December 1948): 48. *Air Force Missile* 3, no. 3 (January 1952): 4, in History of Detachment No. 8, Headquarters Tenth Air Force, Purdue University (1952), IRIS no. 0478686, AFHRA.
29. ARS annual business meeting minutes (14 November 1955), ARS, box 40, von Braun. For the full story on the ARS, see Tom D. Crouch, *Rocketeers and Gentlemen Engineers: A History of the American Institute of Aeronautics and Astronautics* (Reston, VA: AIAA, 2006), 123–27.
30. Minutes (2 November and 1 December 1953, 19 April and 11 November 1955, 24 September and 26 November 1956); and Executive Committee minutes (5 December 1955 and 28 February 1957), AIAA.
31. *Missiles and Rockets* (November 1956): 39, which was the main competitor to the ARS's *Journal of Jet Propulsion*.
32. *Rocket Newsletter* [Southern California Section of the ARS] (January 1956): 1–6. Northrop Aircraft advertisement, *Journal of Jet Propulsion* (June 1956): 510.

33. "ARS Sets Committees," *Journal of Jet Propulsion* (October 1956): 908. Andrew Haley (4 April 1955) to the ARS Policy Committee, ARS, box 42; and minutes of the Board of Directors (10 June 1958), ARS, box 41; both in von Braun. Zucrow voted alone and was defeated.
34. Minutes (3 March 1952), AIAA. L. J. Carter, ed., *The Artificial Satellite: The Proceedings of the Second International Congress on Astronautics* (London: British Interplanetary Society, 1951).
35. Annual business meeting minutes (28 November 1951), AIAA. Crouch, *Rocketeers and Gentlemen Engineers*, 127–37.
36. Minutes (4 February and 3 March 1952), AIAA.
37. Andrew G. Haley, *Rocketry and Space Exploration: The International Story* (New York: Van Nostrand, 1958), 122. *Missile Design and Development* (April 1959): 34.
38. Charles J. Hitch and Roland N. McKean, *The Economics of Defense in the Nuclear Age* (Cambridge: Harvard University Press, 1960), 1–4, chap. 13, 243–44.
39. *Air Research and Development Command* (Long Island City, NY: Federal Procurement Publications, 1957), 7; and *Guided Missiles* (Long Island City, NY: Federal Procurement Publications, 1957), 20.
40. Solberg memorandum to Hovde (30 March 1956), Bruce Reese file (Reese file); and to R. J. Grosh (25 March 1963), C. F. Warner file (Warner file); both in the Department of Mechanical Engineering, Purdue University.
41. M. J. Zucrow, "Report on Progress and Planning" (August 1951), 10, folder 2, box 2; and M. J. Zucrow, "Report on Progress and Planning for Gas Turbines and Jet Propulsion Program at Purdue University" (May 1958), 3, folder 4, box 2; both in Ross.
42. "Report of Industrial Experience of C. F. Warner from June 1953 to September 1954," Warner file.
43. Michael Gorn, "The NACA and its Military Patrons in the Supersonic Era," *Air Power History* (Fall 2001): 18–27.
44. Purdue sent the Army Air Forces copies of its aeronautical engineering theses by January 1948. Collation of Air Technical Intelligence Information, 1–11, IRIS no. 142231, AFHRA.
45. *Aero Digest* (September 1948): 50.
46. Office instruction (4 April 1947) from Maj. Gen. L. C. Craigie, chief, Engineering Division, AMC, Collation of Air Technical Intelligence Information, IRIS no. 142231, AFHRA. Military requests to any NACA laboratory for special research studies required the formidable written approval of the Aeronautical Board Clearance Sub-Committee, this to protect NACA's commitment to "basic research programs."
47. *Purdue Exponent* (4 October 1949): 1. *Aviation* (April 1947): 12–13; and (May 1947): 5. Hugh Dryden memorandum (29 August 1949), Administration, box 501, RG 255.
48. AMC memorandum to NACA Lewis Flight Propulsion Laboratory (23 May 1949), Rocket Propellants, box 502, RG 255.
49. *Thirty-Ninth Annual Report of the National Advisory Committee for Aeronautics, 1953*

(Washington, DC: GPO, 1954), i. *New York Times* (1 April 1946): 13. Alex Roland, *Model Research: The National Advisory Committee for Aeronautic, 1915–1958*, vol. 1 (Washington, DC: GPO, 1985), 252–53.

50. Minutes of the NACA Subcommittee on Propulsion Systems (23 October 1945), Subcommittee, box 2, Virginia Dawson Papers (Dawson). Petrovic, "Design for Decline," 101, 157. William F. Trimble, *Jerome C. Hunsaker and the Rise of American Aeronautics* (Washington, DC: Smithsonian Institution, 2002), 178–83, argues that the aviation industry even pressured to have NACA Lewis cease all basic engine research.
51. "A Survey of Fundamental Problems Requiring Research at the Aircraft Engine Research Laboratory" (May 1946), folder 376, box 501, RG 255.
52. "Report" in John Victory's files (2 October 1947), Administration, box 501; and Abe Silverstein memorandum (25 November 1946), Plans, Programs and Schedules, box 501; both in RG 255. Memorandum from Abe Silverstein, chief of research Lewis to NACA HQ (6 January 1952), NARA Chicago folder, box 4, Raymond R. Sharp Director's Office files. Silverstein interview with Mauer, 74.
53. Trimble, *Jerome C. Hunsaker*, 162–95, 212–13.
54. By agreement with the Executive Committee in December 1944, in Roland, *Model Research*, vol. 1, 252–53. Roland, *Model Research*, vol. 1, 383. Joseph Shortal, *A New Dimension: Wallops Island Flight Test Range* (Washington, DC: NASA, 1978), 16.
55. Petrovic, "Design for Decline," 113–14.
56. Olson interview with Sloop, 11–12.
57. "A Survey of Fundamental Problems Requiring Research at the Aircraft Engine Research Laboratory" (May 1946), 1, 16–18, 29–30, 44–45. Frank Friswold, "A Central Instrumentation System for a Rocket Laboratory" (1950), Friswold folder, box 5, RG 255. John L. Sloop, *Liquid Hydrogen as a Propulsion Fuel, 1945–1959* (Washington, DC: GPO, 1978), 78. NACA and Purdue also cooperated on the "reactivity of acid–fuel systems," the chemical starting of engines, "ejector designs," and combustion oscillations, in tandem with the Naval Rocket Test Station (Lake Denmark, New Jersey). John Sloop, "Rocket Research Program" (1952), Rocket Engine, box 502, RG 255. *Thirty-Eighth Annual Report of the National Advisory Committee for Aeronautics, 1952* (Washington, DC: GPO, 1953), 38.
58. John Sloop, "Aircraft Propulsion Principles" (1949), Case Institute Lectures, box 11, Sloop.
59. "Rocket Motor Cooling," Research Staff Meetings, 1945–1950; and "Test Program for Internal Film Cooling of Rocket Nozzles" (29 July 1946); both in Papers, box 11, Sloop. "Trip to Ohio State University" by John Sloop (14 January 1949), Visits and Visitors, box 507, RG 255.
60. Sloop, *Liquid Hydrogen as a Propulsion Fuel*, 73–93. Virginia P. Dawson, *Engines and Innovation: Lewis Laboratory and American Propulsion Technology* (Washington, DC: NASA, 1995), chap. 4. Virginia P. Dawson, *Ideas into Hardware: A History of the Rocket Engine Test Facility at the NASA Glenn Research Center* (Cleveland: NASA, 2004), 15–24. Robert S.

Arrighi, *Bringing the Future within Reach: Celebrating 75 Years of NASA John H. Glenn Research Center* (Cleveland: NASA, 2016), 43–92.

61. Pinkel, a specialist in heat transfer and cooling, gave two lectures for Purdue's Rocket Lab in 1953–1954: "The Nuclear Powerplant for Aircraft" and "Some Special Stress and Materials Problems in Turbojet Engines." Olson, a fuels and combustion expert for aircraft and missile power plants, spoke to the Purdue ARS in 1954: "Propulsion for Supersonic Flight." Their lectures are in box 8, RG 255.
62. *Wing Tips* [NACA Lewis newsletter] (29 October 1948): 4; and (3 September 1948): 3.
63. Sloop talk for the ASME/SAE (8 November 1948); the Purdue talk was John L. Sloop, "The Role of Chemistry in NACA Research" (March 1952); both are in Papers, box 11, Sloop.
64. "Future Trends in Rocket Propulsion" (10 October 1957), in Papers, box 11, Sloop.
65. *Thirty-Fifth Annual Report of the National Advisory Committee for Aeronautics, 1949* (Washington, DC: GPO, 1949), 35, 60. Besides its national research laboratories (Langley, Ames, and Lewis), NACA was a committee of committees. Four major ones focused on aerodynamics, power plants, aircraft construction, and operating problems. Each had its own set of technical subcommittees, twenty-two in all by 1950, filled with four hundred national experts to advise NACA.
66. "Recommendation of the NACA Subcommittee on Propulsion Systems Analysis" (31 May 1949), based on a 2–3 February 1949 meeting, NACA Subcommittees, box 2, Dawson.
67. Minutes (11 November 1951 and 4 February 1952), AIAA. Minutes of the Board of Directors (10 June 1958), ARS, box 41, von Braun. Hovde to Robert Wilson of Standard Oil (28 December 1949), Research and Development, box 16, Hovde.
68. *Thirty-Eighth Annual Report of the National Advisory Committee for Aeronautics*, 38. Robert Arrighi, "Bruce T. Lundin," in *Memorial Tributes* (Washington, DC: NAE, 2014). My interpretation is based on Dawson, *Engines and Innovation*, 149–57; and Arrighi, *Bringing the Future within Reach*, 115. Petrovic, "Design for Decline," 42–50, 97–108, 142–44.
69. "Proposed Subcommittee on Rocket Engines" (19 October 1950), NACA Subcommittees, box 2, Dawson. Abe Silverstein, chief of research, memorandum to NACA Headquarters (20 April 1950), NACA Aeronautical Research, box 4, Raymond R. Sharp Director's Office files. Sloop, *Liquid Hydrogen as a Propulsion Fuel*, 77.
70. Abe Silverstein, chief of research, Lewis memorandum to NACA Headquarters (26 July 1950), Visits and Visitors, box 507, RG 255.
71. *Thirty-Seventh Annual Report of the National Advisory Committee for Aeronautics, 1951* (Washington, DC: GPO, 1952), 1–2, 26, 50.
72. M. B. Ames, acting assistant director for research, NACA Headquarters, memorandum "Research on Space Flight and Associated Problems" (10 July 1952) to NACA Lewis, NACA Aeronautical Research Projects, box, RG 255. James R. Hansen, *Engineer in Charge: A History of the Langley Aeronautical Laboratory, 1917–1958* (Washington, DC: NASA, 1987), 350–51.

73. *NACA Conference on Supersonic Missile Performance* (Cleveland: Lewis Flight Propulsion Laboratory, 13 March 1952), marked "Secret," declassified in 1962.
74. M. J. Zucrow, "Program Guidance Report and Technical Estimate" (21 October 1952), GMC folder, box 192, RG 330.
75. Members included Henry E. Alquist, NACA; Lt. Col. Langdon F. Ayers, USAF-ARDC; Richard B. Canright, JPL-Caltech; Comdr. K. C. Childers, USN BuAer; B. F. Coffman, USN BuAer; H. F. Dunholter, General Dynamics; R. Bruce Foster, Bell Aircraft; Benson E. Gammon, NACA; Stanley Gendler, RAND; Joseph L. Gray, Office of Chief of Army Ordnance; Paul R. Hill, NACA-Langley; Thomas E. Meyers, North American Aviation; Eugene Miller, Redstone Arsenal; G. E. Moore, General Electric; W. P. Munger, Reaction Motors Inc.; J. R. Patton, ONR; C. C. Ross, Aerojet; C. N. Satterfield, MIT; C. W. Schnare, USAF-ARDC; F. E. Schultz, General Electric; Jack H. Sheets, Curtiss-Wright; Capt. Levering Smith, USN, Naval Ordnance Test Station; R. J. Thompson Jr., M. W. Kellog; John Sloop, NACA Lewis; A. J. Stosick, JPL; R. C. Swann, Redstone Arsenal; F. I. Tanczos, USN BuAer; Paul Winternitz, Reaction Motors Inc.; David A. Young, Aerojet. Thomas E. Myers of North American Aviation replaced Zucrow as chair in 1954, though Zucrow remained a member.
76. Sloop, *Liquid Hydrogen as a Propulsion Fuel*, 75–81, 93. John Sloop, "NACA High Energy Rocket Propellant Research in the Fifties," paper presented at the Panel on Rocketry in the 1950s, AIAA 8th Annual Meeting (Washington, DC, 28 October 1971), John Sloop Materials, box 5, Dawson.
77. Abe Silverstein interview with John Sloop (29 May 1974), 18–20, 92–94, NASA. Walter T. Olson interview with John L. Sloop (11 July 1974), 42, NASA.
78. The first quote is from John Sloop, "Rocket Engine Research Program" (May–June 1953), 12–14, Rocket Engine, box 502, RG 255. The second quote is from Sloop, *Liquid Hydrogen as a Propulsion Fuel*, 75–81.
79. "Conference on Proposed Rocket Facility" (27 March 1952), RETF Design, box 2, Rocket Engine Test Facility Papers.
80. Abe Silverstein, associate director, "Discussion with M. J. Zucrow of Proposed Addition to Rocket Research Facilities," based on the Zucrow-Olson conversation on 13 November, in the memorandum from Lewis to NACA Headquarters (24 November 1952), Rocket Engine, box 502, RG 255.
81. Zucrow quoted in 1963 NASA Authorization. *Hearings before the Subcommittee on Advanced Research and Technology of the Committee on Science and Astronautics, US House of Representatives*, 87th Congress (March and April 1962), 1560. Bruce Lundin letter (1974), folders 3–5, box 9, Zucrow.
82. M. B. Ames, acting assistant director for research, NACA Headquarters, memorandum (10 July 1952). Sloop's handwritten note is in folder 2, box 16, Sloop. On Hunsaker's financial pressures, see Petrovic, "Design for Decline," 125; and Trimble, *Jerome C. Hunsaker*, 202–4.

Hunsaker's comments are in *Fortieth Annual Report of the National Advisory Committee for Aeronautics, 1954* (Washington, DC: GPO, 1955), ix.

83. Sloop, *Liquid Hydrogen as a Propulsion Fuel*, 79–81. Memorandum to the associate director of NACA from Walter Olson on the Panel on Propulsions and Fuels (MPF) of the RDB-GMC, on 30 January 1953, R&D Board, box 503, RG 255. Walter T. Olson, "Lewis at 40: A Reflection," *Lewis News* (25 September 1981): 3.
84. Memorandum of the Technical Advisory Panel on Fuels and Lubricants, Department of Defense (13 January 1955), Department of Defense, box 503, RG 255.
85. Abe Silverstein, associate director, NACA Lewis, memorandum to NACA Headquarters (18 January 1956), Defense Department, box 503, RG 255. Silverstein's communication (18 June 1956), Papers, box 11, Sloop.
86. "Rocket Research Program" (4–5 May 1953) for the Subcommittee on Rocket Engines, NACA Lewis Laboratory, Rocket Engine, box 502, RG 255.
87. *Panel on Science and Technology Fourth Meeting. Hearings before the Committee on Science and Astronautics, US House of Representatives*, 87th Congress (21 and 22 March 1962), 18, 61–62.
88. Executive Committee minutes (22 June 1955), AIAA. *Missiles and Rockets* (October 1956): 39. George S. James, "A Background of Memories of Working with Dr. Wernher von Braun," *Acta Astronautica* 113 (August–September 2015): 212–20.
89. Beryl Williams and Sam Epstein, *The Rocket Pioneers on the Road to Space* (New York: J. Mesner, 1955), 212. Roger D. Launius, "First Steps into Space: Projects Mercury and Gemini," in *Exploring the Unknown: Selected Documents in the History of the U.S. Civil Space Program*, ed. John M. Logsdon, vol. 7, *Human Spaceflight: Projects Mercury, Gemini, and Apollo* (Washington, DC: NASA, 2008), 2–7; and Dwayne Day, "The Von Braun Paradigm," *Space Time Magazine* 33 (November–December 1994): 12–15.
90. *Purdue Exponent* (7 May 1953): 3. *Journal and Courier* (7 May 1953): 5. Elliott Katz interview with Michael G. Smith (29 September 2016), 7, 25, ASC. Personal email communication between Michael G. Smith and John P. Sellers III (18 January 2021). Reese interview with Grimm, 9.
91. Walter Olson interview with Virginia Dawson (16 July 1984), 18, Walter Olson, box 4, Dawson.
92. *Indianapolis Times* (6 November 1960): 10. The photograph of Wernher von Braun with Maurice Zucrow (1953) is in *MEmo: Annual Newsletter of Mechanical Engineering* (1995): 1.
93. Zucrow interview with Eckles, 27.
94. Quotes from Grayson Merrill, *The Reminiscences of Captain Grayson Merrill* (Annapolis: US Naval Institute, 1997), 171, 181; and Stewart interview with Greenberg, 12. Abe Silverstein interview with John Sloop (29 May 1974), 121–34, NASA. Michael Neufeld, *Von Braun: Dreamer of Space, Engineer of War* (New York: Knopf, 2007), 368.
95. Werner von Kirchner to Zucrow (15 December 1974), folders 3–5, box 9, Zucrow. Also see Catrine Clay, *The Good Germans: Resisting the Nazis, 1933–1945* (London: Weidenfeld, 2020).

## CHAPTER 7

1. M. J. Zucrow, "The Engineering Graduate and his Career as a Professional Engineer" (1947), Zucrow Talks, box 1 (addition), Zucrow.
2. *Indianapolis Times* (6 November 1960): 10; and (3 November 1957): 4–5.
3. M. J. Zucrow, "Report on Progress and Planning" (August 1951), 7, folder 2, box 2, Ross. Katz interview with Westwick, 9.
4. Zucrow interview with Eckles, 21–24.
5. M. J. Zucrow, "Report on Progress and Planning" (August 1951), 5–7; and Zucrow, "Report on Progress and Planning" (May 1958), 5–8. Zucrow interview with Eckles, 21–24. Warner, "Thermal Sciences and Propulsion Center," 2. Zucrow letter and PRF funding "Justification" (21 December 1950), folder 12, box 95, von Kármán.
6. *Journal and Courier* (9 December 1965): 9. Hovde to Zucrow (15 December 1974), box 1 (addition), Zucrow.
7. *Journal of the American Rocket Society* (September 1951): 133. Alexander J. Smits and Courtland D. Perkins, "Aerospace Education and Research at Princeton University, 1942–1975," paper presented at the 41st Aerospace Sciences Meeting and Exhibit (Reno, NV, January 2003).
8. Stuart, Leslie, *The Cold War and American Science: The Military–Industrial–Academic Complex at MIT and Stanford* (New York: Columbia University Press, 1993), 14–31. Rebecca Lowen, *Creating the Cold War University: The Transformation of Stanford* (Berkeley: University of California Press, 1997), 121.
9. F. N. Andrews, "Report to the University Senate" (16 March 1970), folder 6, box 7, Special Committee on Sponsored Research (SCSR). G. S. Meikle, confidential report (12 May 1948), Research Director's Calendars, box 2; and "Report of President and Research Director" (9 May 1951), Research Director's Reports, box 3; both in PRF Records.
10. *Engineering News* [published as a supplement to *Purdue Alumnus*] (December 1958): 2; and *Purdue Alumnus* (April 1958): 7; and (May 1957): 17.
11. *Purdue Alumnus* (April 1958): 4–5. *Indianapolis Times* (3 November 1957): 4–5. *New York Times* (3 August 1958): 41.
12. History of Detachment #8, Headquarters Tenth Air Force, Purdue University, 1 January to 31 March 1952, IRIS no. 0478686, AFHRA.
13. *Purdue Alumnus* (January 1954): 5.
14. *Indianapolis Times* (3 November 1957): 4–5; and (6 November 1960): 10. "Prof. M. J. Zucrow," *Purdue Engineer* (March 1960): 50–51. *Panel on Science and Technology First Meeting. Proceedings of the Committee on Science and Astronautics, US House of Representatives*, 86th Congress (4 May 1960), 51–54.
15. Everard Arthur Bonney, Maurice J. Zucrow, and C. W. Besserer, *Aerodynamics, Propulsion, Structures and Design Practice* (Princeton: Van Nostrand, 1956). Maurice J. Zucrow, *Ord-*

*nance Engineering Design Handbook—Ballistic Missile Series—Propulsion and Propellants* (Redstone Arsenal: US Army Ordnance Corps, October 1959).

16. Reese interview with Grimm, 10; and the résumés in the Reese file. *Missile Design and Development* (July 1958): 17.
17. M. J. Zucrow, "Report on Progress and Planning" (May 1958), 5–8, folder 4, box 2, Ross.
18. *Panel on Science and Technology Fourth Meeting. Hearings before the Committee on Science and Astronautics, US House of Representatives*, 87th Congress (21 and 22 March 1962), 89. *Panel on Science and Technology Seventh Meeting. Proceedings of the Committee on Science and Astronautics, US House of Representatives*, 89th Congress (25, 26, and 27 January 1966), 57.
19. I have drawn these important insights, here and in the paragraphs below, from Charles Ehresman's history, "Jet Propulsion Center," in folder 3, box 28, Ehresman; and "Rocket Research," *PRF Horizons* 2, no. 8 (April 1956): 1–2.
20. C. F. Warner and M. J. Zucrow, *An Evaluation of the Heat Transfer Encountered in a Rocket Motor Operating at High Chamber Pressure*, Project Squid Technical Report No. 18 (Washington, DC: US Navy, July 1949); and M. J. Zucrow and Cecil Warner, "The Application of White Fuming Nitric Acid and Jet Engine Fuel (AN-F-58) as Rocket Propellants," *Journal of the American Rocket Society* (September 1950): 139–50.
21. Several of these efforts were mentor–student partnerships. C. H. Trent and M. J. Zucrow, "The Hypergolic Reaction of Dicyclopentadiene with WFNA," *Journal of the American Rocket Society* (May 1951): 129–31; and their "Behavior of Liquid Hydrocarbons with White Fuming Nitric Acid," *Journal of Industrial and Engineering Chemistry* 44, no. 11 (November 1952): 2668–73. M. J. Zucrow and C. M. Beighley, "Experimental Performance of WFNA-JP-3 Rocket Motors at Different Combustion Pressures," *Journal of the American Rocket Society* (June 1952): 323–30.
22. "Rocket Research," *PRF Horizons* 2, no. 8 (April 1956): 1–2. H. L. Solberg, "Nomination of Dr. M. J. Zucrow for Sigma Xi Award" (6 January 1955), Zucrow file.
23. Solberg, "Nomination of Dr. M. J. Zucrow" (6 January 1955).
24. M. J. Zucrow and A. R. Graham, "Some Considerations of Film Cooling for Rocket Motors," *Journal of Jet Propulsion* (June 1957): 650–56. M. J. Zucrow and J. P. Sellers, "Experimental Investigation of Rocket Motor Film Cooling," *ARS Journal* (May 1961): 668–70. C. F. Warner and D. L. Emmons, "Effects of Selected Gas Stream Parameters and Coolant Properties on Liquid Film Cooling," *Journal of Heat Transfer* (May 1964): 271–78.
25. Stanley Veerin Gunn, "The Effects of Several Variables upon the Ignition Lags of Hypergolic Fuels Oxidized by Nitric Acid" (PhD diss.,1953), partially republished in *Journal of the American Chemical Society* 22, no. 1 (January 1952): 33–38. Edward Dobbins, David Charvonia, H. L. Wood, and Charles M. Ehresman also contributed to this work.
26. J. R. Osborn and J. M. Bonnell published their joint work as "Effect of Fuel Composition

on High Frequency Oscillations in Rocket Motors Burning Premixed Hydrocarbon Gases and Air," *ARS Journal* (October 1961): 1397–1401; and their "Combustion Pressure Oscillations in an Unmixed Gas Rocket Motor," *Journal of Spacecraft and Rockets* 3, no. 11 (1966): 1680–82. For relevant dissertations, see James Bottorff, "Carbon Deposition in Gas Generators using WFNA and JP-4" (PhD diss., June 1957). J. M. Murphy, "An Experimental and Analytical Investigation of the Erosive Burning Characteristics of a Non-Homogeneous Solid Propellant" (PhD diss., 1964); and Robert Lain Glick, "The Effect of Acceleration on the Burning Rate of Nonmetallized Composite Propellants" (PhD diss., 1966).

27. Bruno Latour and Steve Woolgar, *Laboratory Life* (Beverly Hills: Sage Publications, 1979).
28. Robert Glick interview with Michael G. Smith (20 January 2017), 24; George Schneiter interview with Michael G. Smith (10 September 2016), 11; both in ASC. For example, Del Robison was the phase leader of the project Effect of High Combustion Pressure on Rocket Motor Performance and Heat Transfer. David Charvonia directed the phase work of George Schneiter.
29. Glick interview with Smith, 27. J. Michael Murphy, "Papa and the Space Shuttle" (2011), 10, in the J. Michael Murphy Papers. Robert Strickler interview with Tracy Grimm (7 October 2015), 12, 21–25, ASC.
30. Strickler interview with Grimm (7 October 2015), 12. Murphy interview with Grimm, 10–11, ASC. Schneiter interview with Smith, 12. Glick interview with Smith, 26–27. Delbert Robison, "Rocket Records: Rocket Fuel Reactions Integrated by Simultaneous Electronic Recording of Rocket Motor Functions," *Instrumentation* 7, no. 3 (1954): 13. George Schneiter, "Purdue's Jet Propulsion Center," *Purdue Engineer* (January 1960): 22–26.
31. Paul Petty interview with Tracy Grimm (18 November 2015), 5–9, ASC. Elliott Katz interview with Michael G. Smith (29 September 2016), 18, ASC.
32. These are from the 1974 testimonies of Bruce Reese, Bob Glick, and Eldon Knuth, folders 3–5, box 9, Zucrow; and Glick interview with Smith, 12–14.
33. Zucrow, "The Engineering Graduate." Dave and Athena Charvonia and Mel L'Ecuyer testimonies (1974), folders 3–5, box 9, Zucrow. Katz interview with Smith, 14. For Zucrow's edits, deleting passive voice and wordiness, see Richard James Rossbach, "The Comparison of Several Methods for Cycle Analysis of Continuous-Flow Air-Burning Engines" (MSME, 1948), at Chaffee.
34. Harvard College, Class of 1922, *Fiftieth Anniversary Report* (Cambridge: Harvard University, 1972), 584–85. A. R. Graham testimony (1974), folders 3–5, box 9, Zucrow. Katz interview with Smith, 22. Schneiter interview with Smith, 10. Strickler interview with Grimm (7 October 2015), 21.
35. Dave and Athena Charvonia testimonies (1974), folders 3–5, box 9, Zucrow. *Indianapolis News* (7 April 1966): 45.

36. H. E. Solberg to J. G. Skifstad (20 November 1956), J. G. Skifstad file, Department of Mechanical Engineering, Purdue University.
37. James Warner interview with Michael G. Smith (17 December 2016), 7–8, ASC. Eldon Knuth, Bernard Hill, Betty Yost, and Jim Kempe, "Prof. M. J. Zucrow," *Purdue Engineer* (November 1948): 35.
38. "'Doc' Zucrow," in *MEmo: Annual Newsletter of Mechanical Engineering* (1995): 2–4. The quotes are from J. Michael Murphy, "Papa and the Space Shuttle," 10; and Glick interview with Smith, 12–13.
39. Warner, "Thermal Sciences and Propulsion Center," 5.
40. Knuth, "Prof. M. J. Zucrow," 35. These are the words of Jesus Christ, to do mercy, in Luke 10:37.
41. "'Doc' Zucrow," in *MEmo*, 2. F. L. Hovde testimony (1974), folders 3–5, box 9, Zucrow. For a list of Zucrow's official PhD students as of 1965, see appendix 3.
42. William Wisely, "Profile of a Successful Career," *Purdue Engineer* (May 1964): 28. Zucrow, "The Engineering Graduate," 2–5. This was the Purdue model: to become part of a "research team working on the fringe of existing knowledge." *Engineering News* (December 1958): 2.
43. See the various advertisements in *Purdue Engineer* (December 1958): 11; and (October 1959): 3. *Missiles and Space* (March–April 1963): 43. *Purdue Scientist* (March 1957): 31.
44. Matthew Wisnioski, *Engineers for Change: Competing Visions of Technology in 1960s America* (Cambridge: MIT Press, 2012), 22–29. Ruth Oldenziel, *Making Technology Masculine: Men, Women, and Modern Machines in America, 1870–1945* (Amsterdam: Amsterdam University Press, 2000).
45. Advertisements in *Missiles Away* (Spring 1957): 35; and *Astronautics and Aeronautics* (August 1967): 18. Solberg to G. A. Hawkins (4 May 1950), George Hawkins file, Department of Mechanical Engineering, Purdue University.
46. *Engineering News* (December 1966): 1; and (December 1959): 3.
47. The courses were English Composition (101) and Fundamentals of Speech Communication (114) for the first year, a choice of literature courses for each semester of the second year, a choice of history and government courses in each semester of the third year, then a "nontechnical elective" in each semester of the fourth year. *Purdue Alumnus* (September–October 1950): 10; and (September–October 1951): 5. "Careers in Education," *Purdue News* 24, no. 2 (October 1952): 36.
48. *Journal of Engineering Education* 34 (May 1944): 594; and 35 (September 1944): 36–44. See also James Kip Finch, "Engineering and the Liberal Arts," *Journal of General Education* 2, no. 4 (July 1948): 301; and Melvin Kranzberg, "Educating the Whole Engineer," *ASEE Prism* 3, no. 3 (November 1993): 26.
49. Robert Eckles, "Liberal Science at Purdue: An Experiment," *Journal of General Education* 3, no. 4 (July 1949): 313–16. *Bulletin of Purdue University* (West Lafayette, IN: Purdue University, April 1946), 317.

50. W. E. Howland, "Operation Egghead," *Campus Copy* (October 1955): 3. C. P. Snow, *The Two Cultures* (London: Cambridge University Press, 1959).
51. F. L. Hovde interview with Robert Topping (1 June 1976), Hovde Interviews, box L, Hovde.
52. Zucrow, "The Engineering Graduate," 2–5. *Panel on Science and Technology Third Meeting. Report of the Committee on Science and Astronautics, US House of Representatives*, 87th Congress (17 April 1961), 95–96.
53. *Panel on Science and Technology Eleventh Meeting. Proceedings of the Committee on Science and Astronautics, US House of Representatives*, 91st Congress (27, 28, and 29 January 1970), 205–6. Zucrow testified here with Daniel Boorstin of the University of Chicago and Harvey Poe of St. John's College.
54. "Professor M. U. Clauser," *Purdue Engineer* (December 1950): 38.
55. *Purdue Exponent* (16 March 1950): 2. *Purdue Alumnus* (November 1950): 6. M. U. Clauser to Harold de Groff (24 February 1951), Correspondence 1951–1965, box 1, Harold de Groff Papers.
56. Zucrow, "Report on Progress and Planning" (August 1951), 2, 9–10.
57. The correspondence controversy is in Reese interview with Grimm, 7. The quotes are from Thompson interview with Smith, 12.
58. Memorandum from Milton Clauser to Members of the Aeronautics Staff (7 April 1952), Zucrow file.
59. "To Whom It May Concern" proclamation (4 April 1952); and "Professor Honored at Surprise Dinner" with forty-five attending (organized by Katz, Beighley, Robison, Gunn, Fisher, Bob Graham, Bill Craft); both in the black scrapbook, box 1 (addition), Zucrow.
60. *Engineering News* (December 1958): 2. Aeronautical Engineering was second from last in a list of eight schools.
61. By 1966, Aeronautical Engineering (then called Aeronautics, Astronautics, and Engineering Sciences) taught turbojets and rockets, as well as aircraft and missile propulsion, at the undergraduate and dual graduate level, in Propulsion I (AE 372), Aircraft Power Plants (AE 472 and 486), and Aerodynamics of Propulsion (AE 572). But otherwise it offered no substantial graduate offerings in jet propulsion.
62. "Comments on Aeronautical Engineering as a Field of Study and a Career" (spring of 1957), unnumbered technical report, 10–19, School of Aeronautical Engineering, ASC. A. F. Grandt, et al., *One Small Step* (West Lafayette, IN: Purdue University Press, 1995), 65, 97.
63. *Purdue Exponent* (23 March 1953): 1. *Journal and Courier* (21 March 1953): 2.
64. *PRF Horizons* 2, no. 4 (December 1955): 2. Strickler interview with Grimm (7 October 2015), 27. Purdue's reforms helped to establish a national trend, as discussed in *Journal of Engineering Education* 46 (September 1955): 25–60.
65. William K. LeBold, Edward C. Thoma, John W. Gillis, and George A. Hawkins, *A Study of the Purdue University Engineering Graduate* (West Lafayette, IN: Purdue University, 1960), 95. *Purdue Engineer* (November 1959): 46. On the national trend for engineering

science, see "Funds for R and D in Engineering Schools, Fiscal Year 1958," *Reviews of Data on Research and Development* 21 (July 1960): 1.

66. Thompson interview with Smith, 5. *Purdue Engineer* (December 1959): 9.

67. Zucrow made these points in *Panel on Science and Technology Fifth Meeting. Proceedings before the Committee on Science and Astronautics, US House of Representatives*, 88th Congress (22 and 23 January 1963), 16–18, 78; and Maurice Zucrow, "What's an Engineer?," *Engineering News* (April 1965): 1–2.

68. "A New Approach to Engineering Design," *Backgrounder from Purdue* [School of Engineering and Mathematical Sciences] (May 1963): 2–4. *Engineering News* (April 1963): 3–4. Edwin Layton, "American Ideologies of Science and Engineering," *Technology and Culture* 17, no. 4 (October 1976): 698–99. Editorial, "What Kind of Ph.D. for Aerospace?," *Astronautics and Aeronautics* (May 1967): 50.

69. *Engineering News* (April 1965): 3–4; (February 1966): 9. Joan Edwards, "Inventiveness: Man's Road to Progress," *Purdue Engineer* (October 1969): 8–9.

70. See one of the first polls and studies in Richard Ritti, "The Purdue Engineering Graduate: His Background, College Experience, and Expectations" (PhD diss., Purdue University, 1957), 1–3. LeBold et al.'s *A Study of the Purdue University Engineering Graduate* (1960) covered nearly four thousand of the graduates between 1911 and 1956. *General Education Program for Engineering Students* (West Lafayette, IN: Purdue University, 1964), folder 1, box 41, Ehresman.

71. John Hancock, "George Andrew Hawkins, 1907–1978," in *Memorial Tributes*, vol. 1 (Washington, DC: NAE, 1979), 112–17. E. A. Walker, et al., *Goals of Engineering Education: Final Report of the Goals Committee* (Washington, DC: ASEE, 1968).

72. LeBold et al., *A Study of the Purdue University Engineering Graduate*, 86–87, 94. D. E. Scheiber and W. K. LeBold, "Humanism and the Purdue Environment: Expectations and Experiences of Freshman Engineers," in the William K. LeBold Papers.

73. Ronald Voigt, "Technology versus Horizontal Man," *Purdue Scientist* (December 1957): 21.

74. Quotes from the film *Road to the Stars*, directed by Bill Adams (North American Aviation, 1957). Robert Stedfeld, "Engineering the Earth Satellite," *Machine Design* (1 September 1956): 82–86. James R. Hansen, *Spaceflight Revolution: NASA Langley Research Center from Sputnik to Apollo* (Washington, DC: NASA, 1995), xxvi–xxviii.

75. Paul Forman, "The Primacy of Science in Modernity, of Technology in Postmodernity, and of Ideology in the History of Technology," *History and Technology* 23, no. 1–2 (March/June 2007): 38.

76. President's Science Advisory Committee, *Strengthening American Science* (Washington, DC: GPO, 1958).

77. Gerard Piel, "Science, Censorship, and the Public Interest," *Science* 125, no. 3252 (April 26, 1957): 793. D. H. Radler, "The New Red Moon," *PRF Horizons* 4, no. 2 (October 1957): 1–4. *PRF Horizons* 4, no. 9 (May 1958): 2–3.

78. Dael L. Wolfle, ed., *Symposium on Basic Research* (Washington, DC: American Association for the Advancement of Science, 1959), xi–xiv, 115, 122–24, 171, 175, 266–68; and included in it is Dwight Eisenhower, "Science: Handmaiden of Freedom," 137–38.
79. *Missile Design and Development* (August 1958): 15; and (December 1959): 7. Harvey M. Sapolsky, *Science and the Navy: The History of the Office of Naval Research* (Princeton: Princeton University Press, 1990), 70–71.
80. Andrew Haley correspondence with the ARS Policy Committee and Wernher von Braun (4 and 11 April 1955), ARS 1952-57, box 42, von Braun. Von Braun to George Sutton, then vice president of the ARS (6 March 1957), ARS 1957, box 40, von Braun.
81. Executive Committee minutes (28 February 1957), AIAA. Minutes of the Board of Directors (10 June 1958), ARS 1958, box 41, von Braun.
82. John Gustafson, "Why Not Astronautical Engineers?," *Astronautics and Aeronautics* (September 1957): 52. See the articles and advertisements in *Missile Design and Development* (August 1958): 26, and (October 1959): 23; *Purdue Alumnus* (November 1958): 1; (Summer 1959): 11, 45; and (February 1960): 1–2.
83. "Plans of the Advisory Committee for Missile and Rocket Amateurs," minutes of the ARS Board of Directors (3 December 1957), ARS 1957, box 40, von Braun.
84. ARS "President's Letter" from George Sutton (17 December 1958), ARS 1958, box 41, von Braun. As the ARS Education Committee retreated from youth rocketry experimentation programs, the Membership Committee (especially von Braun and J. P. Layton) advanced their youth model rocketry initiatives. Minutes (18 November 1959), AIAA.
85. Terry and Beryl Leonard remembered how Zucrow gave their son a "live rocket" for a science fair, in their 1974 letter, folders 3–5, box 9, Zucrow. *Indianapolis Times* (6 November 1960): 10; Homer Hickam, *Rocket Boys* (New York: Random House, 1998), 199, 213. This story was the basis for the movie *October Sky* (Universal Pictures, 1999).
86. *Purdue Alumnus* (May 1960): 3. F. L. Hovde, speech to the Fourth International Conference on the Mechanics and Chemistry of Solid Propellants (23 June 1965), Navy Department, box 37, Hovde Papers.
87. Quoted from Solberg interview with Eckles, 29. The engineering statistic is from LeBold, et al., *A Study of the Purdue University Engineering Graduate*, iii. "Twenty Years in Review," *Campus Copy* (January 1967): 12.
88. *Purdue Alumnus* (September–October 1962): 1.
89. F. L. Hovde correspondence with the Preparedness Investigating Subcommittee, Committee on Armed Services, US Senate (7 January 1958); Hovde remarks to the US Senate (7 January 1958); and Hovde to Mrs. Grace Oswalt (5 December 1957); all in Science and Technology, box 10, Hovde Papers.
90. F. L. Hovde, chair of the Army Scientific Advisory Panel, to Wilbur Brucker, secretary of the Army (30 October 1957), Army Scientific Advisory Panel, box 24, Hovde Papers. "Sobering Remarks," *Journal and Courier* (12 November 1957): 6.

91. D. H. Radler, "Science and the Climate of Opinion," *PRF Horizons* 4, no. 3 (January 1958): 1–2. The "Purdue Opinion Poll" in spring of 1956 was part of Purdue sociologist H. H. Remmers's massive sixteen-year study of American teenagers, "stratified by sex, grade, and geographical region." Remmers was director of Purdue's Division of Educational Reference (1928–1963). The March 1958 poll is in *PRF Horizons* 5, no. 2 (October 1958): 1–2.
92. Walter Hirsch, "The Image of the Scientist in Science Fiction," *American Journal of Sociology* (March 1958): 506–12.
93. J. J. C. letter (26 November 1957) and G. L. O. (7 December 1957), Science and Technology, box 10, Hovde.
94. *Purdue Engineer* (January 1958): 32–33. *PRF Horizons* 5, no. 2 (October 1958): 1–2.
95. *Purdue Alumnus* (Summer 1961): 28.
96. Murphy interview with Grimm, 18.
97. E. A. Bonni and M.Dzh. Zukrou, *Aerodinamika. Reaktivnye Dvigateli* (Moscow: Gosizdat, 1960). Ia. M. Paushkin, *Khimiia reaktivnykh topliv* (Moscow: ANSSSR, 1962). In just one premier case, the Soviets republished Joe D. Hoffman and H. Doyle Thompson, "Optimum Thrust-Nozzle Contours for Gas-Particle Flows," *AIAA Journal* 5, no. 10 (October 1967), 1886–87, in *Voprosy raketnoi tekhniki / Problems of Rocket Technology* 3 (1967): 49–63. These notices and quotes are all from memorandums in the J. D. Hoffman and H. D. Thompson files, Department of Mechanical Engineering, Purdue University.
98. Werner von Kirchner to Zucrow (15 December 1974), folders 3–5, box 9, Zucrow.
99. Maurice J. Zucrow, *Aircraft and Missile Propulsion*, which appeared as *Volume 1: Thermodynamics of Fluid Flow and Application to Propulsion Engines* (New York: Wiley, 1958); and *Volume 2: The Gas Turbine Power Plant, the Turboprop, Turbojet, Ramjet, and Rocket Engines* (New York: Wiley, 1958). Quoted from the reviews in *Missile Design and Development* (May 1958): 33; and (December 1958): 47.
100. Fuhs testimonial (4 November 1974) in folders 3–5, box 9, Zucrow; and his review in *ARS Journal* (June 1959): 464–75. Fuhs (MSME Purdue, PhD in ME from Caltech) was chief scientist at the Aero-Propulsion Lab at Wright-Patterson AFB, then professor of engineering (and department chair of Aeronautics and Mechanical Engineering), Naval Postgraduate School.
101. John Sloop, "Review of Rocket Research at Lewis" (9 February 1954), 1–6, sent to NACA Headquarters, Rocket Engine Research, box 502, RG 255.
102. "Resolution Concerning Increased NACA Effort Adopted by the Subcommittee on Rocket Engines" (28 November 1955), NASA Glenn Research Center. Abe Silverstein letter, Rocket Engine Research Program 1952-55, box 10, Raymond R. Sharp Director's Office files. Olson interview with Sloop," 48–49. Walter Olson, "A Suggested Policy and Course of Action for NACA with Regard to Rocket Engine Propulsion" (6 May 1955), 21–22, part of a communication to Eugene Emme (31 January 1972), Propulsion file, NASA.
103. John L. Sloop, *Liquid Hydrogen as a Propulsion Fuel, 1945–1959* (Washington, DC: GPO, 1978), 79, 83–86, with scaling up to up to forty thousand pounds of thrust.

104. "Rocket Seminar" memorandum (4 March 1957), in Papers, box 11, Sloop. John Sloop Materials, box 5, Dawson. "Minutes of Research Planning Council Meeting of 6 September 1957," Research Groups folder, box 298, RG 255.
105. Walter T. Olson interview with John L. Sloop (11 July 1974), 27, 41–52, NASA.
106. Bruce Lundin, "Some Remarks on a Future Policy and Course of Action for the NACA" (9 December 1957, but possibly as early as 25 October), Lewis Aeronautical Research Projects and Programs, box 502, RG 255.
107. Quoted from John L. Sloop, et al., "Performance and Missions," NACA Flight Propulsion Conference at Lewis Flight Propulsion Laboratory (21–22 November 1957), 161–75, originally marked secret, Papers, box 11, Sloop. For notes and discussions about Lundin's proposal at the meetings of the NACA Lewis Research Planning Council, see its minutes of 4 November and 5 December 1957, Research Groups; also see the 2 December discussion of Olson's 15 Point Plan, "Minutes of Meeting of Research Planning Council" (5 December 1957), Research Groups; both in box 298, RG 255.
108. By 2 December 1957, Silverstein had included Olson's "A Suggested Policy and Course of Action for NACA on Space Flight" into his edited copy of Bruce Lundin's "Some Remarks on a Future Policy and Course of Action for the NACA," now titled "Lewis Laboratory Opinion of a Future Policy and Course of Action for the NACA." The Silverstein edits are in the copy at the NASA Glenn Research Center, with thanks to Robert Arrighi, who has called the Lundin–Silverstein intervention the "basic template" for NASA. Robert S. Arrighi, *Bringing the Future within Reach: Celebrating 75 Years of NASA John H. Glenn Research Center* (Cleveland: NASA, 2016), 115–24.
109. Silverstein interview with Mauer, 61. Virginia P. Dawson, *Engines and Innovation: Lewis Laboratory and American Propulsion Technology* (Washington, DC: NASA, 1995), 159–63. Sloop, *Liquid Hydrogen as a Propulsion Fuel*, 180–85. Lundin's report was the model for "A National Research Program for Space Technology: A Staff Study of the NACA" (14 January 1958).
110. Richard J. Barber Associates, *The Advanced Research Projects Agency, 1958–1974* (Washington, DC: DOD, 1975), I-4 to I-8.
111. Silverstein interview with Mauer, 70–78, 94. Erica Karr, "Nasa Expands," *Missiles and Rockets* (6 April 1959): 25–26.
112. Shirley Alley, Editorial, *Purdue Scientist* (October 1957): 9; and (December 1957): 10–11. *PRF Horizons* 5, no. 6 (February 1959): 1–2. "Final Report on Research Grant NSF-G7278" (25 September 1959), folder 28, box 11, PRF Records.
113. "Project Seeks Dust of Moon," *New York Times* (4 November 1957): 4. *Purdue Alumnus* (December 1957): 3; (April 1958): 4–5; and (September–October 1955): 1.
114. *Indianapolis Times* (3 November 1957): 4–5. *New York Times* (3 August 1958): 41.
115. *Purdue Alumnus* (September–October 1960): 4–5.

## CHAPTER 8

1. Krafft A. Ehricke, prologue to *Space Flight Report to the Nation*, ed. by Jerry Grey and Vivian Grey (New York: Basic Books, 1962), 54–57. Hercules Powder Company, "Beating the Egg," *Space/Aeronautics* (March 1962): 117.
2. *Missiles Away* 4, no. 4 (Winter 1957): 43.
3. Remarks by Dorothy M. Simon on 18 May 1959, 86th Congress, *Congressional Record*, vol. 105, part 6, 8282–84. Krafft Ehricke, "The Anthropology of Astronautics," *Astronautics* (November 1957): 29.
4. Dwight Eisenhower, "Farewell Address to the Nation" (17 January 1961).
5. Alex Roland, *The Military–Industrial Complex* (Washington, DC: AHA, 2005), 3–9. James Ledbetter, *Unwarranted Influence: Dwight D. Eisenhower and the Military-Industrial Complex* (New Haven: Yale University Press, 2011). William I. Hitchcock, *The Age of Eisenhower: America and the World in the 1950s* (New York: Simon & Schuster, 2018), 109–10, 507–9.
6. Eisenhower, "Farewell Address to the Nation."
7. National Academy of Sciences, *Federal Support of Basic Research in Institutions of Higher Learning* (Washington, DC: GPO, 1964), 8, 38, 46, 69–73, 86–100, 107–8. Linda Weiss, *America Inc.? Innovation and Enterprise in the National Security State* (Ithaca: Cornell University Press, 2014), 32–33.
8. National Science Foundation, *Government University Relationships in Federally Sponsored Research and Development* (Washington, DC: GPO, 1958). Nathan Pusey, "Harvard and the Federal Government," *New Scientist* 12, no. 256 (12 October 1961): 100.
9. Archie Palmer, *University Research and Patent Policies, Practices and Procedures* (Washington, DC: NAS, 1962), 1, 19–24.
10. *Reviews of Data on Research and Development* 19 (April 1960): 1–7. Clarence Danhof, *Government Contracting and Technological Change* (Washington, DC: Brookings Institution, 1968), 307, 332–33. Elliott V. Converse III, *Rearming for the Cold War, 1945–1960* (Washington, DC: Department of Defense, 2012), 36–37.
11. National Academy of Sciences, *Federal Support of Basic Research*, 102–8. *Principles for Determining Costs Allocable to Research and Development and Educational Services under Grants and Contracts with Educational Institutions*, Circular A-21 (Washington, DC: GPO, 1958).
12. Bruce Reese, "Overhead Rates" memorandum (1 April 1969), in Reese file.
13. "Prof. M. J. Zucrow," *Purdue Engineer* (March 1960): 50–51. *Panel on Science and Technology Fourth Meeting. Hearings before the Committee on Science and Astronautics, US House of Representatives*, 87th Congress (21 and 22 March 1962), 82. Peter F. Drucker, *Landmarks of Tomorrow* (New York: Harper Brothers, 1957), xi, 10–11, 91.
14. *Panel on Science and Technology Third Meeting. Report of the Committee on Science and Astronautics, US House of Representatives*, 87th Congress (17 April 1961), 19–25.

15. *National Science Policy. Hearings of the Committee on Science and Astronautics, US House of Representatives,* 91st Congress (July, August, and September of 1970), 716–21.
16. *Missile Design and Development* (November 1959): 4.
17. P. R. Trumpler, "Is Sponsored Research Destroying Our Universities?," *Journal of Engineering Education* 51, no. 8 (April 1961): 618–25.
18. *Report to the President on Government Contracting for Research and Development,* S. Doc. 94, 87th Congress (May 1962). Aaron L. Friedberg, *In the Shadow of the Garrison State: America's Anti-Statism and Its Cold War Grand Strategy* (Princeton: Princeton University Press, 2000), 334.
19. *PRF Horizons* 7, no. 8 (April 1961): 1–2; 9, no. 10 (June 1963): 1–3; and 10, no. 9 (May 1964): 2–4.
20. *Purdue Alumnus* (April 1966): 11.
21. *Purdue Alumnus* (February 1962): 4; and (March 1964): 3. Draft "Electronic Research Center," Frederick Hovde to George Simpson, NASA (4 December 1963), NASA, box 8, Hovde.
22. Special Projects Office, US Navy, *An Introduction to the PERT/COST System for Integrated Project Management* (Washington, DC: GPO, 1961). Office of Programs, NASA, *PERT Handbook* (Washington, DC: GPO, 1961). Air Force Systems Command, *PERT-Time System Description Manual,* vol. 1 (Washington, DC: USAF, 1963). For context, see Booz, Allen, and Hamilton, *Review of Navy R&D Management* (Washington, DC: Department of the Navy, 1976), 82, 138–40, 347–52; and Nancy Petrovic, "Design for Decline: Executive Management and the Decline of NASA" (PhD diss., University of Maryland, 1982), 165–66.
23. Fremont Kast and James Rosenzweig, eds., *Science, Technology, and Management* (New York: McGraw-Hill, 1962), 1–2; and one of its chapters, Lt. Gen. F. S. Bessom, "Project Management within the Army Materiel Command," 90–105.
24. Seymour Melman, *Pentagon Capitalism: The Political Economy of War* (New York: McGraw-Hill, 1970), 2–10, 97–98. Walter A. McDougall, *The Heavens and the Earth: A Political History of the Space Age* (New York: Basic, 1985), 8.
25. Ken Hechler and Albert E. Eastman, *The Endless Space Frontier: A History of the House Committee on Science and Astronautics* (San Diego: Univelt, 1982).
26. *Panel on Science and Technology Second Meeting. Report of the Committee on Science and Astronautics, US House of Representatives,* 86th Congress (17 September 1960), 58.
27. *Panel on Science and Technology Fifth Meeting. Proceedings before the Committee on Science and Astronautics, US House of Representatives,* 88th Congress (22 and 23 January 1963), 58–59, 72, 89–91.
28. *Panel Third Meeting,* 75.
29. *Panel Fourth Meeting,* 57, 97. *Panel Third Meeting,* 51.
30. *Panel on Science and Technology First Meeting. Proceedings of the Committee on Science and*

*Astronautics, US House of Representatives*, 86th Congress (4 May 1960), 52–54. *Panel Second Meeting*, 23. *Panel Fourth Meeting*, 88.

31. *Panel on Science and Technology Seventh Meeting. Proceedings of the Committee on Science and Astronautics, US House of Representatives*, 89th Congress (25, 26, and 27 January 1966), 134.
32. *Panel First Meeting*, 51–54, 60, 63. Zucrow's "beans" remark became a theme for this meeting. He also set the tone with the same theme at the panel's third meeting (*Panel Third Meeting*, 23–32, 49–50).
33. *Panel Fourth Meeting*, 82. For an earlier reference to "the cupboard of basic research," see Milton Rosen, "A Down-to-Earth View of Spaceflight," *Navigation* 3, no. 9 (September 1953): 318.
34. For these metaphors, see *Panel First Meeting*, 18; *Panel Third Meeting*, 65; and *Panel Fifth Meeting*, 47.
35. "Prof. M. J. Zucrow," *Purdue Engineer* (March 1960): 50–51. *Indianapolis Times* (6 November 1960): 10. *Panel Third Meeting*, 63.
36. "Statement on ARS Views of US National Space Program" (23 February 1960), AIAA. Hugh Dryden, "NASA Mission and Long-Range Plan," in *NASA–Industry Program Plans Conference, July 28–29, 1960* (Washington, DC: NASA, 1960), 6–9.
37. Grey and Grey, eds., *Space Flight Report to the Nation*. Jerry Grey, Princeton University's nuclear propulsion expert, was one of the key organizers. *Purdue Exponent* (6 October 1961): 3. "Free Enterprise v. the Moon," *Time* (20 October 1961). The event, along with Zucrow's role, were surveyed in *Astronautics* (December 1961): 31, whose front cover was Don Trembath's painting, "Space Trajectories," also the front cover for this book.
38. *Missiles and Space* (March 1962): 10.
39. Martin Summerfield to ARS president William Pickering (27 August 1962); and minutes of the Board of Directors (8–9 September 1962); both in ARS Minutes, box 13, G. Edward Pendray Papers (Pendray).
40. G. E. Pendray to Martin Summerfield (20 June 1962), General Correspondence, box 8, von Braun. Pendray to Harry Guggenheim (15 July 1962), Aviation and Rocketry, box 1, Guggenheim Foundation Records.
41. *Missile Design and Development* (December 1959): 6. John Sherill interview with Wernher von Braun, "The Day He Stopped Dreaming," *Guideposts* (October 1960): 1–5.
42. Wernher von Braun, "Changing Patterns in Leadership of a Research Team," in *Proceedings of the Sixteenth National Conference on the Administration of Research* (Denver: University of Denver, 1963), 63–71.
43. *Panel Third Meeting*, 61–65, 79.
44. See John M. Logsdon, *The Decision to Go to the Moon: Project Apollo and the National Interest* (Chicago: University of Chicago Press, 1970).
45. John F. Kennedy, "Special Message to the Congress on Urgent National Needs" (25 May

1961). David Halberstam, *The Best and the Brightest* (New York: Random House, 1972). James Webb, *Space Age Management: The Large-Scale Approach* (New York: McGraw-Hill, 1969). Senator Warren Magnuson's comments (in 1962) in Kast and Rosenzweig, *Science, Technology and Management*, 1.

46. *Panel Fourth Meeting*, 19–53. M. J. Zucrow, "Space Propulsion Engines: Their Characteristics and Problems," *American Scientist* 50, no. 3 (September 1962): 409–35.
47. Zucrow witnessed a nuclear reactor engine test at the Jackass Flats, Nevada, Test Site in September of 1961. *Purdue Exponent* (6 October 1961): 3.
48. Raymond Bisplinghoff, "Introduction to the Advanced Research and Technology Programs," in *NASA–Industry Program Plans Conference, February 11–12, 1963* (Washington, DC: NASA, 1963), 121–25. See also Ernst Stuhlinger and Gustav Mesmer, eds., *Space Science and Engineering* (New York: McGraw-Hill, 1965), 243; and Maxwell Hunter, *Thrust into Space* (New York: Holt, Rinehart and Winston, 1966).
49. David G. Elliott's, "Developing Electric-Propulsion Power Plants," *Astronautics and Aerospace Engineering* 1, no. 5 (June 1963): 82–87; his "Two-Fluid Magnetohydrodynamic Cycle for Nuclear-Electric Power Conversion," *ARS Journal* (June 1962): 924–28; and his "Magnetohydrodynamic Power Systems," *Journal of Spacecraft and Rockets* 4, no. 7 (July 1967): 842–46.
50. *Panel Fifth Meeting*, 43–55.
51. See Frank Rom's letter to Zucrow (6 November 1974), in folders 3–5, box 9, Zucrow. For context, see Mark D. Bowles, *Science in Flux: NASA's Nuclear Program at Plum Brook Station, 1955–2005* (Washington, DC: NASA, 2006).
52. C. F. Warner letter to R. J. Grosh (14 October 1963), Warner file.
53. Chandler C. Ross, "Life at Aerojet-General University: A Memoir" (1981), 58–85, folder 1, box 21, Ehresman.
54. Visit of John L. Sloop to Rocketdyne Division, Rockwell International (24 April 1974), 13, 32–54; R. W. Graham interview with Sandra Johnson (20 September 2005); both at NASA. Robert Kraemer and Vince Wheelock, *Rocketdyne: Powering Humans into Space* (Reston: AIAA, 2005), 139–45. Gunn was also the founding chair of the ARS Nuclear Propulsion Committee and won the Schreiber-Spence Award for Distinguished Contributions to Space Nuclear Power (1990). Also see Stanley Gunn, "Nuclear Propulsion: A Historical Perspective," *Space Policy* 17 (2001): 291–98.
55. The syllabus and files are in folders 2–3, box 41, Ehresman. Murthy, also known as B. G., was then a visiting research professor.
56. Junior Class of 1969, Air Force ROTC, "Project METRO: Manned Environment for Technology and Research in Orbit," submitted for AS 360 (May 1969), folder 12, box 2, Ross. Ross was an Outstanding Basic Cadet in the USAF ROTC, 1967–1968.
57. I have culled the facts and quotes here and in the previous paragraph from *Purdue Alumnus*

(September–October 1962): 23; (January 1963): 4–5; and (May 1965): 3–7, 26. *Journal and Courier* (28 January 1967): 1–6.

58. M. J. Zucrow sabbatical report (24 January 1959 to 1 June 1959) in Zucrow file.
59. Zucrow sabbatical report. Office of Technology Utilization, *A Study of NASA University Programs* (Washington, DC: NASA, 1968), 68–72.
60. "NASA–University Relationships," in *A Review of Space Research*, by Space Sciences Board (Washington, DC: NAS, 1962), 12–25. SUP lasted until 1970.
61. T. L. K. Smull, *The Nature and Scope of the NASA University Program* (Washington, DC: NASA, 1965); and W. Henry Lambright, *Launching NASA's Sustaining University Program* (Syracuse, NY: The Inter-University Case Program, 1969).
62. *Engineering News* (December 1965): 4.
63. *Purdue Alumnus* (January 1973): 10. Brian Jirout, "Farming from Space: Landsat and the Development of Agricultural Surveillance during the Cold War," *Quest* 24, no. 4 (2017): 3–15.
64. *Engineering News* (December 1961): 2.
65. Warner to R. J. Grosh (5 November and 5 December 1963), Warner file.
66. "Final Report to NASA on Facilities Grant NsG(F)-14" (10 May 1965), box 22, Pendray.
67. The Princeton dedication speeches are in box 22, Pendray.
68. Robert A. Caro, *The Years of Lyndon Johnson*, vol. 4, *The Passage of Power* (New York: Knopf, 2012), 559–60.
69. As the MOU summarized, "Research is being conducted on the ignition of liquid bipropellants, erosion burning of solid propellants, heat transfer to gases under conditions of high temperature gradients, heat transfer characteristics of liquid propellants, film cooling of heated surfaces, heat transfer to a gas bubbling through a liquid, combustion pressure oscillations in rocket motors, solid propellant basic combustion phenomena, combustion of hybrid propellants, two-phase flow (gases in liquid), flow of gases containing solid particles in supersonic nozzles, gas dynamics of thrust vector control by secondary injection, application of three dimensional characteristics to supersonic flow in nozzles, heat transfer from a hot plasma forming a tube arc, the application of optical spectroscopy to the analysis of plasmas, the fundamentals of accelerating an air column by means of charged colloidal particles, the dynamics of annular two-phase flow in nozzles, diffusion of two-phase flows, mass transfer in annular two-phase flow, and heat transfer to a rotating disk." Memorandum of understanding, Research Facilities Grant NsG(f)-21 (April 1964), Zucrow Labs Research Proposals, box 39, Ehresman. Office of Technology Utilization, *A Study of NASA University Programs*, 58. W. Henry Lambright, *Powering Apollo: James E. Webb of NASA* (Baltimore: Johns Hopkins University Press, 2000), 136–37.
70. Office of Technology Utilization, *A Study of NASA University Programs*, 56. F. N. Andrews, "Report to the University Senate" (16 March 1970), folder 6, box 7, Special Committee on Sponsored Research (SCSR).

71. Eugene Manganiello, quoted in Charles Ehresman's bound notebook "High Pressure Combustion Facility" (September 1964 to March 1965), 10; C. M. Ehresman to Zucrow, "Trip Report, NASA Coordination Meeting" (20–21 January 1965), Zucrow Labs Facilities; both in box 31, Ehresman.
72. Strickler interview with Grimm (7 October 2015), 9, 15–16, ASC. C. M. Ehresman memorandum "NASA Lewis Functional Design Review and Approval of High Pressure Combustion Research Facility" (5 November 1964), Zucrow Labs Facilities, box 31, Ehresman. Among the Rocket Lab students who interned or later worked at Aerojet were Clair Beighley, Dell E. Robison, J. P. Sellers, James Bottorff, Joe Hoffman, and H. Doyle Thompson.
73. Strickler interview with Grimm (7 October 2015), 11–12. C. M. Ehresman, "Design Consideration for a High Pressure Combustion Research Facility" (4 January 1965), 6–8, Facilities, box 32, Ehresman.
74. Ehresman, "Design Consideration," 5–6, 9–10. He used the *PERT General Information Manual* (White Plains: IBM Corporation, 1962).
75. Strickler interview with Grimm (7 October 2015), 20–21. "High Pressure Combustion Research Facility Projected Execution Plan" (17 September 1964), Zucrow Labs Facilities, box 31; Memorandum from Purdue Development Planning (17 January 1964), Zucrow Labs Research Proposals, box 39; both in Ehresman.
76. Zucrow interview with Eckles, 22–29. L. J. Freehafer, vice president and treasurer, memorandum to Zucrow (3 November 1964), with copies to President Hovde and campus administrators, Zucrow Labs Facilities, box 31, Ehresman.
77. Dick Smith, "Purdue's Jet Propulsion Center," Indiana–Purdue Football Program (November 1964), 7, in the black scrapbook, box 1 (addition), Zucrow.
78. Ehresman letter to Babcock and Wilcox (29 September 1964), Propellant Storage System; and his "Trip Report" (12 May 1965), Zucrow Labs Facilities; both in box 31, Ehresman.
79. I have drawn these facts from Charles Ehresman's brief history, "Jet Propulsion Center," in folder 3, box 28, Ehresman; and C. M. Ehresman, "The Design, Construction, and Operation of Purdue University's New Combustion Research Laboratory," in *ISA-70 Silver Jubilee, International Conference and Exhibit* (26–29 October 1970), 648–70.
80. Priscilla Decker, "Pioneer in Jet Propulsion," *Campus Copy* [Purdue University] (March 1965): 8–10. *Journal and Courier* (1 April 1966): 13. Ehresman bound notebook "High Pressure Combustion Facility," 57.
81. John Sloop, "Summary of Remarks," dedication of the JPC Library and Office Building and High Pressure Lab (5 April 1966), Papers, box 12, Sloop.
82. Office of Technology Utilization, *A Study of NASA University Programs*, 2, 12, 15, 20. W. H. Siegfried, "Space Colonization: Benefits for the World," *American Institute of Physics Conference Proceedings* 654, no. 1270 (2003): 1277.
83. Zucrow letter to the Board of Trustees (7 November 1958), and Solberg's letter to Dean Hawkins (26 May 1958), both in Zucrow file.

84. Warner to R. J. Grosh (5 November 1963), Warner file. Thompson interview with Smith, 14. M. J. Zucrow recommendation for Reese to dean G. A. Hawkins (16 September 1965), Zucrow file.
85. R. J. Grosh to Dean Hawkins (26 December 1962), Zucrow file.
86. *Journal and Courier* (9 December 1965): 9. *Purdue Exponent* (16 December 1965): 2.
87. News clippings from the *Terre Haute Star* and *Tribune* and Huntington *Herald Press*, *Gary Post-Tribune*, and *Vincennes Sun Commercial* (1965), Zucrow file.
88. Joe D. Hoffman and H. Doyle Thompson, "A General Method for Determining Optimum Thrust-Nozzle Contours for Gas-Particle Flows," AIAA Paper 66-538, AIAA Second Propulsion Joint Specialist Conference, Colorado Springs (13–17 June 1966). *Huntsville News* (17 May 1967), a clipping from the black scrapbook, box 1 (addition), Zucrow. *Astronautics and Aeronautics* (August 1967): 33. William Bollay (1966), in folders 3–5, box 9, Zucrow.
89. Zucrow interview with Eckles, 21–22. P. W. McFaddin to G. A. Hawkins (24 May 1966), Ehresman file.
90. "Graduate Study in Mechanical Engineering, Purdue University," folder 8, box 41, Ehresman. For the array of courses offered by the Rocket Lab between 1966 and 1970, see appendix 3.
91. Among the research topics and graduate students combined from 1966 and 1967 were the following: Liquid-film cooling: D. L. Crabtree, R. A. Gater. High combustion pressure: T. F. Larsen, T. W. Carpenter, J. W. Converse, G. R. Johnson, A. W. Brecheisen, and N. J. Barsic. Thrust nozzles: M. P. Scofield. Combustion studies: R. L. Strickler, D. W. Netzer (instability), C. A. Bryce. Solid rockets: S. D. Kershner, R. L. Derr, P. J. Goede, R. J. Burick, B. W. Farquhar, D. J. Norton. And some new approaches: J. W. O. Anderson on energy transfers, G. C. Trenker and R. Poulsen on air-augmented rockets, R. D. Guhse on injections into supersonic streams, V. H. Ransom on the analysis of nozzle flow in scramjets, W. L. Allan on applying "attitude and trajectory control by external jets," and W. F. Hassel on magnetically accelerated colloidal particles. "1966 Review of Research" (5–6 April 1966)," "1967 Review of Research" (11–12 April 1967)," and "Graduate Student Assignments" (spring semester 1966–1967), folder 4, box 41, Ehresman.
92. See the brochures and internal memorandums in folders 4, 5, and 8, box 41, Ehresman, especially the memorandum "Why Graduate Students *Don't* Come to Purdue" (11 May 1967) in folder 4.
93. Advertisements in *Purdue Alumnus* (May 1959): 1; and *Rocket Newsletter* [Southern California Section of the ARS] (January 1956): 3.
94. Philp Abelson, "The Midwest in the Scientific Revolution," *PRF Horizons* 11, no. 8 (April 1965): 1. Zucrow's comments in *Panel Third Meeting*, 20.
95. "Why Graduate Students *Don't* Come to Purdue."
96. Bruce Reese memorandum "Visit to Forrestal Research Center" (27 October 1966), Trip Reports, folder 28, box 31, Ehresman. Len Caveny, "Martin Summerfield and His Princeton University Propulsion and Combustion Laboratory," paper presented at the 47th AIAA

Joint Propulsion Conference & Exhibit (San Diego, CA, 31 July 2011), 14, 38–42, recognized Summerfield's hardware approach, if framed within his physics expertise.

97. Small appointments booklet, Nov. 68–Dec. 69, "Purdue University," box 1 (addition), Zucrow.

98. *Astronautics and Aeronautics* (February 1967): 65. *Panel on Science and Technology Eighth Meeting. Government, Science, and International Policy. Proceedings of the Committee on Science and Astronautics, US House of Representatives*, 90th Congress (24, 25, and 26 January 1967), 160–64.

99. "Science and Public Policy," *PRF Horizons* 11, no. 6 (February 1965): 1; and 11, no. 9 (May 1965): 2.

100. Boyd R. Keenan, ed., *Science and the University* (New York: Columbia University Press, 1966).

101. Kurt Stehling, "In Print," *Astronautics and Aeronautics* (October 1968), 98. C. S. Draper, "Technology, Engineering, Science and Modern Education," *Leonardo* 2 (1969): 147–53; and Herbert Fox, "Alienation in the Space Industry," *Astronautics* (August 1969): 17. Norman Mailer, *Of a Fire on the Moon* (Boston: Little, Brown, 1970), 376–77.

102. Neil Armstrong interview (1966), 13, NASA.

103. Neil M. Maher, *Apollo in the Age of Aquarius* (Cambridge: Harvard University Press, 2019).

104. Bruce L. R. Smith, *American Science Policy Since World War II* (Washington, DC: Brookings Institution, 1990), 50. Elizabeth Berman, *Creating the Market University: How Academic Science Became an Economic Engine* (Princeton: Princeton University Press, 2011), 35–37.

105. *Purdue Exponent* (16 December 1965): 2. "President Frederick Hovde," *Purdue Alumnus* (Summer 1971): 9–11.

106. *New York Times* (7 May 1969): 32. *Purdue Engineering Magazine* 65, no. 8 (November 1968): 8–10.

107. Matthew Wisnioski, *Engineers for Change: Competing Visions of Technology in 1960s America* (Cambridge: MIT Press, 2012), 163, 173–85.

108. Warren E. Howland, "Engineering Education for Social Leadership," *Technology and Culture* 10, no. 1 (January 1969): 3; and 11, no. 2 (April 1970).

109. Joseph Haberer, ed., *Technology and the Future of Man* (West Lafayette, IN: Purdue University, 1973), 3, 9–11, 21, 30, 45; and his "The Two Cultures Syndrome," *Science, Technology and Public Policy News* [Purdue University] 24 (February 1975): 1–2.

110. From Trip Reports, folder 28, box 31, Ehresman. Purdue News Bureau release (24 November 1967), Reese file.

111. As head of the school, with Mel L'Ecuyer's collaboration, Reese helped integrate its upper-level undergraduate and graduate coursework with the offerings of the Department of Mechanical Engineering. Reese's Airbreathing Jet Propulsion (ME 651), for example, became a core offering in both graduate student curriculums.

112. Kuhn letter to Frederick Hovde, president of Purdue University (2 July 1970), folder 6, box 7; and "Review of Research at Other Universities," folder 4, box 3, Special Committee on Sponsored Research (SCSR).

113. "DOD-Sponsored Research" (6 May 1970); and address of R. G. Jahn to students (4 May 1970), folder 7, box 4, SCSR.
114. Summerfield memorandum (5 October 1970), folder 7, box 4; and Summerfield memorandum to the committee (13 November 1970), folder 14, box 3, SCSR.
115. Formal report "Review of Research at Other Universities," folder 4, box 3, SCSR.
116. Kuhn to Frederick Hovde (2 July 1970); F. N. Andrews to Thomas Kuhn (28 July 1970); and accompanying report "Purdue Research Foundation—Division of Sponsored Programs" (13 May 1970), all from folder 6, box 7, SCSR.
117. David Beers, *Blue Sky Dream: A Memoir of America's Fall from Grace* (New York: Doubleday, 1996), 131–35. See the statistics in Hechler and Eastman, *The Endless Space Frontier*, 252. William Cowdin interview with Tracy Grimm (11 November 2015), ASC.
118. Thomas P. Hughes, *Rescuing Prometheus: Four Monumental Projects That Changed Our World* (New York: Pantheon, 1998), 171–74. See also Jennifer S. Light, *From Warfare to Welfare: Defense Intellectuals and Urban Problems in Cold War America* (Baltimore: Johns Hopkins University Press, 2003).
119. *Panel on Science and Technology Tenth Meeting: Proceedings of the Committee on Science and Astronautics, US House of Representatives*, 91st Congress (4, 5, and 6 February 1969), 75, 91. *Panel Seventh Meeting*, 156, 168.
120. Minutes of the meetings on 18–19 November 1965 and 4–6 May 1966 (at NACA Lewis, Cleveland), Air Breathing Propulsion Systems, box 514, RG 255. Zucrow was also a member of the Committee on Airbreathing Engines of the SAE (1965) and the Subcommittee on Aircraft Gas Turbines of the ASME (1963–1965). Also see M. J. Zucrow and S. N. B. Murthy, "Jet Propulsion and Aircraft Propellers," in *Standard Handbook for Mechanical Engineers*, 7th ed., ed. Theodore Baumeister (New York: McGraw-Hill, 1967), 11-112 to 11-139.
121. Minutes of the meetings on 22–23 April 1965 and 3–4 November 1966, Air Breathing Propulsion Systems, box 514, RG 255. *Panel on Science and Technology Sixth Meeting. Proceedings before the Committee on Science and Astronautics, US House of Representatives*, 89th Congress (26 and 27 January 1965), 21–24, 30, 73, 81–86. The separate quote is from Zucrow interview with Eckles, 28.
122. M. J. Zucrow, "The Engineering Graduate and his Career as a Professional Engineer" (1947), 2, Zucrow Talks, box 1 (addition), Zucrow.
123. Robert J. Gordon, *The Rise and Fall of American Growth: The U.S. Standard of Living since the Civil War* (Princeton: Princeton University Press, 2017); and Monika Gisler and Didier Sornette, "Exuberant Innovations: The Apollo Program," *Society* 46, no. 1 (2009), 55–68.
124. Clair Beighley (28 March 1966), folders 3–5, box 9, Zucrow.
125. Glen A. Robertson and Darryl W. Webb, "The Death of Rocket Science in the 21st Century," *Physics Procedia* 20 (2011): 319–30.
126. Alexander Welsh, "The Whiteness of the Moon," *New Republic* (28 September 1963): 13.
127. Herman Kahn, et al., *The Next 200 Years: A Scenario for America and the World* (New York: William Morrow, 1976), 6–7, 56–57. Henry Etzkowitz and Loet Leydesdorff, "The Endless

Transition: A 'Triple Helix' of University-Industry-Government Relations," *Minerva* 36 (1998): 203–8. Rudi Beichel and D. W. Culver, "Rocket Propulsion for the Next Forty Years" (September–October 1998), in box 28, Ehresman. Victor Ransom's letter to Zucrow (1974), in folders 3–5, box 9, Zucrow.

## EPILOGUE

1. State of California road test score sheet (11 January 1966), Maurice Zucrow folder, box 1 (addition), Zucrow.
2. *Purdue University Bulletin* (Lafayette: Purdue University, 1971), 4–5, 73.
3. These items are in the black scrapbook, box 1 (addition), Zucrow.
4. Harvard College, Class of 1922, *Fiftieth Anniversary Report* (Cambridge: Harvard University, 1972), 584–85.
5. Manila folder titled "Four (4) Talks Given by Dr. Zucrow," box 1 (addition), Zucrow.
6. This was his answer to John McCormack, congressman from Boston, in *Panel on Science and Technology First Meeting. Proceedings of the Committee on Science and Astronautics, US House of Representatives*, 86th Congress (4 May 1960), 52–54.
7. M. J. Zucrow, "Sponsored Research—Shall We Permit It to Undermine Our Educational Objectives," Brigham Young University (22 April 1961), manila folder titled "Four (4) Talks Given by Dr. Zucrow," box 1 (addition), Zucrow.
8. "Energy Management of Munitions Plant Modernization" (February 1975), Recorday appointment book, blue folder, box 1 (addition), Zucrow.
9. Maurice Zucrow and Joe D. Hoffman, *Gas Dynamics*, vol. 1 (New York: Wiley, 1976), 72.
10. From the small booklet calendars titled "Appointments Nov. 68–Dec. 69," "Appointments 1970," and "Appointments 1971," brown folder, box 1 (addition), Zucrow.
11. *Journal of Engineering Education* 51, no. 1 (October 1960): 26.
12. John L. McDaniel, US Army Missile Command, Redstone Arsenal, "Contributions of Dr. Maurice Zucrow to Army Missile Command Programs" (14 November 1968), Zucrow file.
13. McDaniel, "Contributions of Dr. Maurice Zucrow to Army Missile Command Programs."
14. "Nomination of M. J. Zucrow for Honorary Doctorate in Engineering" (1969), Zucrow file.
15. Louis Martin Sears, "The Engineer and the Historian," *Mechanical Engineering* (July 1947): 581–84. Donald Putt interview with John L. Sloop (30 April 1974), 40, NASA.
16. Frank Malina to Andrew Haley (9 March 1955), Personal Correspondence, box 2, Frank Malina Papers. Martin Summerfield to Eugene Emme (15 June 1965), Biographical Material on Martin Summerfield, NASA.
17. Ernst Stuhlinger and Gustav Mesmer's *Space Science and Engineering* (New York: McGraw-Hill, 1965) ignored Zucrow, though Stuhlinger cited his own work nine times. George P. Sutton's *History of Liquid Propellant Rocket Engines* (Reston: AIAA, 2006) also ignored Zucrow, though Sutton cited his own work twenty-four times.

18. Maurice J. Zucrow, *Engineering Design Handbook: Elements of Aircraft and Missile Propulsion* (Redstone Arsenal: US Army Materiel Command, 1969), xxxvii–xxxix. This is a revised 750-page edition of his 98-page *Propulsion and Propellants* (1959).
19. Charles Ehresman to Fred Durant, Walter Rocket binder, box 24, Ehresman.
20. See Milt Rosen's comments in "Rocketry in the 1950s," AIAA Panel Discussion (28 October 1971), IRIS no. 00907747, AFHRA.
21. Douglas Bradley and Katherine P. VanHooser, "Space Shuttle Main Engine: The Relentless Pursuit of Improvement," paper presented at the AIAA Space 2011 Conference and Exposition (Long Beach, CA, 27–29 September 2011).
22. Robert E. Biggs, *Space Shuttle Main Engine: The First Twenty Years and Beyond* (San Diego: AAS, 2008), 7; and his "Engineering the Engine: The Space Shuttle Main Engine," in *Space Shuttle Legacy*, ed. Roger Launius et al. (Reston: AIAA, 2013): 77–110.
23. Bradley and VanHooser, "Space Shuttle Main Engine."
24. "Zucrow, Prof. Maurice Joseph," *American Men of Science: Biographical Directory, Volume 1965* (New York: Bowker, 1967), 6084. Harvard College, *Fiftieth Anniversary Report*, 584–85. Maurice Zucrow, "What's an Engineer?," *Engineering News* (April 1965): 1–2.
25. Reese memorandum to graduate students (15 February 1967), folder 12, box 32, Ehresman. Thompson interview with Smith, 15–16.
26. The letters are in folders 3–5, box 9, Zucrow.
27. Stewart interview with Eckles, 11.
28. Robert B. Eckles, *The Dean: A Biography of A. A. Potter* (West Lafayette, IN: Purdue University, 1974), 86–87.
29. Zucrow interview with Eckles, 26–27.
30. "Gold Plated," folder 9, box 66, Ehresman.
31. Frank Incropera, "Request for Name Change of TSPC" (20 February 1996), Zucrow Memorial Lab Rename, box 27; and Ehresman letter (1999), folder 4, box 28; both in Ehresman. Zucrow Labs acquired a new sign in 2018.
32. Stephen Heister, William E. Anderson, Timothée L. Pourpoint, and R. Joseph Cassady, *Rocket Propulsion* (New York: Cambridge University Press, 2019).
33. Robert Lucht, ed., *Zucrow Labs, 2016–2017 Annual Report* (West Lafayette, IN, Purdue University, 2017).
34. Jerry Ross, *Spacewalker: My Journey in Space and Faith as NASA's Record-Setting Frequent Flyer* (West Lafayette, IN: Purdue University Press, 2013), 47. "Miracles," Wellesley Foundation (4 September 1988), folder 7, box 3, Warner.
35. Jerry and Karen Ross interview with Tracy Grimm (31 August 2012), 11, ASC. Jerry L. Ross, "A Study of Fluid Injection into a Subsonic Air Flow" (MSME thesis, Purdue University, 1972).

# ARCHIVAL SOURCES

IN THE ARCHIVAL CITATIONS IN THE NOTES, THE AUTHOR, TITLE, AND DATE ARE followed by the name of the folder, number of the box, and record group number or name (with abbreviations).

Air Force Historical Records, Maxwell Air Force Base (AFHRA)
- USAF Collections

American Institute of Aeronautics and Astronautics (AIAA)
- Minutes and Reports, Board of Director's Files, American Rocket Society

California Institute of Technology Archives and Special Collections (Caltech)
- Interviews and Oral Histories
- Theodore von Kármán Papers
- Charles Lauritsen Papers

Huntington Library, San Marino, California (Huntington)
- Interviews and Oral Histories

Library of Congress, Manuscripts Division
- Vannevar Bush Papers (Bush)
- Guggenheim Foundation Records
- Frank Malina Papers
- Bernard Schriever Papers
- Nathan Twining Papers (Twining)
- Wernher von Braun Papers (von Braun)

NASA Glenn Research Center, Cleveland
- Virginia Dawson Papers (Dawson)
- Raymond R. Sharp Director's Office Files
- Rocket Engine Test Facility Papers
- Abe Silverstein Papers
- John Sloop Papers (Sloop)

NASA History Office, Washington, DC (NASA)
- Topic and Subject Folders

National Air and Space Museum, Steven F. Udvar-Hazy Center
- Andrew Haley Papers (Haley)
- Richard Porter Papers (Porter)
- Robert Truax Papers

National Archives and Records Administration at College Park
- RG 38, Office of the Chief of Naval Operations
- RG 72, Bureau of Aeronautics, US Navy
- RG 156, Office of the Chief of Ordnance
- RG 227, Office of Scientific Research and Development
- RG 330, Office of the Secretary of Defense
- RG 342, US Air Force Commands, Activities, and Organizations

National Archives and Records Administration at Chicago
- RG 255, National Advisory Committee on Aeronautics

Princeton University, Mudd Manuscript Library
- Astrophysical Sciences Papers
- G. Edward Pendray Papers
- Special Committee on Sponsored Research (SCSR)

Purdue University Archives and Special Collections (ASC)
- Harold de Groff Papers
- Charles M. Ehresman Papers (Ehresman)
- George A. Hawkins Papers
- Frederick L. Hovde Papers (Hovde)
- Warren E. Howland Papers
- Horton B. Knoll Papers (Knoll)
- William K. LeBold Papers
- G. Stanley Meikle Papers
- J. Michael Murphy Papers
- Andrey A. Potter Papers (Potter)
- Jerry Ross Papers (Ross)
- R. B. Stewart Papers (Stewart)
- M. J. Zucrow Papers (Zucrow)
- Purdue Research Foundation Records

The following interviews were conducted by Michael G. Smith as part of the Maurice J. Zucrow Laboratories Oral History Project:
- Raymond Cohen, 6 January 2017
- Robert Glick, 20 January 2017
- Elliott Katz, 29 September 2016
- George Schneiter, 10 September 2016
- H. Doyle Thompson, 17 January 2017
- James Warner, 17 December 2016

Purdue University, Maurice J. Zucrow Laboratories, Chaffee Hall (Chaffee)
- Project Squid Records, 1948–1952
- Dissertations and Reports

Purdue University, Department of Mechanical Engineering, Main Office Files
    George Hawkins
    Joe D. Hoffman
    Bruce Reese (Reese file)
    J. G. Skifstad
    H. Doyle Thompson
    Cecil Warner (Warner file)
    Maurice J. Zucrow (Zucrow file)
Tufts University Digital Collections and Archives
    Student Army Training Corps (SATC) Records
UCLA University Archives
    Record Series 587, Engineering, Science, and Management War Training (ESMWT)
Western Reserve Historical Society in Cleveland
    Abe Silverstein Papers

# INDEX

Image insert page numbers are italicized and identified as *p1*, *p2*, *p3*, and so forth.

T

## X



## Y



## Z

# ABOUT THE AUTHOR

MICHAEL G. SMITH IS A PROFESSOR OF HISTORY AT PURDUE UNIVERSITY, WHERE he has taught Russian history and aerospace history since 1996. He is also the author of *Rockets and Revolution: A Cultural History of Early Spaceflight* (2014) and *Language and Power in the Creation of the USSR* (1998).

Printed in the USA
CPSIA information can be obtained
at www.ICGtesting.com
CBHW072225180324
5549CB00005B/205